PHOTOSYNTHESIS

PHOTOSYNTHESIS
CONCEPTS AND APPLICATIONS

Charles Furlow

Kruger Brentt
Publishers
2023

Kruger Brentt Publishers UK. LTD.
Company Number 9728962

Regd. Office: 68 St Margarets Road, Edgware, Middlesex HA8 9UU

ISBN: 9781787150454

For information on all our publications visit our website at http://krugerbrentt.com/

PREFACE

Most life on Earth depends on photosynthesis. The process is carried out by plants, algae, and some types of bacteria, which capture energy from sunlight to produce oxygen (O2) and chemical energy stored in glucose (a sugar). Herbivores then obtain this energy by eating plants, and carnivores obtain it by eating herbivores. Photosynthesis is important to living organisms because it is the number one source of oxygen in the atmosphere. Without photosynthesis, the carbon cycle could not occur, oxygen-requiring life would not survive and plants would die. Green plants and trees use photosynthesis to make food from sunlight, carbon dioxide and water in the atmosphere: It is their primary source of energy. The importance of photosynthesis in our life is the oxygen it produces. Without photosynthesis there would be little to no oxygen on the planet. Photosynthesis uses light energy from the sun and carbon dioxide and water in the atmosphere to make food for plants, trees, algae and even some bacteria. It releases oxygen as a byproduct. The chlorophyll in these living organisms, which also contributes to their green hues, absorbs the sunlight and combines it with carbon dioxide to convert these compounds into an organic chemical called adenosine triphosphate (ATP). ATP is crucial in the relationship between energy and living things, and is known as the "energy currency for all life."

Humans are reliant on photosynthesis to produce the food that we eat, as a source of energy to create heat, light and electricity and for many of the materials that we make and use in our everyday lives. All living organisms that photosynthesize contain chlorophyll. Chlorophyll is an enzyme that allows water and carbon dioxide to be recombined into sugars and oxygen. These sugars are then used to fuel cellular processes and allow for the growth and reproduction of an organism. The importance of photosynthesis to man cannot be understated. Humans would not exist without photosynthesis and we continue to rely upon it for survival. All civilizations in human history were created using energy sourced from photosynthesis. When the process of photosynthesis is seriously impeded - either from extensive deforestation, drought or crop disease - this can lead to famine and societal collapse. These issues have been associated with the fall of empires and civilizations, including the Mayans in Central America, the Khmer in Southeast Asia and the Akkadians in the Middle East. The Irish potato famine is another famous example of the societal effects of reduced photosynthetic production. Humans are totally reliant on photosynthesis for the production of food, and we depend on photosynthesis to produce almost all the food products we consume. The fruits and vegetables in our diets directly depend on the photosynthesis process for life, and the animals we consume (in the form of meat, eggs, milk, and so on) depend on plants for their survival.

The present b ook contains thirteen chapters covering all related disciplines. These chapters inc lude Introduction to Photosynthesis, Photosynthesis in Plants, Energy Conversion I n Photosynthesis, The Photosynthetic Pigments, Photosynthetic Energy Conversion, Utilization of Biological Solar Energy and Development of Artificial Photosynthes is in Combating Climate Change, Photosynthesis and Respiration, Photosynthes is in Bacteria, Photosynthetic Alterations of Virus Infected Plants, Photosynthes is in Algae: Potential and Challenges, Metabolite Transport Across the Chloroplast Envelope Of C_3-Plants, Control of Photosynthesis, Allocation and Partitioning By Sugar Regulated Gene Expression and Starch Metabolism in Leaves Related terminology is given at the end for ready reference. This book provides a comprehensive introduction of on the complex sequence of photosynthetic energy transduction and carbon fixation. It is writte n in a lucid style with theoretical discussions and being supplemented with illustrations, and tables for easy understanding of the subject. The book is also supported with a compr ehensive glossary and index in the end. This book is an essential read for students and researchers interested in photosynthesis, plant science, plant physiology and climate change.

We are grate ful to all those persons as well as various books, manuals, periodicals, magazines, journals etc. that helped in the preparation of this book. In spite of the best efforts, it is possible that some errors may have occurred into the compilation and editing of the book. Further queries, constructive suggestions and criticisms for the improvement of the book are always welcome and shall be thankfully acknowledged.

Charles Furlow

CONTENTS

1

INTRODUCTION TO PHOTOSYNTHESIS

1.1 INTRODUCTION: AN OVERVIEW

Photosynthesis is the process by which plants, some bacteria and some protistans use the energy from sunlight to produce glucose from carbon dioxide and water. This glucose can be converted into pyruvate which releases adenosine triphosphate (ATP) by cellular respiration. Oxygen is also formed.

Photosynthesis may be summarised by the word equation:

$$\text{carbon dioxide} + \text{water} \xrightarrow[\text{chlorophyll}]{\text{sunlight}} \text{glucose} + \text{oxygen}$$

The conversion of usable sunlight energy into chemical energy is associated with the action of the green pigment chlorophyll. Chlorophyll is a complex molecule. Several modifications of chlorophyll occur among plants and other photosynthetic organisms. All photosynthetic organisms have chlorophyll a. Accessory pigments absorb energy that chlorophyll a does not absorb. Accessory pigments include chlorophyll b (also c, d, and e in algae and protistans), xanthophylls, and carotenoids (such as beta-carotene). Chlorophyll a absorbs its energy from the violet-blue and reddish orange-red wavelengths, and little from the intermediate (green-yellow-orange) wavelengths.

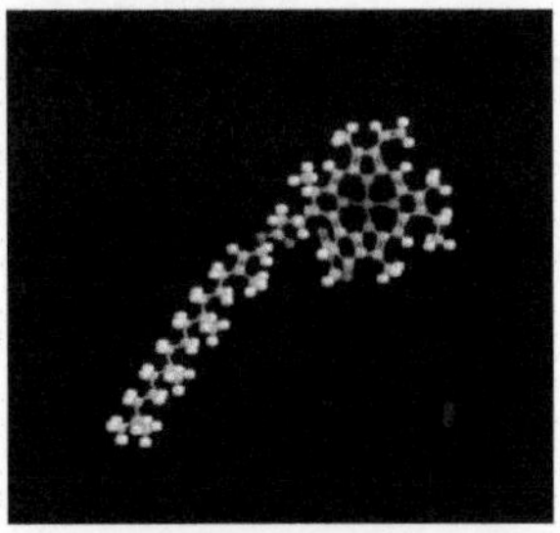

Chlorophyll

All chlorophylls have:

- A lipid-soluble hydrocarbon tail (c20h39 -)
- A flat hydrophilic head with a magnesium ion at its centre; different chlorophylls have different side-groups on the head

The tail and head are linked by an ester bond.

1.2 LEAVES AND LEAF STRUCTURE

Plants are the only photosynthetic organisms to have leaves (and not all plants have leaves). A leaf may be viewed as a solar collector crammed full of photosynthetic cells. The raw materials of photosynthesis, water and carbon dioxide, enter the cells of the leaf, and the products of photosynthesis, sugar and oxygen, leave the leaf. Water enters the root and is transported up to the leaves through specialized plant cells known as xylem vessels. Land plants must guard against drying out and so have evolved specialized structures known as stomata to allow gas to enter and leave the leaf. Carbon dioxide cannot pass through the protective waxy layer covering the leaf (cuticle), but it can enter the leaf through the stoma (the singular of stomata), flanked by two guard cells. Likewise, oxygen produced during photosynthesis can only pass out of the leaf through the opened stomata. Unfortunately for the plant, while these gases are moving between the inside and outside of the leaf, a great deal of water is also lost. Cottonwood trees, for example, will lose 100 gallons (about 450 dm3) of water per hour during hot desert days.

1.3 THE STRUCTURE OF THE CHLOROPLAST AND PHOTOSYNTHETIC MEMBRANES

The thylakoid is the structural unit of photosynthesis. Both photosynthetic prokaryotes and eukaryotes have these flattened sacs/vesicles containing photosynthetic chemicals. Only eukaryotes have chloroplasts with a surrounding membrane. Thylakoids are stacked like pancakes in stacks known collectively as grana. The areas between grana are referred to as stroma. While the mitochondrion has two membrane systems, the chloroplast has three, forming three compartments.

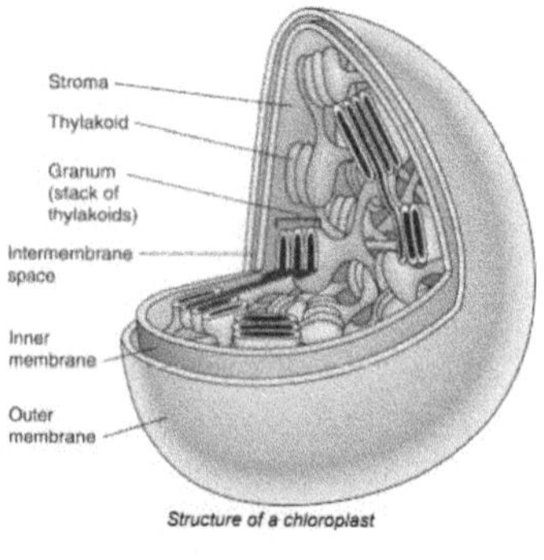

Structure of a chloroplast

1.4 STAGES OF PHOTOSYNTHESIS

When chlorophyll a absorbs light energy, an electron gains energy and is 'excited'. The excited electron is transferred to another molecule (called a primary electron acceptor). The chlorophyll molecule is oxidized (loss of electron) and has a positive charge. Photoactivation of chlorophyll a results in the splitting of water molecules and the transfer of energy to ATP and reduced nicotinamide adenine dinucleotide phosphate (NADP).

The chemical reactions involved include:

- Ccondensation reactions - responsible for water molecules splitting out, including phosphorylation (the addition of a phosphate group to an organic compound)
- Ooxidation/reduction (redox) reactions involving electron transfer

1.4.1 Photosynthesis is a Two-Stage Process

The light dependent reactions, a light-dependent series of reactions which occur in the grana, and require the direct energy of light to make energy-carrier molecules that are used in the second process:

- Light energy is trapped by chlorophyll to make ATP (photophosphorylation)
- At the same time water is split into oxygen, hydrogen ions and free electrons:

$$2H_2O \rightarrow 4H^+ + O_2 + 4e^- \text{ (photolysis)}$$

- The electrons then react with a carrier molecule nicotinamide adenine dinucleotide phosphate (NADP), changing it from its oxidised state (NADP+) to its reduced state (NADPH):

$$NADP^+ + 2e^- + 2H^+ \rightarrow NADPH + H^+$$

The light-independent reactions, a light-independent series of reactions which occur in the stroma of the chloroplasts, when the products of the light reaction, ATP and NADPH, are used to make carbohydrates from carbon dioxide (reduction); initially glyceraldehyde 3-phosphate (a 3-carbon atom molecule) is formed.

1.4.2 The Light-Dependent Reactions

When light energy is absorbed by a chlorophyll molecule its electrons gain energy and move to higher energy levels in the molecule (photoexcitation). Sufficient energy ionises the molecule, with the electron being 'freed' leaving a positively charged chlorophyll ion. This is called photoionisation. In whole chloroplasts each chlorophyll molecule is associated with an electron acceptor and an electron donor. These three molecules make up the core of a photosystem. Two electrons from a photoionised chlorophyll molecule are transferred to the electron acceptor. The positively charged chlorophyll ion then takes a pair of electrons from a neighbouring electron donor such as water.

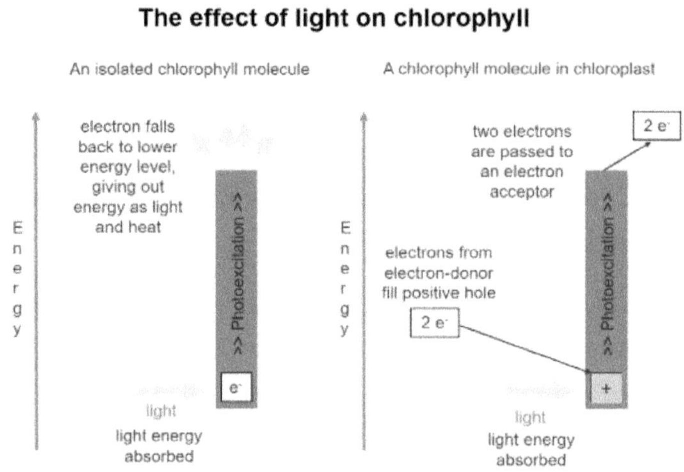

An electron transfer system (a series of chemical reactions) carries the two electrons to and fro across the thylakoid membrane. The energy to drive these processes comes from two photosystems:

- Photosystem II (PSII) (P680)
- Photosystem I (PSI) (P700)

It may seem confusing, but PSII occurs before PSI. It is named because it was the second to be discovered and hence named second.

The energy changes accompanying the two sets of changes make a Z shape when drawn out. This is why the electron transfer process is sometimes called the Z scheme. Key to the scheme is that sufficient energy is released during electron transfer to enable ATP to be made from ADP and phosphate.

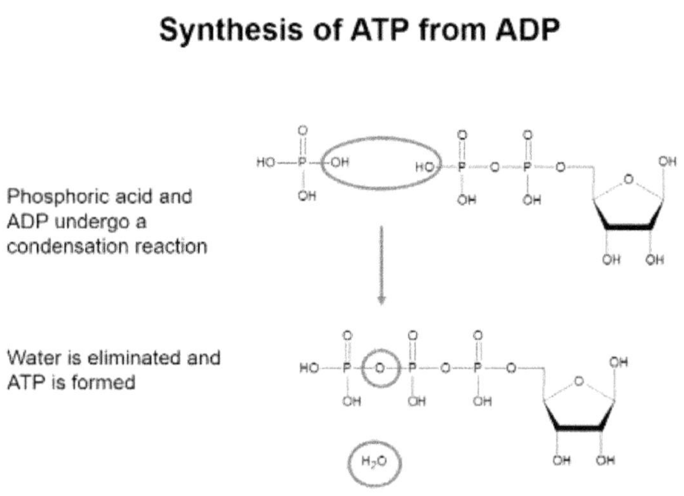

A condensation reaction has led to phosphorylation.

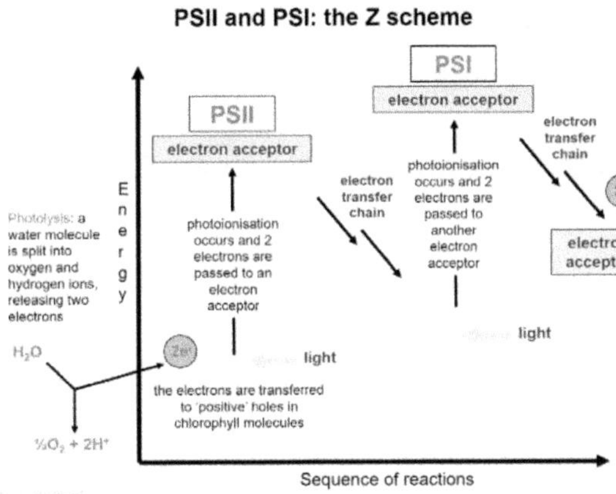

1.4.3 Non-Cyclic Phosphorylation (The Z Scheme)

Both adenosine triphosphate (ATP) and NADPH are produced.

In the first photosystem (Photosystem II, PSII):

- Photoionisation of chlorophyll transfers excited electrons to an electron
- Photolysis of water (an electron donor) produces oxygen molecules, hydrogen ions and electrons, and the latter are transferred to the positively-charged chlorophyll
- The electron acceptor passes the electrons to the electron transport chain; the final acceptor is photosystem psi
- Further absorbed light energy increases the energy of the electrons, sufficient for the reduction of NADP+ to NADPH

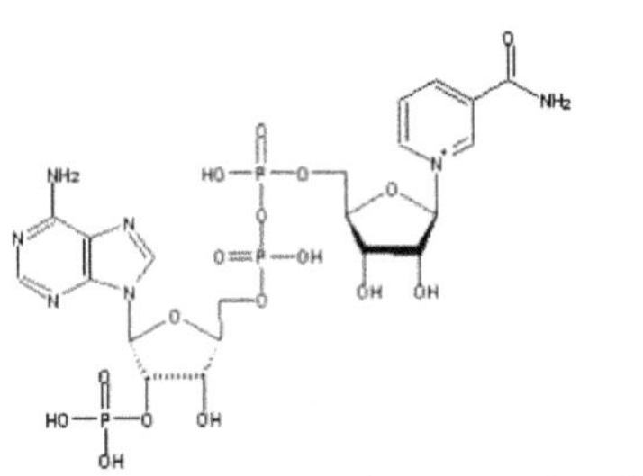

The oxidised form of nicotinamide adenine dinucleotide phosphate (NADP⁺)

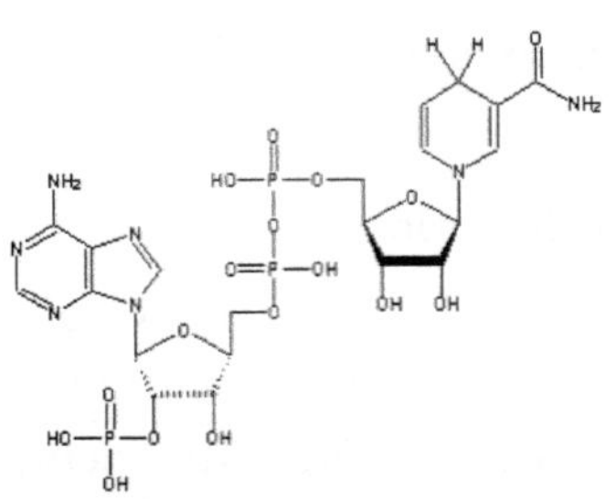

The reduced form of nicotinamide adenine dinucleotide phosphate (NADPH)

1.4.4 Chemiosmosis And Atp Synthesis

The components of non-cyclic phosphorylation are found in the thylakoid membranes of the chloroplast. Electrons passing through the transport chain provide energy to pump H^+ ions from the stroma, across the thylakoid membrane into the thylakoid compartment. H^+ ions are more concentrated in the thylakoid compartment than in the stroma. We say there is an electrochemical gradient. H^+ ions diffuse from the high to the low regions of concentration. This drives the production of ATP.

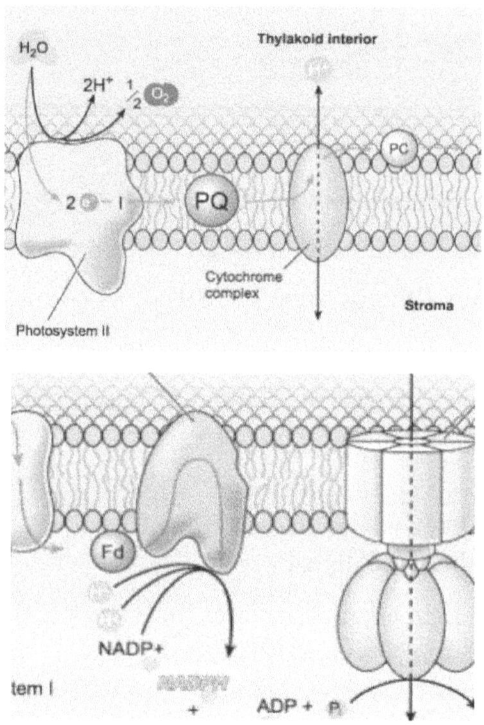

1.4.5 Cyclic Phosphorylation

The net effect of non-cyclic phosphorylation is to pass electrons from water to NADP. Energy released enables the production of ATP. But much more ATP is needed to drive the light-independent reactions. This extra energy is obtained from cyclic phosphorylation. This involves only Photosystem I which generates excited electrons. These are transferred to the electron transport chain between PSII and PSI, rather than to NADP+ and so no NADPH is formed. The cycle is completed by electrons being transported back to PSI by the electron transport system.

1.4.6 The Light-Independent Reactions

In the Light-Independent Process (the Dark reaction) carbon dioxide from the atmosphere (or water for aquatic/marine organisms) is captured and modified by the addition of hydrogen to form carbohydrates. The incorporation of carbon dioxide into organic compounds is known as carbon fixation. The energy for this comes from the first phase of the photosynthetic process. Living systems cannot directly utilize light energy, but can, through a complicated series of reactions, convert it into C-C bond energy that can be released by glycolysis and other metabolic processes.

Carbon dioxide combines with a five-carbon sugar, ribulose 1,5-biphosphate (RuBP). A six-carbon sugar forms but is unstable. Each molecule breaks down to form two glycerate 3-phosphate (GP) molecules.

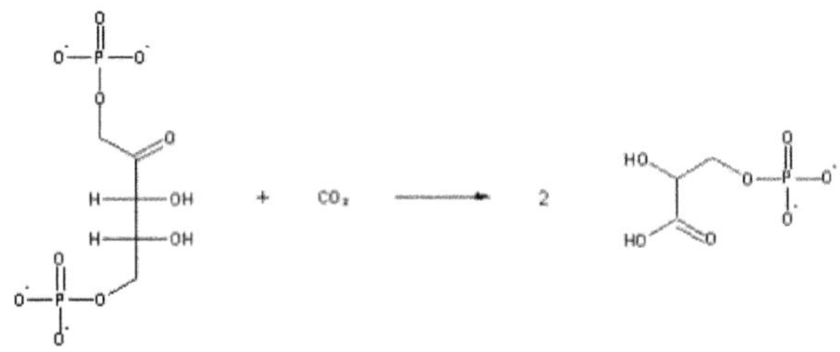

These glycerate 3-phosphate (GP) molecules are phosphorylated by ATP into glycerate diphosphate molecules.

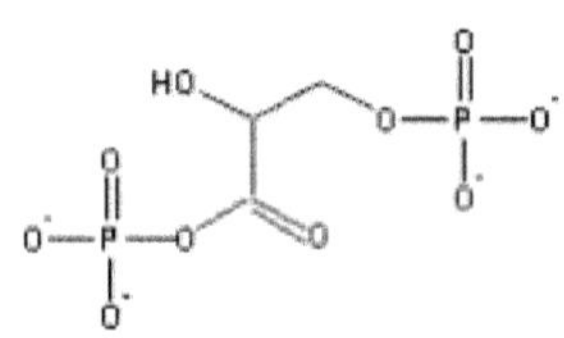

These are reduced by NADPH to two molecules of glyceraldehyde 3-phosphate (GALP).

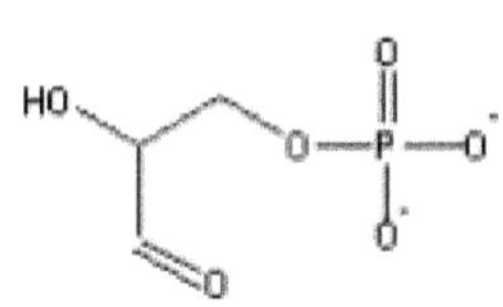

Of each pair of GALP molecules produced

- One molecule is the initial end product of photosynthesis; it is quickly converted to glucose and other carbohydrates, lipids or amino acids
- One molecule forms RuBP through a series of chemical reactions

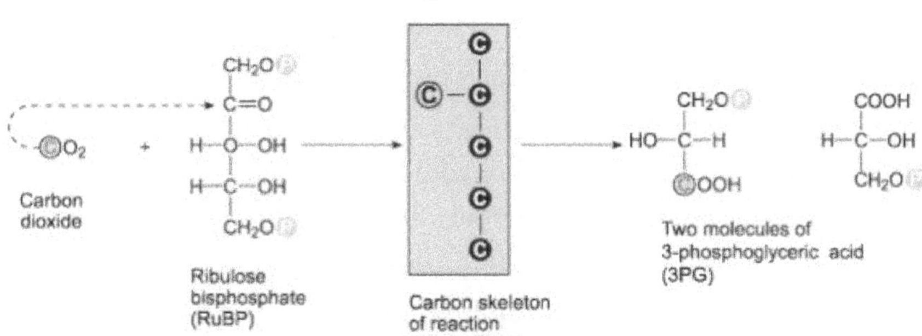

1.4.7 The First Steps in the Calvin Cycle

The first stable product of the Calvin Cycle is phosphoglycerate (PGA), a 3-C chemical. The energy from ATP and NADPH energy carriers generated by the photosystems is used to phosphorylate the PGA. Eventually there are 12 molecules of glyceraldehyde phosphate (also known as phosphoglyceraldehyde or PGAL, a 3-C), two of which are removed from the cycle to make a glucose. The remaining PGAL molecules are converted by ATP energy to reform six RuBP molecules, and thus start the cycle again.

1.5 SUMMARY OF STAGES OF PHOTOSYNTHESIS

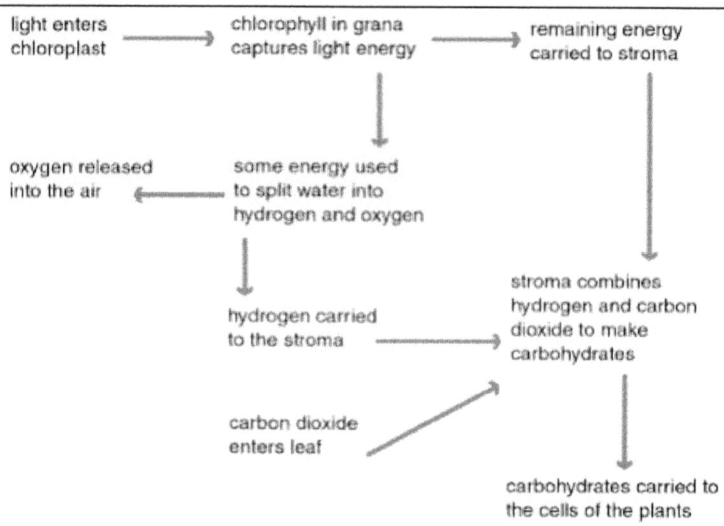

1.6 FACTORS AFFECTING THE RATE OF PHOTOSYNTHESIS

The main factors are light intensity, carbon dioxide concentration and temperature, known as limiting factors.

- As light intensity increases, the rate of the light-dependent reaction, and therefore photosynthesis generally, increases proportionately. As light intensity is increased however, the rate of photosynthesis is eventually limited by some other factor. Chlorophyll a is used in both photosystems. The wavelength of light is also important. PSI absorbs energy most efficiently at 700 nm and PSII at 680 nm. Light with a high proportion of energy concentrated in these wavelengths will produce a high rate of photosynthesis.
- An increase in the carbon dioxide concentration increases the rate at which carbon is incorporated into carbohydrate in the light-independent reaction and so the rate of photosynthesis generally increases until limited by another factor.
- Photosynthesis is dependent on temperature. It is a reaction catalysed by enzymes. As the enzymes approach their optimum temperatures the overall rate increases. Above the optimum temperature the rate begins to decrease until it stops.

2

PHOTOSYNTHESIS IN PLANTS

2.1 INTRODUCTION

Life on earth ultimately depends on energy derived from sun. Photosynthesis is the only process of biological importance that can harvest this energy.

- Literally photosynthesis means 'synthesis using light'. Photosynthetic organisms use solar energy to synthesize carbon compound that cannot be formed without the input of the energy.
- Photosynthesis (Photon = Light, Synthesis = Putting together) is an anabolic, endergonic process by which green plant synthesize carbohydrates (initially glucose) requiring carbon dioxide, water, pigments and sunlight. In other words, we can say that photosynthesis is transformation of solar energy/radiant energy/ light energy (ultimate source of energy for all living organisms) into chemical energy.

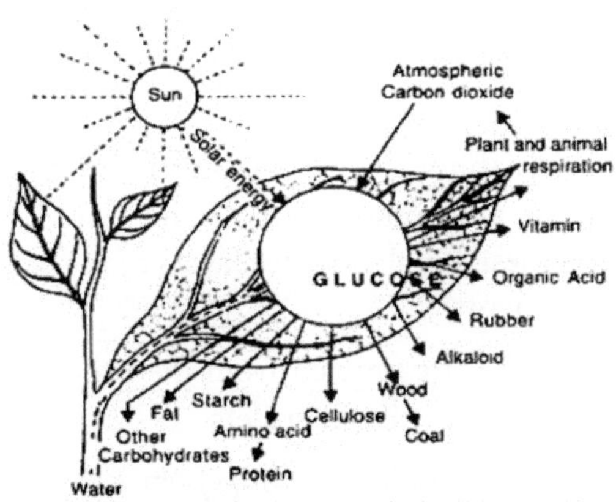

Fig. 6.1. Diagram showing the formation of different organic components from glucose which was also formed by the process of photosynthesis.

2.2 SIMPLE GENERAL EQUATION OF PHOTO SYNTHESIS IS AS FOLLOWS

$$6CO_2 + 6H_2O \xrightarrow[\text{Chlorophyll}]{\text{Light}} C_6H_{12}O_6 + 6O_2$$

According to Van Neil and Robert Hill, oxygen liberated during photosynthesis comes from water and not from carbon dioxide.

Thus, the overall correct biochemical reaction for photosynthesis can be written as:

$$6CO_2 + 12H_2O \xrightarrow[\text{Chlorophyll}]{\text{Sun Light}} C_6H_{12}O_6 + 6O_2 \uparrow + 6H_2O$$

Some photosynthetic bacteria use hydrogen donor other than water. Therefore, photosynthesis is also defined as the anabolic process of manufacture of organic compounds inside the chlorophyll containing cells from carbon dioxide and hydrogen donor with the help of radiant energy.

2.3 SIGNIFICANCE OF PHOTOSYNTHESIS

- Photosynthesis is the most important natural process which sustains life on earth.
- The process of photosynthesis is unique to green and other autotrophic plants. It synthesizes organic food from inorganic raw materials.
- All animals and heterotrophic plants depend upon the green plants for their organic food, and therefore, the green plants are called producers, while all other organisms are known as consumers.
- Photosynthesis converts radiant or solar energy into chemical energy. The same gets stored in the organic food as bonds between different atoms. Photosynthetic products provide energy to all organisms to carry out their life activities (all life is bottled sunshine).
- Coal, petroleum and natural gas are fossil fuels which have been produced by the application of heat and compression on the past plant and animal parts (all formed by photosynthesis) in the deeper layers of the earth. These are extremely important source of energy.
- All useful plant products are derived from the process of photosynthesis, e.g., timber, rubber, resins, drugs, oils, fibers, etc.
- It is the only known method by which oxygen is added to the atmosphere to compensate for oxygen being used in the respiration of organisms and burning of organic fuels. Oxygen is important in (a) efficient utilization and complete

breakdown of respiratory substrate and (b) formation of ozone in stratosphere that filters out and stops harmful UV radiations in reaching earth.

- Photosynthesis decreases the concentration of carbon dioxide which is being added to the atmosphere by the respiration of organisms and burning of organic fuels. Higher concentration of carbon dioxide is poisonous to living beings.
- Productivity of agricultural crops depends upon the rate of photosynthesis. Therefore, scientists are busy in genetically manipulating the crops.

2.4 MAGNITUDE OF PHOTOSYNTHESIS

- Only 0.2% of light energy falling on earth is utilized by photosynthetic organisms. The total carbon dioxide available to plants for photosynthesis is about 11.2 x 1014 tonnes. Out of this only 2.2 x 1013 tonnes are present in the atmosphere @ 0.03%. Oceans contain 11 x 1014 (110,000 billion) tonnes of carbon dioxide.
- About 70 to 80 billion tonnes of carbon dioxide are fixed annually by terrestrial and aquatic autotrophs and it produces near about 1700 million tonnes of dry organic matter. Out of these 10% (170 million tonnes) of dry matter is produced by land plants and rest by ocean (about 90%). This is an estimate by Robinowitch (1951),According to more recent figures given by Ryther and Woodwell (1970) only 1/3 of total global photosynthesis can be attributed to marine plants.

2.5 HISTORICAL BACKGROUND

320 B.C. Two great Greek philosophers and scientists **Aristotle** and **Theophrastus** thought plants obtain nourishment directly Min the soil.

1648. The first studies on photosynthesis may be attributed to **Van Helmont** a flemish botanist demonstrated that plants snake their own food (organic materials; and do not simpl) absorb them from the soil. Van Ilelmont weighed a pot filled with soil and planted a willow me m it and demonstrated that the tree gained 164 pounds in five years but the soil weighed only two ounces less. Van Helmont concluded that the rest of substance Came from the water he had added

1727 **Stephen Hales** for the first I ore recognized the importance of air and light in the nourish-ment of plants. He commented "the plants very prolxibly draw through their leaves some part of their nourishment from the air" and he further added "may not light also contribute to the ennobling principle, of vegetables", Further advances were not possible yet because different types of gases of air were not yet distinguished. It was not until 1754 that **Joseph Black** distinguished fixed air or carbon dioxide.

1772 **Joseph Priestley** discovered oxygen. Priestley reported that plants were capable of purifying air winch had been "injured" by the respiration of animals or in the burning of candles. In his classic experiment in which he showed that a mouse placed under a bell-jar survived longer if a plant of mint was placed together with it Priestley called the air produced by burning of candle as Phlogiston which is noxious for a mouse and said that plants convert it into Dephlogiston. In subsequent experiments Priestley obtained irratic results. probably he was not aware that light was essential in the process of purification of air by green plants.

1779 **Jan Ingen-Housz** discovered the role of light and green pans of the plants in purifying noxious air.

1783 **Lavoisler** Identified the purifying principle produced by green plants in sunlight as oxygen (dephlogistonl and the noxious air produced by the burning of candle as carbon dioxide (phlogiston).

1782 **To Jean Senebier** showed that carbon dioxide accelerated the process of photosynthesis and

1788 these results were made quantitative by **De Saussure.**

1804 **De Saussure** realised the importance of water in this process. Ile in the increase in dry weight of a plant and showed that it was equal to the amount of Carbondioxide and water consumed. Thus the broad outlines of the photosynthesis appealed to be Carbon dioxide plus water plus light energy yields oxygen and organic matter

1845 **Von Mayer** discovered that green plants convert solar energy into the chemical energy of organic matter.

1845 **Liebig** pointed out that the organic matter was synthesized from CO_2 by green plants. and water was also used In the process of photosynthesis.

1862 **Julius Von Sachs** and his students have the credit of laying the foundation of modern views on photosynthesis. They demonstrated that starch was the end product of photosynthesis.

1888 **Engelmann** plotted the action spectrum of photosynthesis.

1891 Jean B.J.D. Boussingault undertook detailed study of photosynthesis. By now the basic metabolism of green plants was understood and the relationship between the process of photosynthesis producing food and that of respiration which oxidised food stuffs had become clear.

1937 **Robert Hill** demonstrated that in the presence of sun light, water and a suitable hydrogen acceptor. Isolated chloroplasts released oxygen. even if carbon dioxide was absent. thus Hits experiments proved that In photosynthesis oxygen Is released from water.

1941 **Ruben. Hassid and Kamen** . using radioactive oxygen (O^{18}) also proved that the source of oxygen in photosynthesis was water.

1954 **Arnon. Allen and Whatley.** using $C^{14}O_2$ demonstrated the fixation of CO_2 by isolated chloroplasts.

1954 **Melvin Calvin** and his coworkers discovered various reactions involved in the conversion of carbon dioxide into carbohydrates using $C^{14}O_2$. These reactions are collectively known as **C_3-Cycle** or **Calvin Cycle**. Ile was awarded Nobel Prize for this work in 1961.

1967 M.D. Hatch and **C.R. Slack** discovered a new series of reactions involved in photosynthetk CO, assimilation in many tropical plants. In these plants the first stable compound in photosynthetic process is a 4.carbon compound Imalatel. hence, this series of reactions Is known as C_4-cycle.

1985 **Huber, Michel and Dissenhofer** crystallized the photosynthetic reaction centre of Rhodobacter (bacteria) and investigated its structure using X-ray crystallography technique. They were awarded Nobel Prize for this work in 1988

2.6 FUNCTIONAL RELATIONSHIP BETWEEN LIGHT AND DARK REACTIONS

- During photosynthesis water is oxidized and carbon dioxide is reduced, but where in the overall process light energy intervenes to drive the reaction. However, it is possible to show that photosynthesis consists of a combination of light-requiring reactions (the "light reactions") and non-light requiring reactions (the "dark reactions").
- It is now clear that tall the reactions for the incorporation of CO_2 into organic materials (i.e., carbohydrate) can occur in the dark (the "dark reactions"). The reactions dependent on light (the "light reactions") are those in which radiant energy is converted into chemical energy.
- According to Arnon, the functional relationship between the "light" and "dark" reactions can be established by examining the requirements of the dark reactions. The "dark reactions" comprise a complex cycle of enzyme-mediated reactions (the Calvin Cycle) which catalyzes the reduction of carbon dioxide to sugar. This cycle requires reducing power in the form of reduced nicotinamide adenine dinucleotide phosphate (NADPH) and chemical energy in the form of adenosine triphosphate (ATP).
- The reduced NADP (NADPH) and ATP are produced by the "light reactions". It is thus possible to divide a description of photosynthesis into those reactions associated with the Calvin cycle and the fixation of carbon dioxide, and those reactions (i.e., capture of light by pigments, electron transport, photophosphorylation) which are directly driven by light.

2.7 SITE OF PHOTOSYNTHESIS

- Chloroplast (Fig. 2) in green plants constitute the photosynthetic apparatus and act as site of photosynthesis. Chloroplasts of higher plants are discoid or ellipsoidal in shape measuring 4 —6 μ in length and 1—2 μ in thickness. It is a double membranous cytoplasmic organelle of eukaryotic green plant cells. The thickness of the two membranes including periplastidial space is approximately 300Å.

- Ground substance of chloroplast is filled with a hydrophilic matrix known as stroma. It contains cp-DNA (0.5%), RNA (2—3%), Plastoribosome (70S), enzymes for carbon dioxide assimilation, proteins (50—60%), starch grains and osmophilic droplets, vitamin E and K, Mg, Fe, Mn, P, etc. in traces. In stroma are embedded a number of flattened membranous sacs known as thylakoids. Photosynthetic pigments occur in thylakoid membranes.
- Aggregation of thylakoids to form stacks of coin like structures known as granna. A grannum consists near about 20 — 30 thylakoids. Each thylakoid encloses a space known as loculus. The end of disc shape thylakoid is called as margin and the area where the thylakoids membranes are appressed together is called partition.
- Some of the granna lamella are connected with thylakoids of other granna by stroma lamella or fret membranes. Thylakoid membrane and stroma lamella both are composed of lipid and proteins. In photosynthetic prokaryotes (blue-green algae and Bacteria) chloroplast is absent. Chromatophore is present in photosynthetic bacteria and photosynthetic lamellae in blue-green algae.

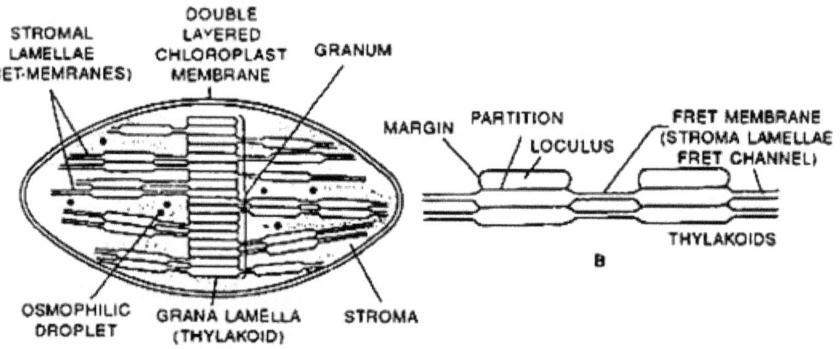

2.8 MECHANISM OF PHOTOSYNTHESIS

Photosynthesis is an oxidation reduction process in which water is oxidized and carbon dioxide is reduced to carbohydrate.

Blackmann (1905) pointed out that the process of photosynthesis consists of two phases:

- Light reaction or Light phase or Light-dependent phase or Photochemical phase
- Dark reaction or Dark phase or Light independent phase or Biochemical phase.

During light reaction, oxygen is evolved and assimilatory power (ATP and NADPH2) are formed. During dark reaction assimilatory power is utilized to synthesize glucose.

- Oxygenic photosynthesis (with evolution of (O_2) takes place in green eukaryotes and cyanobacteria (blue-green algae).
- An oxygenic photosynthesis (without the evolution of (O_2) takes place in photosynthetic bacteria.

$$CO_2 + H_2O \xrightarrow[\text{Chlorophyll}]{\text{Light}} \text{Sugar} + \text{Oxygen}$$

$$CO_2 + H_2S \xrightarrow[\text{Chlorophyll}]{\text{Light}} \text{Sugar} + \text{Sulphur or other oxidised compounds or some other}$$
inorganic compounds.

2.9 PHOTOSYNTHETIC PIGMENTS

Photosynthetic pigments are substances that absorb sunlight and initiate the process of photosynthesis. **Photosynthetic pigments are grouped into 3 categories:**

2.9.1 Chlorophyll

These are green coloured most abundant photosynthetic pigments that play a major role during photosynthesis. Major types of chlorophylls are known to exist in plants and photosynthetic bacteria viz., Chlorophyll a, b, c, d and e, Bacteriochlorophyll a, b and g, and Chlorobium chlorophyll (Bacterio viridin).

The structure of chlorophyll was first studied by Wilstatter, Stoll and Fischer in 1912. Chemically a chlorophyll molecule consists of a porphyrin head (15 x 15Å) and phytol tail (20Å). Porphyrin consists of tetrapyrrole rings and central core of Mg. Phytol tail is side chain of hydrocarbon. It is attach to one of the pyrrole ring. This chain helps the chlorophyll molecules to attach with thylakoid membrane.

Out of various types of chlorophyll, chlorophyll a and chlorophyll b are the most important for photosynthetic process. Chlorophyll a is found in all photosynthetic plants except photosynthetic bacteria. For this reason it is designated as Universal Photosynthetic Pigment or Primary Photosynthetic Pigment.

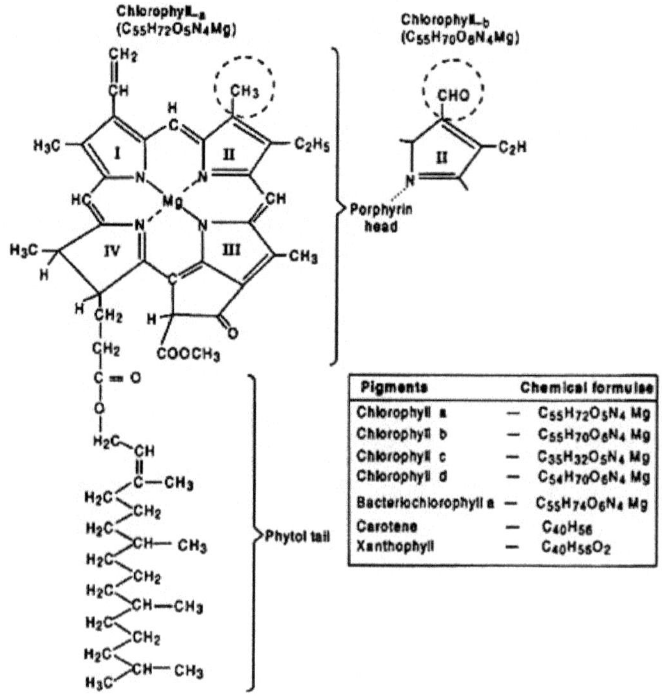

Pigments		Chemical formulae
Chlorophyll a	—	$C_{55}H_{72}O_5N_4Mg$
Chlorophyll b	—	$C_{55}H_{70}O_6N_4Mg$
Chlorophyll c	—	$C_{35}H_{32}O_5N_4Mg$
Chlorophyll d	—	$C_{54}H_{70}O_6N_4Mg$
Bacteriochlorophyll a	—	$C_{55}H_{74}O_6N_4Mg$
Carotene	—	$C_{40}H_{56}$
Xanthophyll	—	$C_{40}H_{56}O_2$

Structure of Chlorophyll a.

2.9.2 Carotenoids

These are yellow, red or orange colour pigments embedded in thylakoid membrane in association with chlorophylls but their amount is less. These are insoluble in water and precursor of Vitamin A. These are of two of types viz., Carotene and Xanthophyll (Carotenol/Xanthol).

Carotenes are pure hydrocarbons, red or orange in colour and their chemical formula is – C40H56 Some of the common carotenes are -α, β, γ and δ carotenes, Phytotene, Neurosporene, Lycopene (Red pigment found in ripe tomato). β—carotene on hydrolysis gives Vitamin A.

$$C_{40}H_{56} + 2H_2O \xrightarrow{\text{carotenase}} 2C_{20}H_{29}OH \text{ (Vitamin A)}$$

Xanthophylls are yellow coloured oxygen containing carotenoids and are most abundant in nature. The ratio of xanthophyll to carotene in nature is 2:1 in young leaves. The most common xanthophyll in green plant is Lutein (C40H56O2) and it is responsible for yellow colour in autumn foliage. Both carotene and xanthophylls are soluble in organic solvents like chloroform, ethyl ether, carbondisulphide etc.

2.9.3 Phycobilins (Biliproteins)

These are water soluble pigments and are abundantly present in algae, and also found in higher plants. There are two important types of phycobilins-Phycoerythrin (Red) and

Phycocyanin (Blue). Like chlorophyll, these pigments are open tetrapyrrole but do not contain Mg and Phytol chain.

PHOTOSYNTHETIC PIGMENTS : THEIR DISTRIBUTION AND LIGHT ABSORPTION

Pigment	Distribution	Light absorption (in nm)
1. Chlorophylls		
Chlorophyll-a	All photosynthesizing plants except bacteria.	429,410,660
Chlorophyll-b	Higher plants and green algae.	453,430,642
Chlorophyll-c	Diatoms, dinoflagellates and brown algae.	645
Chlorophyll-d	In some red algae	-
Chlorophyll-e	In *Tribonema* and zoospores of *Vaucheria*	-
Bacteriochlorophyll-a	Purple and green bacteria	-
Bacteriochlorophyll-b	In a strain of purple bacterium *Rhodopseudomonas.*	-
Bacteriochlorophyll-c, d & e (Chlorobium chlorophyll or Bacterioviridin)	Green bacteria	
Bacteriochlorophyll-g	Heliobacteria	
2. Carotenoids		
Carotenes	Mostly in algae and higher plants	
Xanthophylls (Carotenols)	Mostly in algae and higher plants	
3. Phycobilins		
Phycoerythrins	In blue-green and red algae	490,546,576
Phycocyanins	In blue-green and red algae	618

Nature of Light: The source of light for photosynthesis is sunlight. Sun Light is a form of energy (solar energy) that travels as a stream of tiny particles. Discrete particles present in light are called photons. They carry energy and the energy contained in a photon is termed as quantum. The energy content of a quantum is related to its wave length. Shorter the wave length, the greater is the energy present in its quantum. Depending upon the wave length electromagnetic spectrum comprises cosmic rays, gamma rays, X-rays,-UV rays, visible spectrum, infra-red rays, electric rays and radio waves. The visible spectrum ranges from 390 nm to 760 nm (3900 – 7600A), however, the plant life is affected by wave length ranging from 300 – 780 nm. Visible spectrum can be resolved into light of different colours i.e., violet (390-430 nm), blue or indigo (430-470 nm), blue green (470-500 nm), green (500 – 580 nm), yellow (580 – 600 nm), orange (600 – 650 nm), orange red (650 – 660 nm) and red (660 – 760 nm). Red light above 700 nm is called far red. Radiation shorter than violet are UV rays (100 – 390 nm). Radiation longer than those of red are called infra-red (760 – 10,000 nm).

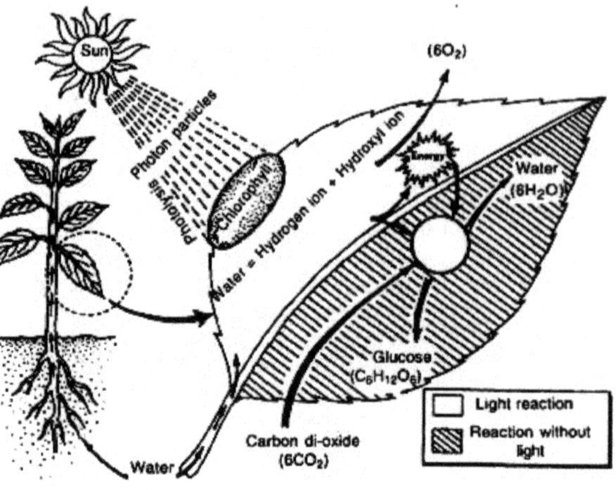

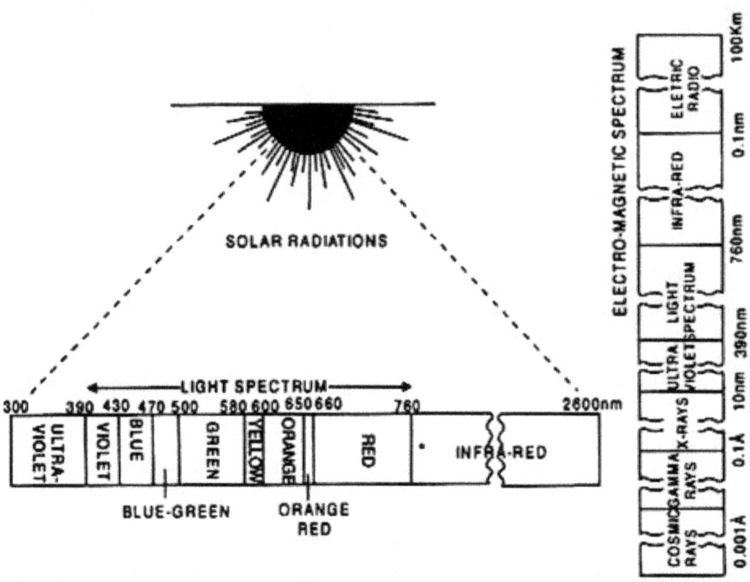

A ray of light falling upon a leaf behaves in 3 different ways. Part of it is reflected, a part transmitted and a part absorbed. The leaves absorb near about 83% of light, transmit 5% and reflect 12%. From the total absorption, 4% light is absorbed by the chlorophyll. Engelmann (1882) performed an experiment with the freshwater, multicellular filamentous green alga spirogyra. In a drop of water having numerous aerobic bacteria, the alga was exposed to a narrow beam of light passing through a prism. The bacteria after few minutes aggregated more in that regions which were exposed to blue and red wave length. It confirms that maximum oxygen evolution takes place in these regions due to high photosynthetic activities.

Absorption Spectrum: All photosynthetic organisms contain one or more organic pigments capable of absorbing visible radiation which will initiate the photochemical reactions of photosynthesis. When the amount of light absorbed by a pigment is plotted as a function of wave length, we obtain absorption spectrum (Fig. 4). It varies from pigment to pigment. By passing light of specific wave length through a solution of a substance and measuring the fraction absorbed, we obtain the absorption spectrum of that substance. Each type of molecules has a characteristic absorption spectrum, and measuring the absorption spectrum can be useful in identifying some unknown substance isolated from a plant or animal cell.

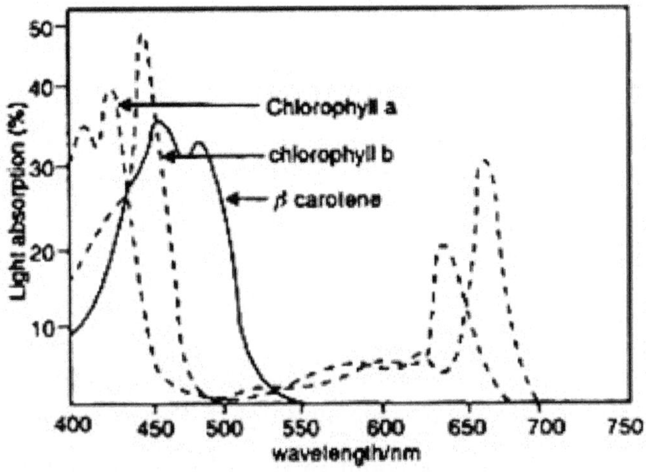

Action Spectrum: It represents the extent of response to different wave lengths of light in photosynthesis. It can also be defined as a measure of the process of photosynthesis when a light of different wave lengths is supplied but the intensity is the same. For photochemical reactions involving single pigment, the action spectrum has same general shape as the absorption spectrum of that pigment, otherwise both are quite distinct (Fig. 5).

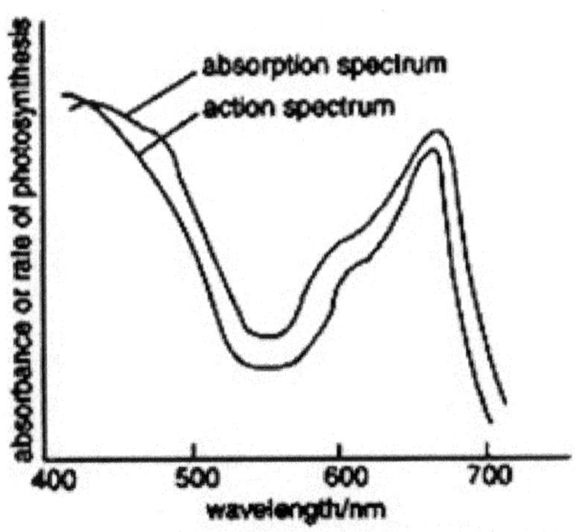

Quantum Requirement and Quantum Yield: The solar light comes to earth in the form of small packets of energy known as photons. The energy associated with each photon is called Quantum. Thus, requirement of solar light by a plant is measured in terms of number of photons or quanta.

The number of photons or quanta required by a plant or leaf to release one molecule of oxygen during photosynthesis is called quantum requirement. It has been observed that in most of the cases the quantum requirement is 8. It means that 8 photons or quantum's are required to release one molecule of oxygen. The number of oxygen molecules released per photon of light during photosynthesis is called Quantum yield. If the quantum requirement is 8 then quantum yield will be 0.125 (1/8).

2.9.4 Photosynthetic Unit Or Quantasome

- It is defined as the smallest group of collaborating pigment molecules necessary to affect a photochemical act i.e., absorption and migration of a light quantum to trapping centre where it promotes the release of an electron.
- Emmerson and Arnold (1932) on the basis of certain experiments assumed that about 250 chlorophyll molecules are required to fix one molecule of carbon dioxide in photosynthesis. This number of chlorophyll molecules was called the chlorophyll unit but the name was subsequently changed to photosynthetic unit and later it was designated as Quantasome by Park and Biggins (1964).
- The size of a quantasome is about 18 x 16 x l0nm and found in the membrane of thylakoids. Each quantasome consists of 200 – 240 chlorophyll (160 Chlorophyll a and 70 – 80 Chlorophyll b), 48 carotenoids, 46 quinone, 116 phospholipids, 144 diagalactosyl diglyceride, 346 monogalactosyl diglyceride, 48 sulpholipids, some sterols and special chlorophyll molecules (P680 and P700).
- 'P' is pigment, 680 and 700 denotes the wave length of light these molecule absorb. Peso and P700 constitute the reaction centre or photo centre. Other accessory pigments and chlorophyll molecules are light gatherers or antenna molecules. It capture solar energy and transfer it to the reaction centre by resonance transfer or inductive resonance.

2.10 PHOTOLUMINESCENCE:

It is the phenomenon of re-radiation of absorbed energy. It is of two types:

(1) Fluorescence and

(2) Phosphorescence

The normal state of the molecule is called as ground state or singlet state. When an electron of a molecule absorbs a quantum of light it is raised to a higher level of energy a state called Excited Second Singlet State. From first singlet state excited electron may return to the ground state either losing its extra energy in the form of heat or by losing energy in the form of radiant energy. The later process is called fluorescence. The substance which can emit back the absorbed radiations is called fluorescent substance. All photosynthetic pigments have the property of fluorescence.

The excited molecule also losses its electronic excitation energy by internal conversion and comes to another excited state called triplet state. From this triplet state excited molecule may return to ground state in three ways-by losing its extra energy in the form of heat, by losing extra energy in the form of radiant energy is called phosphorescence. The electron carrying extra energy may be expelled from the molecule and is consumed in some other chemical reactions and a fresh normal electron returns to the molecule. This mechanism happens in chlorophyll a (Universal Photosynthetic Pigment).

2.11 EMERSON RED DROP EFFECT AND ENHANCEMENT EFFECT

R. Emerson and Lewis (1943) while determining the quantum yield of photosynthesis in Chlorella by using monochromatic light of different wave lengths noticed a sharp decrease in quantum yield at wave length greater than 680 mμ. This decrease in quantum yield took place in the far red part of the spectrum i.e., the curve shows quantum yield drops dramatically in the region above 680 nm (Red region). This decline in photosynthesis is called Red drop effect (Emerson's first experiment).

Emerson and his co-workers (1957) found that the inefficient far red light in Chlorella beyond 680nm could be made fully efficient if supplemented with light of short wave length. The quantum yield from the two combined beams was found to be greater than the effect of both beams when used separately. This enhancement of photosynthesis is called Emerson Enhancement Effect (Emerson's second experiment) (Fig. 6).

Rate of oxygen evolution in combined beam – Rate of oxygen evolution in red beam/ Rate of oxygen evolution in far red beam

E = Emerson effect.

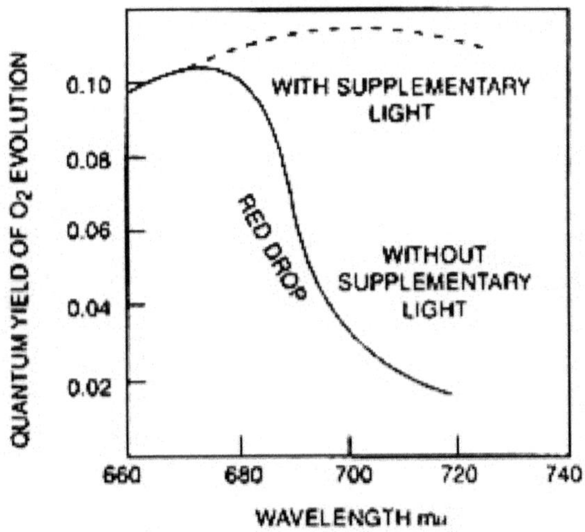

2.12 LIGHT TRAPPING CENTRES (PSI & PSII)

The discovery of red drop effect and the Emerson's enhancement effect concluded in a new concept about the role played bychlorophyll-a and accessary pigments in photosynthesis that photosynthesis involves two distinct photochemical processes. These processes are associated with two groups of photosynthetic pigments called as Pigment system I (Photoact I or Photosystem I) and Pigment system II (Photoact II or Photosystem II).

Each pigment system consists of a central core complex and light harvesting complex (LHC). LHC comprises antenna pigments associated with proteins (viz.., antenna complex). Their main function is to harvest light energy and transfer it to their respective reaction centre. The core complex consists of reaction centre associated with proteins and also electon donors and acceptors.

Wave length of light shorter than 680 nm affect both the pigment systems while wave length longer than 680 nm affect only pigment system I. PSI is found in thylakoid membrane and stroma lamella. It contains pigments chlorophyll a 660, chlorophyll a 670, chlorophyll a 680, chlorophyll a 690, chlorophyll a 700. Chlorophyll a 700 or P700 is the reaction centre of PS I. PS II is found in thylakoid membrane and it contains pigments as chlorophyll b 650, chlorophyll a 660, chlorophyll a 670, chlorophyll a 678, chlorophyll a 680 – 690 and phycobillins.

$P_{680\text{-}690}$ is the reaction centre of PS II. Chlorophyll a content is more in PS I than PS II. Carotenoids are present both in PS II and PS I. PS I is associated with both cyclic and non-cyclic photophosphorylation, but PS II is associated with only non-cyclic photophosphorylation.

Both the pigment systems are believed to be inter-connected by a third integral protein complex called cytochrome b – f complex. The other intermediate components of electron transport chain viz., PQ (plasto quinone) and PC (plastocyanin) act as mobile electron carriers between two pigment systems. PS I is active in both red and far red light and PS II is inactive in far red light (Fig. 6.7).

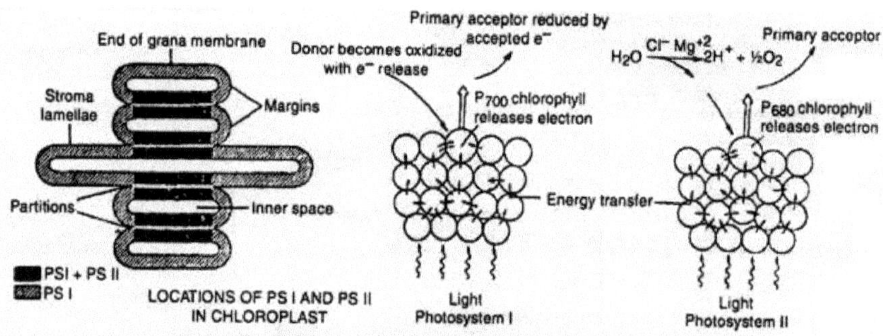

Fig. 6.7. Harvesting of light by two photosytems– PS-I & PS-II

2.13 EVIDENCE IN SUPPORT OF TWO PHASES OF PHOTOSYNTHESIS

2.13.1 Physical Separation of Chloroplast into Granna and Stroma Fraction

It is now possible to separate granna and stroma fraction of chloroplast. If light is given to granna fraction in the presence of suitable hydrogen acceptor and in complete absence of carbon dioxide then assimilatory power, ATP and NADPH2, are produced. If these assimilatory powers are given to stroma fraction in the presence of carbon dioxide and absence of light then carbohydrate is synthesized.

2.13.2 Temperature Coefficient (Q_{10}) Q_{10}

is the ratio of the rate of reaction at a given temperature and a temperature 10°C lower. Q_{10} value of photosynthesis is found to be two or three (for dark reaction) when photosynthesis is fast, but Q_{10} is one (for light reaction) when photosynthesis is slow.

2.13. 3 Evidence from Intermittent Light

Warburg observed that when intermittent light (flashes of light) of about 1/16 seconds were given to green algae (Chlorella vulgaris and Scenedesmus obliquus), the photosynthetic yield per second was higher as compared to the continuous supply of same intensity of light. This confirms that one phase of photosynthesis is independent of light.

2.13.4 Evidence from Carbon dioxide in Dark

It comes from tracer technique by the use of heavy carbon in carbon dioxide (C14O2). The leaves which were first exposed to light have been found to reduce carbon dioxide in the dark It indicates that carbon dioxide is reduced to carbohydrate in dark and it is purely a biochemical phase.

2.14 LIGHT REACTION (PHOTOCHEMICAL PHASE)

Light Reaction

Light reaction or photochemical reaction takes place in thylakoid membrane or granum and it is completely dependent upon the light. The raw materials for these reactions are pigments, water and sunlight. It can be discussed in the following three steps:

- Excitation of chlorophyll
- Photolysis of water
- Photophosphorylation

2.14.1 Excitation of Chlorophyll

It is the first step of light reaction. When P680 or P700 (special type of chlorophyll a) of two pigment systems receives quantum of light then it becomes excited and releases electrons.

$$\text{Chlorophyll} \xrightarrow{\text{Light}} \text{Chl}^{+} \text{ (oxidised)} + e^{-} \text{ (electron)}$$

2.14.2 Photolysis of Water and Oxygen Evolution (Hill Reaction)

Before 1930 it was thought that the oxygen released during photosynthesis comes from carbon dioxide. But for the first time Van Neil discovered that the source of oxygen evolution is not carbon dioxide but H_2O. In his experiment Neil used green sulphur bacteria which do not release oxygen during photosynthesis. They release sulphur. These bacteria require H_2S in place of H_2O.

The idea of Van Neil was supported by R. Hill. Hill observed that the chloroplasts extracted from leaves of Stellaria media and Lamium album when suspended in a test tube containing suitable electron acceptors (Potassium feroxalate or Potassium fericyanide), Oxygen evolution took place due to photochemical splitting of water.

The splitting of water during photosynthesis is called Photolysis of water. Mn, Ca, and CI ions play prominent role in the photolysis of water. This reaction is also known as Hill reaction. To release one molecule of oxygen, two molecules of water are required.

$$4H_2O \xrightarrow{\text{Light}} 4H^{+} + 4(OH^{-})$$

$$4(OH) \longrightarrow 4(OH) + 4e^{-}$$

$$4(OH) \longrightarrow 2H_2O + O_2 \uparrow$$

$$2H_2O \longrightarrow 4H^{+} + 4e^{-} + O_2 \uparrow$$

The evolution of oxygen from water was also confirmed by Ruben, Randall, Hassid and Kamen (1941) using heavy isotope (O18) in green alga Chlorella. When the photosynthesis is allowed to proceed with H2O18 and normal CO2, the evolved oxygen contains heavy isotope. If photosynthesis is allowed to proceed in presence of CO218 and normal water then heavy oxygen is not evolved.

$$6CO_2 + 12H_2O^{18} \xrightarrow[\text{chlorophyll}]{\text{Sunlight}} C_6H_{12}O_6 + 6H_2O + 6O_2^{18}$$

$$6CO_2^{18} + 12H_2O \xrightarrow[\text{chlorophyll}]{\text{Sunlight}} C_6H_{12}O_6^{18} + 6H_2O^{18} + 6O_2$$

Thus, the fate of different molecules can be summarized as follows

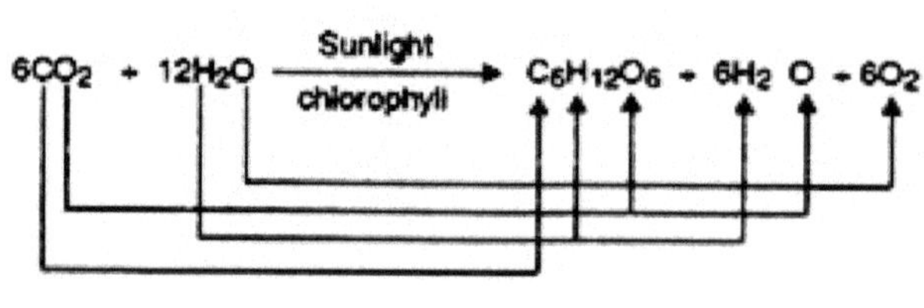

2.14.3 Photophosphorylation

Synthesis of ATP from ADP and inorganic phosphate (pi) in presence of light in chloroplast is known as photophosphorylation. It was discovered by Arnon et al (1954).

Photophosphorylation is of Two Types

- Cyclic photophosphorylation
- Non-cyclic photophosphorylation.

Cyclic Photophosphorylation (Fig.)

It is a process of photophosphorylation in which an electron expelled by the excited photo Centre (PSI) is returned to it after passing through a series of electron carriers. It occurs under conditions of low light intensity, wavelength longer than 680 nm and when C_O2 fixation is inhibited. Absence of C_O2 fixation results in non requirement of electrons as $NADPH_2$ is not being oxidized to NAD^{P+}. Cyclic photophosphorylation is performed by photosystem I only. Its photo Centre P700 extrudes an electron with a gain of 23 kcal/mole of energy after absorbing a photon of light (hv).

After losing the electron the photo Centre becomes oxidized. The expelled electron passes through a series of carriers including X (a special chlorophyll molecule), FeS, ferredoxin, plastoquinone, cytochrome b- f complex and plastocyanin before returning to photo Centre. While passing between ferredoxin and plastoquinone and/or over the cytochrome complex, the electron loses sufficient energy to form ATP from ADP and inorganic phosphate.

Halobacteria or halophile bacteria also perform photophosphorylation but ATP thus produced is not used in synthesis of food. These bacteria possess purple pigment bacteriorhodopsin attached to plasma membrane. As light falls on the pigment, it creates a proton pump which is used in ATP synthesis.

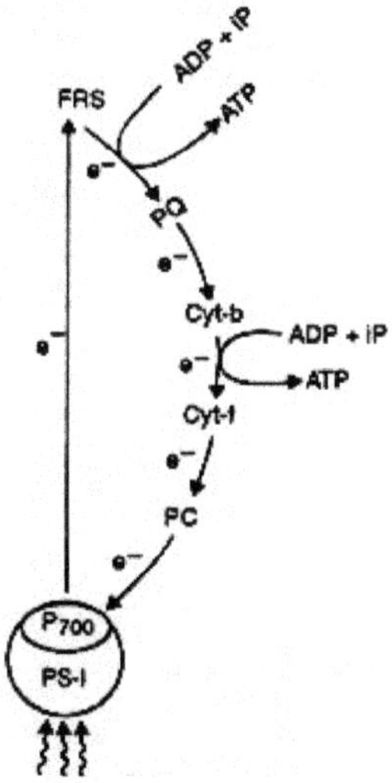

Noncyclic Photophosphorylation (Z-Scheme) (Fig.)

It is the normal process of photophosphorylation in which the electron expelled by the excited photo Centre (reaction centre) does not return to it. Non-cyclic photophosphorylation is carried out in collaboration of both photo system I and II. (Fig. 6.9). Electron released during photolysis of water is picked up by reaction centre of PS-II, called P680. The same is extruded out when the reaction centre absorbs light energy (hv). The extruded electron has an energy equivalent to 23 kcal/mole.

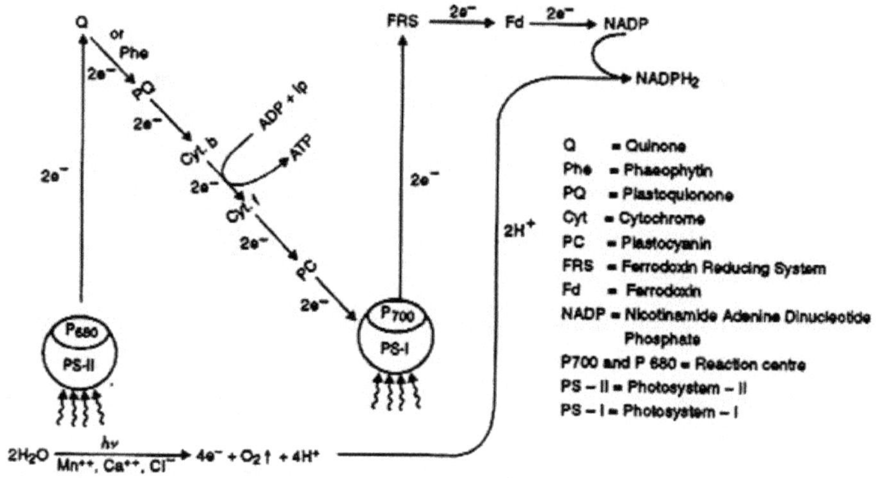

It passes through a series of electron carriers— Phaeophytin, PQ, cytochrome b- f complex and plastocyanin. While passing over cytochrome complex, the electron loses sufficient energy for the synthesis of ATP. The electron is handed over to reaction centre P700 of PS-I by plastocyanin. P700 extrudes the electron after absorbing light energy.

The extruded electron passes through FRS ferredoxin, and NADP -reductase which combines it with NADP+ for becoming reduced through H+ releasing during photolysis to form NADPH2. ATP synthesis is not direct. The energy released by electron is actually used for pumping H+ ions across the thylakoid membrane. It creates a proton gradient. This gradient triggers the coupling factor to synthesize ATP from ADP and inorganic phosphate (Pi).

2.15 CHEMIOSMOTIC HYPOTHESIS

How actually ATP is synthesized in the chloroplast?

The chemiosmotic hypothesis has been put forward by Peter Mitchell (1961) to explain the mechanism. Like in respiration, in photosynthesis too, ATP synthesis is linked to development of a proton gradient across a membrane. This time these are membranes of the thylakoid. There is one difference though, here the proton accumulation is towards the inside of the membrane, i.e., in the lumen. In respiration, protons accumulate in the inter-membrane space of the mitochondria when electrons move through the ETS.

Let us understand what causes the proton gradient across the membrane. We need to consider again the processes that take place during the activation of electrons and their transport to determine the steps that cause a proton gradient to develop (Figure 9).

- Since splitting of the water molecule takes place on the inner side of the membrane, the protons or hydrogen ions that are produced by the splitting of water accumulate within the lumen of the thylakoids.
- As electrons move through the photosystems, protons are transported across the membrane. This happens because the primary accepter of electron which is located towards the outer side of the membrane transfers its electron not to an electron carrier but to an H carrier. Hence, this molecule removes a proton from the stroma while transporting an electron. When this molecule passes on its electron to the electron carrier on the inner side of the membrane, the proton is released into the inner side or the lumen side of the membrane.
- The NADP reductase enzyme is located on the stroma side of the membrane. Along with electrons that come from the acceptor of electrons of PS I, protons are necessary for the reduction of $NADP^+$ to NADPH+ H^+ .These protons are also removed from the stroma. Hence, within the chloroplast, protons in the stroma decrease in number, while in the lumen there is accumulation of protons. This creates a proton gradient across the thylakoid membrane as well as a measurable decrease in pH in the lumen.

Why are we so interested in the proton gradient?

This gradient is important because it is the breakdown of this gradient that leads to release of energy. The gradient is broken down due to the movement of protons across the membrane to the stroma through the trans membrane channel of the F0 of the ATPase. The ATPase enzyme consists of two parts: one called the F0 is embedded in the membrane and forms a trans-membrane channel that carries out facilitated diffusion of protons across the membrane. The other portion is called F1 and protrudes on the outer surface of the thylakoid membrane on the side that faces the stroma.

The break down of the gradient provides enough energy to cause a conformational change in the F1 particle of the ATPase, which makes the enzyme synthesis several molecules of energy-packed ATP. Chemiosmosis requires a membrane, a proton pump, a proton gradient and ATPase. Energy is used to pump protons across a membrane, to create a gradient or a high concentration of protons within the thylakoid lumen.

ATPase has a channel that allows diffusion of protons back across the membrane; this releases enough energy to activate ATPase enzyme that catalyzes the formation of ATP. Along with the NADPH produced by the movement of electrons, the ATP will be used immediately in the biosynthetic reaction taking place in the stroma, responsible for fixing $CO_{2,}$ and synthesis of sugars.

Where are the ATP and NADPH Used?

We have seen that the products of light reaction are ATP, NADPH and O_2. Of these O2 diffuses out of the chloroplast while ATP and NADPH are used to drive the processes leading to the synthesis of food, more accurately, sugars. This is the biosynthetic phase of photosynthesis.

This process does not directly depend on the presence of light but is dependent on the products of the light reaction, i.e., ATP and NADPH, besides CO_2 and H_2O. You may wonder how this could be verified; it is simple: immediately after light becomes unavailable the biosynthetic process continues for some time, and then stops. If then, light is made available, the synthesis starts again.

Can we hence say that calling the biosynthetic phase as the dark reaction is a misnomer?

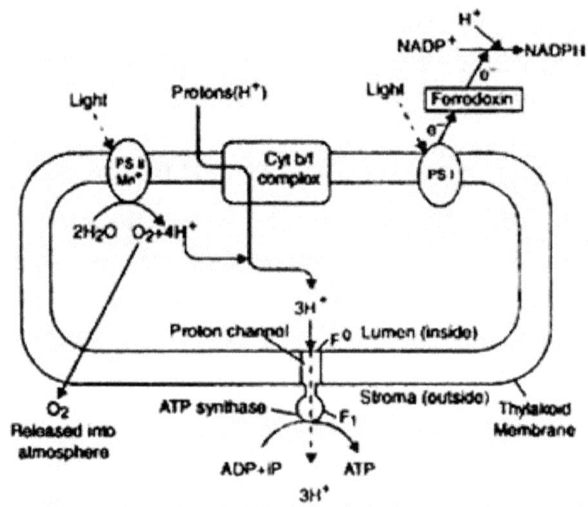

Dark Reaction (Biosynthetic Phase)-The Second Phase of Photosynthesis

- The pathway by which all photosynthetic eukaryotic organisms ultimately incorporate CO_2 into carbohydrate is known as carbon fixation or photosynthetic carbon reduction (PCR.) cycle or dark reactions. The dark reactions are sensitive to temperature changes, but are independent of light hence it is called dark reaction, however it depends upon the products of light reaction of photosynthesis, i.e., $NADPH_2$ and ATP.
- The carbon dioxide fixation takes place in the stroma of chloroplasts because it has enzymes essential for fixation of CO_2 and synthesis of sugar. Dark reaction is the pathway by which CO_2 is reduced to sugar. Since CO_2 is an energy poor compound; its conversion to an energy-rich carbohydrate involves a sizable jump up the energy ladder. This is accomplished through a series of complex steps involving small bits of energy.
- The CO_2 assimilation takes place both in light and darkness when the substrates $NADPH_2$ and ATP are available. Because of the need for $NADPH_2$ as a reductant and ATP as energy equivalent, CO_2 fixation is closely linked to the light reactions. During evolution three different ecological variants have evolved with different CO_2 incorporation mechanism: C_3, C_4 and CAM plants.

2.16 CALVIN OR C3 CYCLE OR PCR (PHOTOSYNTHETIC CARBON REDUCTION CYCLE)

It is the basic mechanism by which CO_2 is fixed (reduced) to form carbohydrates. It was proposed by Melvin Calvin. Calvin along with A.A. Benson, J. Bassham used radioactive isotope of carbon (C^{14}) in Chlorella pyrenoidosa and Scenedesmus oblique's to determine the sequences of dark reaction. For this work Calvin was awarded Nobel prize in 1961. To synthesize one glucose molecule Calvin cycle requires $6CO_2$, 18 ATP and 12 $NADPH_2$.

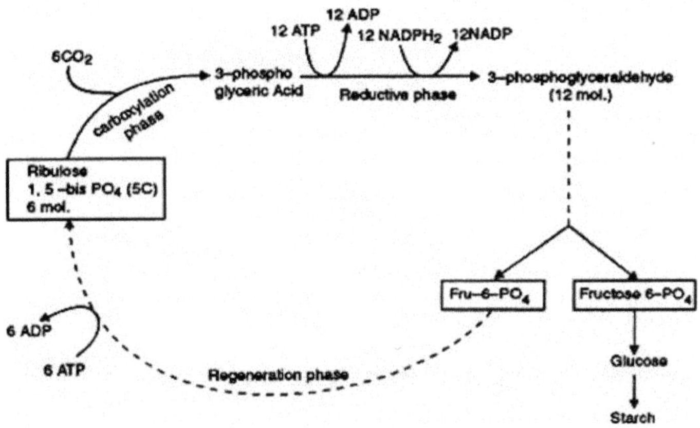

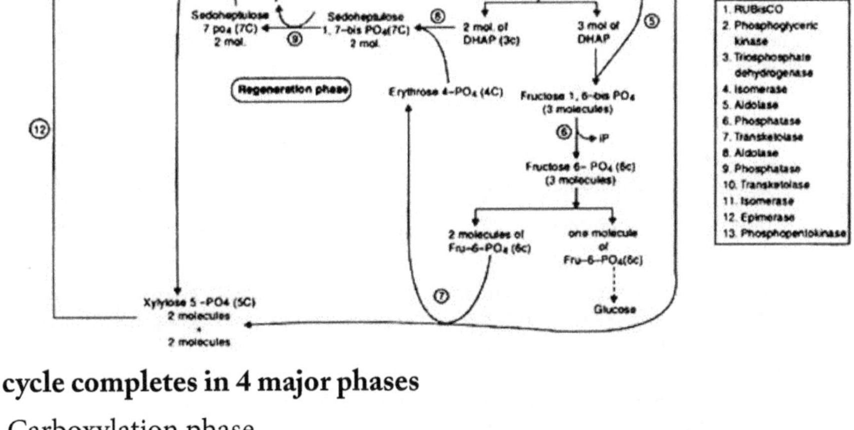

Calvin cycle completes in 4 major phases

- Carboxylation phase
- Reductive phase
- Glycolytic reversal phase (sugar formation phase)

- Regeneration phase

2.16.1 Carboxylation Phase

CO_2 enters the leaf through stomata. In mesophyll cells, CO_2 combines with a phosphorylated 5-carbon sugar, called Ribulose bisphosphate (or RuBP). This reaction is catalyzed by an enzyme, called RUBISCO. The reaction results in the formation of a temporary 6 carbon compound (2-carboxy 3-keto 1,5-biphosphorbitol) Which breaks down into two molecules of 3-phosphoglyceric acid (PGA) and it is the first stable product of dark reaction (C_3 Cycle).

$$\underset{(6\ \text{mols})}{\text{Ribulose-1, 5- biphosphate}} + 6CO_2 + 6H_2O \xrightarrow[Mg^{++}]{\text{RUBISCO}} \underset{(12\ \text{mols})}{\text{3-phosphoglyceric acid (3C)}}$$

2.16.2 Reductive Phase

The PGA molecules are now phosphorylated by ATP molecule and reduced by NADPH2 (product of light reaction known as assimilatory power) to form 3-phospho-glyceraldehyde (PGAL).

$$\underset{(12\ \text{mols})}{\text{3-Phosphoglyceric acid}} + 12\ ATP \xrightarrow{\text{Phosphoglyceric kinase}} \underset{(12\ \text{mols})}{\text{1, 3-diphosphoglyceric acid}} + 12\ ADP$$

$$\underset{(12\ \text{mols})}{\text{1, 3-diphosphoglyceric acid}} + 12\ NADPH_2 \xrightarrow[\text{isomerase}]{\text{Triose phosphate dehydrogenase}} \underset{(12\ \text{mols})}{\text{Glyceraldehyde-3-phosphate}} + 12\ NADP^+ + 12H_3PO_4$$

2.16.3 Glycolytic Reversal (Formation of Sugar) Phase

Out of two mols of 3-phosphoglyceraldehyde one mol is converted to its isomer 3-dihydroxyacetone phosphate.

$$\underset{(1\ \text{mol})}{\text{3-phosphoglyceraldehyde}} \xrightarrow[\text{isomerase}]{\text{Phosphotriose}} \underset{(1\ \text{mol})}{\text{3-dihydroxyacetone phosphate}}$$

One mol of 3-dihydroxyacetone phosphate reacts with 1 mol of 3-phosphoglycraldehyde to form one molecule of fructose-1, 6-biphosphate.

$$\underset{(1\ \text{mol})}{\text{3-Phosphoglyceraldehyde}} + \underset{(1\ \text{mol})}{\text{Dihydroxyacetone phosphate}} \xrightarrow{\text{Aldolase}} \underset{(1\ \text{mol})}{\text{Fructose-1, 6-biphosphate (6C)}}$$

One mol of fructose-6-phosphate and one mol of phosphoric acid is released from one mol of fructose-1, 6-biphosphate with the help of the enzyme phosphatase with utilization of one mol of H_2O.

$$\underset{(1\ \text{mol})}{\text{Fructose-1, 6-biphosphate}} \xrightarrow[H_2O]{\text{Phosphatase}} \underset{(1\ \text{mol})}{\text{Fructose-6-phosphate}} + H_3PO_4$$

Fructose-6-phosphate is converted to fructose-1-phosphate which forms glucose-1-phosphate.

$$\underset{(1\ \text{mol})}{\text{Fructose-6-phosphate}} \longrightarrow \underset{(1\ \text{mol})}{\text{Fructose-1-phosphate}} \longrightarrow \underset{(1\ \text{mol})}{\text{Glucose-1-phosphate}}$$

Fructose-6-phosphate and glucose-1-phosphate condense to form sucrose.

$$\underset{(1\ \text{mol})}{\text{Fructose-6-phosphate}} + \underset{(1\ \text{mol})}{\text{Glucose-1-phosphate}} \xrightarrow{\text{Phosphatase}} \underset{(1\ \text{mol})}{\text{Sucrose}}$$

From sucrose are produced other carbohydrate, *e.g.*, starch etc.

2.16.4 Regeneration Phase

Regeneration of Ribulose-5-phosphate (Also known as Reductive Pentose Phosphate Pathway) takes place through number of biochemical steps.

(*i*) 3-phosphoglyceraldehyde (4 mols) $\xrightarrow[\text{isomerase}]{\text{Triose Phosphatase}}$ Dihydroxyacetone phosphate (4 mols)

(*ii*) Dihydroxyacetone phosphate (2 mols) +3-phosphoglyceraldehyde (2 mols) $\xrightarrow{\text{Aldolase}}$ Fructose-1, 6-diphosphate (2 mols)

(*iii*) Fructose-1, 6-diphosphate (2 mols) $+2H_2O$ $\xrightarrow{\text{Phosphatase}}$ Fructose-6-phosphate (2 mols) $+ 2H_3PO_4$

(*iv*) Fructose-6-phosphate (2 mols) +3-phosphoglyceraldehyde (2 mols) $\xrightarrow{\text{Transketolase}}$ Xylulose-5-phosphate (2 mols) + Erythrose-4-phosphate (2 mols)

(*v*) Erythrose-4-phosphate (2 mols) + Dihydroxyacetone phosphate (2 mols) $\xrightarrow{\text{Transaldolase}}$ Sedoheptulose-1, 7-diphosphate (7C) (2 mols)

(*vi*) Sedoheptulose-1,7-diphosphate (2 mols) $+ H_2O$ $\xrightarrow{\text{Phosphatase}}$ Sedoheptulose-7-phosphate (2 mols) $+ 2H_3PO_4$

(*vii*) Sedoheptulose-7-phosphate (2 mols) + 3-phosphoglyceraldehyde (2 mols) $\xrightarrow{\text{Transketolase}}$ Ribose-5-phosphate (2 mols) + Xylulose-5-phosphate (2 mols)

(*viii*) Ribose-5-phosphate (2 mols) $\xrightarrow{\text{Phosphopentose isomerase}}$ Ribulose-5-phosphate (2 mols)

(*ix*) Xylulose-5-phosphate (4 mols) $\xrightarrow{\text{Phosphopentose epimerase}}$ Ribulose-5-phosphate (4 mols)

(*x*) Ribulose-5-phosphate (6 mols) + 6ATP $\xrightarrow{\text{Phosphopentokinase}}$ Ribulose-1, 5-biphosphate (6 mols) + 6ADP

2.17 SUMMARY OF PHOTOSYNTHESIS

- Light Reaction takes place in thylakoid membrane or granum

$$24H_2O \longrightarrow 24OH^- + 24H^+$$
$$18ADP + 18P_i \longrightarrow 18ATP$$
$$12NADP^+ + 24H^+ \longrightarrow 12NADPH_2$$
$$24OH^- - 24e \longrightarrow 12H_2O + 6O_2$$
$$24OH \longrightarrow 12H_2O + 6O_2$$

- Dark Reaction (C_3 cycle) takes place in stroma of chloroplast.

$$6CO_2 + 18ATP + 12NADPH_2 \xrightarrow{\text{Enzymes}} C_6H_{12}O_6 + 6H_2O + 18ADP + 18iP + 12NADP$$

$$\text{Addition of A \& B}: 6CO_2 + 24H_2O \xrightarrow[\text{Chlorophyll}]{\text{Light}} C_6H_{12}O_6 + 18H_2O + 6O_2$$

or

$$6CO_2 + 18ATP + 12NADPH_2 \xrightarrow{\text{Enzymes}} C_6H_{12}O_6 + 6H_2O + 18ADP + 18iP + 12NADP$$

$$\text{Addition of A \& B}: 6CO_2 + 12H_2O \xrightarrow[\text{Chlorophyll}]{\text{Light}} C_6H_{12}O_6 + 6H_2O + 6O_2$$

2.18 C_4 CYCLE (HSK PATHWAY OR HATCH SLACK AND KORTSCHAsK CYCLE)

- C_4 cycle may also be referred as the di-carboxylic acid cycle or the β-carboxylation pathway or Hatch and Slack cycle or cooperative photosynthesis. For a long time,

C_3 cycle was considered to be the only photosynthetic pathway for reduction of CO_2 into carbohydrates. Kortschak, Hartt and Burr (1965) reported that rapidly photosynthesizing sugarcane leaves produced a 4^{-C} compound like aspartic acid and malic acid as a result of CO_2 – fixation.

- It was later supported by M. D. Hatch and C. R. Slack (1966) and they reported that a 4^{-C} compound oxaloacetic acid (OAA) is the first stable product in CO_2 reduction process. This pathway was first reported in members of family Poaceae like sugarcane, maize, sorghum, etc. (tropical grasses), but later on the other subtropical plant like Atriplex spongiosa (Salt bush), Dititaria samguinolis, Cyperus rotundus, Amaranthus etc. So, the cycle has been reported not only in the members of Graminae but also among certain members of Cyperaceae and certain dicots.

2.19 STRUCTURAL PECULIARITIES OF C_4 PLANTS (KRANZ ANATOMY)

C_4 plants have a characteristic leaf anatomy called Kranz anatomy (Wreath anatomy – German meaning ring or Helo anatomy). The vascular bundles in C_4 plant leaves are surrounded by a layer of bundle sheath cells that contain large number of chloroplast. Dimorphic (two morphologically distinct type) chloroplasts occur in C_4 plants (Fig.).

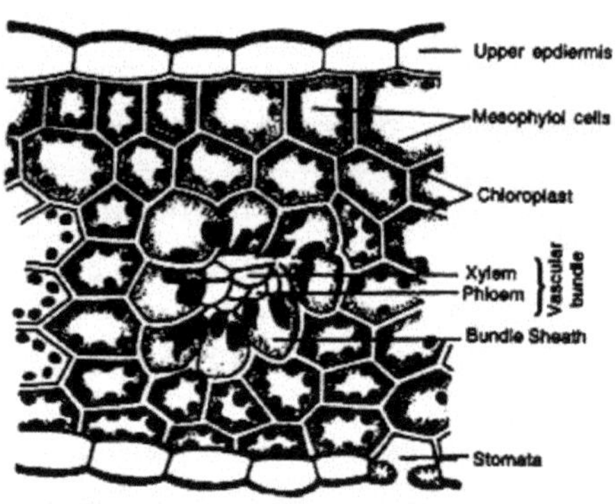

Fig. 6.13 T.S. of monocot plant leaf showing Kranz type of anatomy

In Mesopyll cell

- Chloroplast is small in size
- Well-developed grannum and less developed stroma.
- Both PS-II and PS-I are present.
- Non cyclic photophosphorylation takes place.
- ATP and $NADPH_2$ produces.
- Stroma carries PEPCO but absence of RuBisCO.
- CO_2 acceptor is PEPA (3C) but absence of RUBP
- First stable product OAA (4C) produces.

In Bundle sheath Cell

- Size of chloroplast is large
- Stroma is more developed but granna is poorly developed.
- Only PS-I present but absence of PS-II
- Non Cyclic photophosphorylation does not takes place.
- Stroma carries RuBisCO but absence of PEPCO.
- CO_2 acceptor RUBP (5c) is present but absence of PEPA (3C)
- C^{3-}cycle takes place and glucose synthesies.
- To carry out C^{3-}cycle both ATP and $NADPH_2$ comes from mesophyll cell chloroplast.

Carbon dioxide from atmosphere is accepted by Phosphoenol pyruvic acid (PEPA) present in stroma of mesophyll cell chloroplast and it converts to oxaloacetic acid (OAA) in the presence of enzyme PEPCO (Phosphoenolpyruvate carboxylase). This 4-C acid (OAA) enters into the chloroplast of bundle sheath cell and there it undergoes oxidative decarboxylation yielding pyruvic acid (3C) and CO_2.

The carbon dioxide released in bundle sheath cell reacts with RuBP (Ribulose 1, 5 bisphosphate) in presence of RUBISCO and carry out Calvin cycle to synthesize glucose. Pyruvic acid enters mesophyll cells and regenerates PEPA. In C_4 cycle two carboxylation reactions take place.

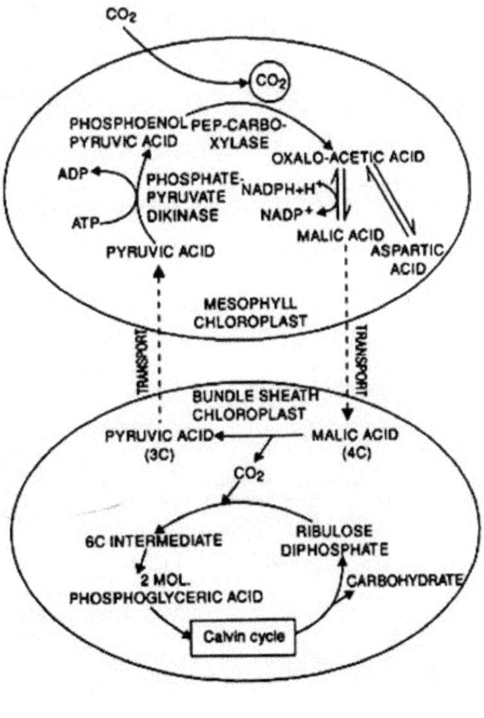

Reactions taking place in mesophyll cells are stated below: (1st carboxylation)

Step–I

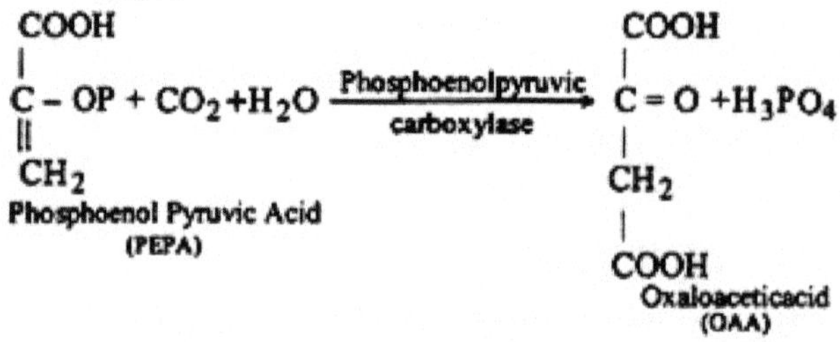

Step-II

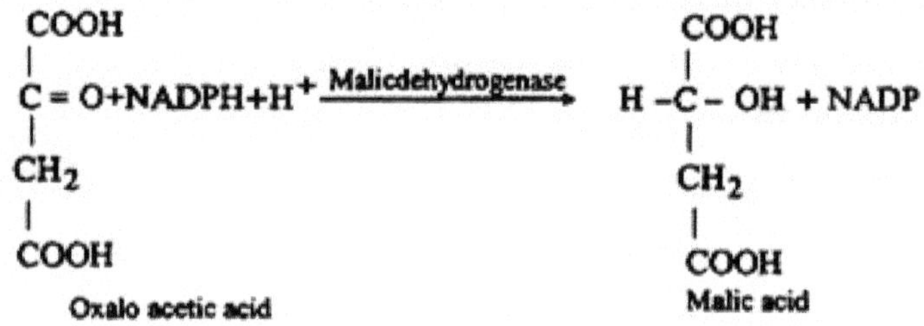

Step-III

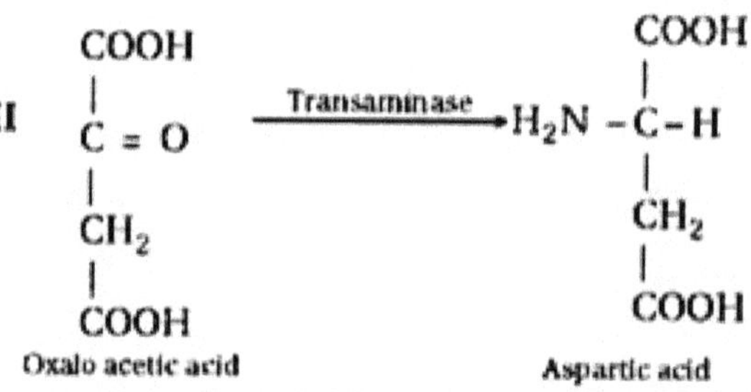

Reactions taking place in bundle sheath cells : (2nd carboxylation)

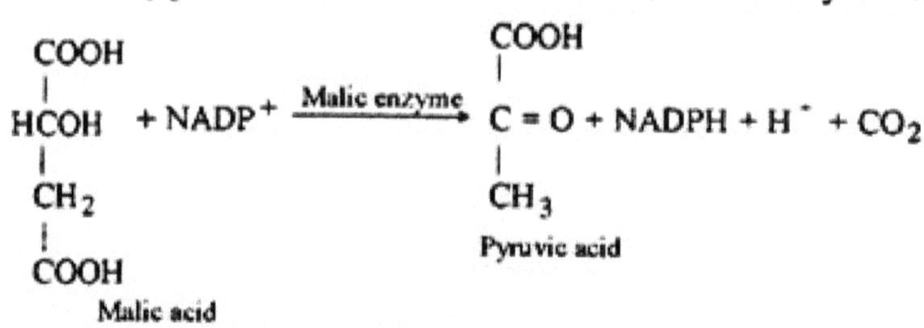

Reaction taking place in mesophyll cells :

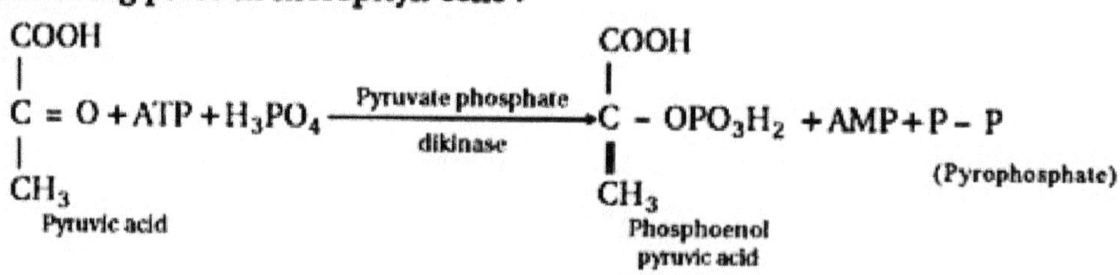

C_4 plants are better photosynthesizes. There is no photorespiration in these plants. To synthesize one glucose molecule, it requires 30 ATP and 12 $NADPH_2$.

Significance of C_4 Cycle

- C_4 plants have greater rate of carbon dioxide assimilation than C_3 plants because PEPCO has great affinity for CO_2 and it shows no photorespiration resulting in higher production of dry matter.
- C_4 plants are better adapted to environmental stress than C3 plants.
- Carbon dioxide fixation by C_4 plants requires more ATP than C_3 plants for conversion of pyruvic acid to PEPA.
- Carbon dioxide acceptor in C_4 plant is PEPA and key enzyme is PEPCO.
- They can very well grow in saline soils because of presence of C4 organic acid.

2.20 CRASSULACEAN ACID METABOLISM (CAM PATHWAY)

It is a mechanism of photosynthesis which occurs in succulents and some other plants of dry habitats where the stomata remain closed during the daytime and open only at night. The process of photosynthesis is similar to that of C_4 plants but instead of spatial separation of initial PEPcase fixation and final Rubisco fixation of CO_2, the two steps occur in the same cells (in the stroma of mesophyll chloroplasts) but at different times, night and day, e.g., Sedum, Kalanchoe, Opuntia, Pineapple (Fig. 13). (CAM was for the first time studied and reported by Ting (1971).

2.20.1 Characteristics of CAM Plants

- Stomatal movement is scoto-active.
- Presence of monomorphic chloroplast.
- Stroma of chloroplast carries both PEPCO and RUBISCO.
- Absence of Kranz anatomy.
- It is more similar to C4 plants than C3 plants.
- In these plants pH decreases during night and increases during day time.

2.20.2 Mechanism of Cam Pathway

PHASE I. During night

Stomata of Crassulacean plants remain open at night. Carbon dioxide is absorbed from outside. With the help of Phosphoenol pyruvate carboxylase (PEPCO) enzyme the CO2 is immediately fixed, and here the acceptor molecule is Phosphoenol pyruvate (PEP).

$$\text{PEP} + HCO_3^- (CO_2 + H_2O) \xrightarrow{\text{PEPCO}} \text{Oxaloacteic acid (OAA)} + H_3PO_4$$

$$\text{Oxaloacetic acid} + \text{NADPH} \xrightarrow{\text{Dehydrogenase}} \text{Malic Acid} + \text{NADP}^+$$

Malic acid is the end product of dark fixation of CO_2. It is stored inside cell vacuole.

PHASE II:

During day time the stomata in Crassulacean plants remain closed to check transpiration, but photosynthesis does take place in the presence of sun light. Malic acid moves out of the cell vacuoles. It is de-carboxylated with the help of malic enzyme. Pyruvate is produced. It is metabolized.

$$\text{Malic Acid (Malate)} + NADP^+ \xrightarrow[\text{Enzyme}]{\text{Malic}} \text{Pyruvate} + CO_2 + NADPH_2$$

The CO_2 thus released is again fixed through Calvin Cycle with the help of RUBP and RUBISCO. This is a unique feature of these succulent plants where they photosynthesis without wasting much of water. They perform acidification or dark fixation of CO_2 during night and de-acidification during day time to release carbon dioxide for actual photosynthesis.

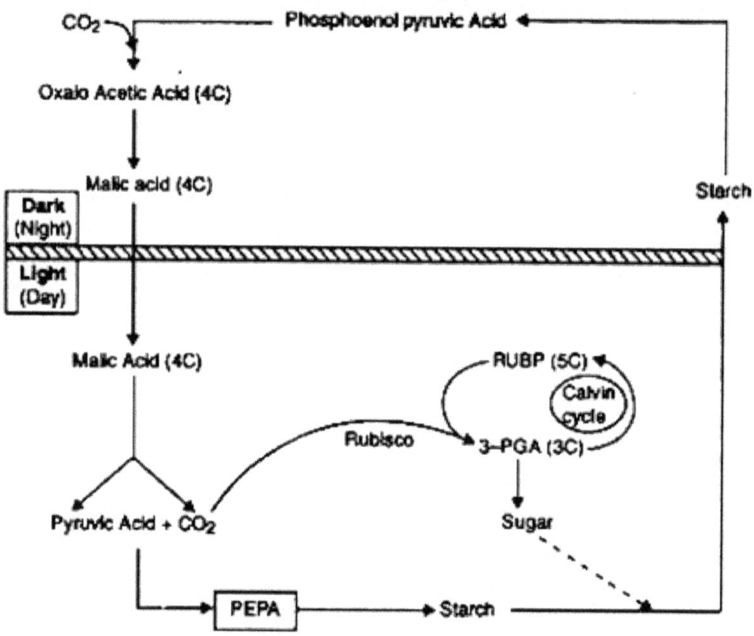

2.20.3 Ecological Significance of CAM Plants

These plants are ecologically significant because they can reduce rate of transpiration during day time, and are well adapted to dry and hot habitats.

- The stomata remain closed during the day and open at night when water loss is little due to prevailing low temperature.
- CAM plants have parenchyma cells, which are large and vacuolated. These vacuoles are used for storing malic and other acids in large amounts.

- CAM plants increase their water-use efficiency, and secondly through its enzyme PEP carboxylase, they are adapted to extreme hot climates.
- CAM plants can also obtain a CO_2 compensation point of zero at night and in this way accomplish a steeper gradient for CO_2 uptake compared to C_3 plants.
- They lack a real photosynthesis during daytime and the growth rate is far lower than in all other plants (with the exception of pineapple).

2.21 PHOTORESPIRATION OR C_2 CYCLE OR GLYCOLATE CYCLE OR PHOTOSYNTHETIC CARBON OXIDATION CYCLE

Photorespiration is the light dependent process of oxygenation of RUBP (Ribulose biphosphate) and release of carbon dioxide by photosynthetic organs of the plant. Otherwise, as we know, photosynthetic organs release oxygen and not CO_2 under normal situation.

Occurrence of photorespiration in a plant can be demonstrated by

- Decrease in the rate of net photosynthesis when oxygen concentration is increased from 2-3 to 21%.
- Sudden increased evolution of CO_2 when an illuminated green plant is transferred to dark.

Photorespiration is initiated under high O_2 and low CO_2 and intense light around the photosynthesizing plant. Photorespiration was discovered by Dicker and Tio (1959), while the term "Photorespiration" was coined by Krotkov (1963). Photorespiration should not be confused with photo- oxidation. While the former is a normal process in some green plants, the latter is an abnormal and injurious process occurring in extremely intense light resulting in destruction of cellular components, cells and tissues.

On the basis of photorespiration, plants can be divided into two groups

- Plants with photorespiration (temperate plants) and plants without photorespiration (tropical plants).

Site of Photorespiration

Photorespiration involves three cell organelles, viz., chloroplast, peroxisome and mitochondria for its completion. Peroxisome, the actual site of photorespiration, contains enzymes like glycolate oxydase, glutamate glyoxalate aminotransferase, peroxidase and catalase enzymes.

2.21.1 Mechanism of Photorespiration

- We Know That The Enzyme Rubisco (Ribulose Biphosphate Carboxylase oxygenase) catalyzes the carboxylation reaction, where CO_2 combines with RuBP for calvin cycle (dark reaction of photosynthesis) to initiate. But this enzyme RUBISCO, under intense light conditions, has the ability to catalyse the combination of O_2 with RuPB, a process called oxygenation.

- In other words the enzyme RUBISCO can catalyse both carboxylation as well as oxygenation reactions in green plants under different conditions of light and O_2/CO_2 ratio. Respiration that is initiated in chloroplasts under light conditions is called photorespiration. This occurs essentially because of the fact that the active site of the enzyme RUBISCO is the same for both carboxylation and oxygenation (Fig. 6.16).
- The oxygenation of RuBP in the presence of O_2 is the first reaction of photorespiration, which leads to the formation of one molecule of phosphoglycolate, a 2 carbon compound and one molecule of phosphoglyceric acid (PGA). While the PGA is used up in the Calvin cycle, the phosphoglycolate is dephosphorylated to form glycolate in the chloroplast (Fig. 16).

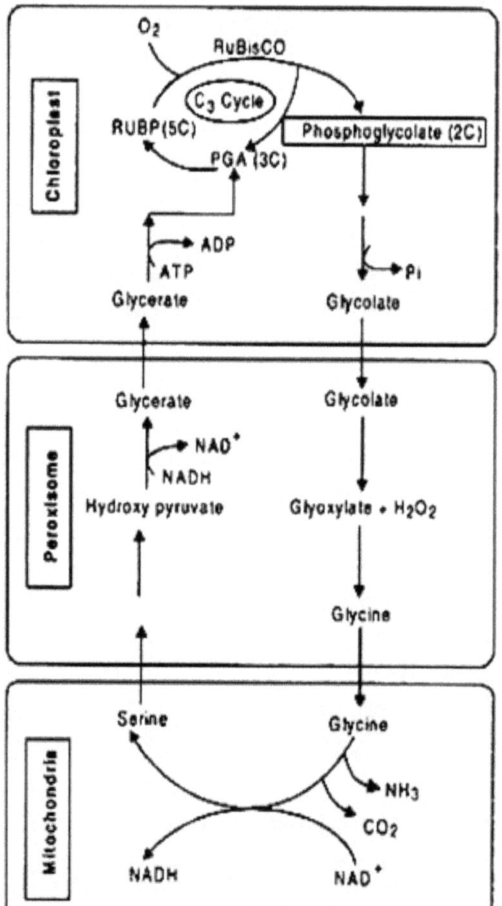

$$RuBP + O_2 \xrightarrow[\text{Oxygenase}]{\text{RuBP}} PGA + Phosphoglycolate$$

- From the chloroplast, the glycolate is diffused to peroxisome, where it is oxidised to glyoxylate. In the peroxisome, the glyoxylate is used to form the amino acid, glycine. Glycine enters mitochondria where two molecules of glycine (4 carbons) give rise to one molecule of serine (3 carbon) and one CO_2 (one carbon).
- The serine is taken up by the peroxisome, and through a series of reactions, is converted to glycerate. The glycerate leaves the peroxisome and enters the chloroplast, where it is phosphorylated to form PGA. The PGA molecule enters the calvin cycle to make carbohydrates, but one CO_2 molecule released in mitochondria during photorespiration has to be re-fixed.
- In other words, 75% of the carbon lost by oxygenation of RuBP is recovered, and 25% is lost as release of one molecule of CO_2. Because of the features described above, photorespiration is also called photosynthetic carbon oxidation cycle.

$$\text{Phosphoglycolate} + H_2O \xrightarrow{\text{Phosphatase}} \text{Glycolate} + H_3PO_4$$
$$\text{Glycolate} + O_2 \xrightarrow{\text{Oxidase}} \text{Glyoxylate} + H_2O_2$$
$$2H_2O_2 \xrightarrow{\text{Catalase}} 2H_2O + O_2$$
$$\text{Glyoxylate} + \text{Glutamate} \xrightarrow{\text{Transaminase}} \text{Glycine} + \alpha\text{-ketoglutarate}$$
$$2\ \text{Glycine} + NAD^+ \xrightarrow{\text{Decarboxylase}} 1\ \text{serine} + CO_2 + NH_3 + NADH$$

$$\text{Serine} + \alpha\text{-ketoglutarate}^* \xrightarrow{\text{Transaminase}} \text{Hydroxypyruvate} + \text{Glutamate}$$
$$\text{Hydroxypyruvate} + NADH \xrightarrow{\text{Reductase}} \text{Glycerate} + NAD^+$$
$$\text{Glycerate} + ATP \xrightarrow{\text{Kinase}} \text{3-phosphoglycerate} + ADP$$

2.21.2 Minimization of Photorespiration (C4 and CAM Plants)

- Since photorespiration requires additional energy from the light reactions of photosynthesis, some plants have mechanisms to reduce uptake of molecular oxygen by Rubisco. They increase the concentration of CO_2 in the leaves so that Rubisco is less likely to produce glycolate through reaction with O_2.
- C_4 plants capture carbon dioxide in cells of their mesophyll (using an enzyme called PEP carboxylase), and they release it to the bundle sheath cells (site of carbon dioxide fixation by Rubisco) where oxygen concentration is low.
- The enzyme PEP carboxylase is also found in other plants such as cacti and succulents who use a mechanism called Crassulacean acid metabolism or CAM in which PEP carboxylase put aside carbon at night and releases it to the photosynthesizing cells during the day.
- This provides a mechanism for reducing high rates of water loss (transpiration) by stomata during the day. This ability to avoid photorespiration makes these plants more hardy than other plants in dry conditions where stomata are closed and oxygen concentration rises.

2.22 FACTORS AFFECTING PHOTOSYNTHESIS

Photosynthesis is affected by both environmental and genetic (internal) factors. The environmental factors are light, CO_2, temperature, soil, water, nutrients etc. Internal or genetic factors are all related with leaf and include protoplasmic factors, chlorophyll contents, structure of leaf, accumulation of end product etc. **Some of the important factors are discussed below:**

2.22.1 Concept of Cardinal Values

The metabolic processes are influenced by a number of factors of the environment. The rate of a metabolic process is controlled by the magnitude of each factor. Sachs (1860) recognized three critical values, the cardinal values or points of the magnitude of each factor. These are minimum, optimum and maximum. The minimum cardinal value is that magnitudes of a factor below which the metabolic process cannot proceed.

Optimum value is the one at which the metabolic process proceeds at its highest rate. Maximum is that magnitude of a factor beyond which the process stops. At magnitudes below and above the optimum, the rate of a metabolic process declines till minimum and maximum values are attained.

2.22.2. Principle of Limiting Factors

Liebig (1843) proposed law of minimum which states that the rate of a process is limited by the pace (rapidity) of the slowest factor. However, it was later on modified by Blackman (1905) who formulated the "principle of limiting factors". It states that when a metabolic process is conditioned as to its rapidity by a number of separate factors, the rate of the process is limited by the pace (rapidity) of the slowest factor. This principle is also known as "Blackman's Law of Limiting Factors."

A metabolic process is conditioned by a number of factors. The slowest factor or the limiting factor is the one whose increase in magnitude is directly responsible for an increase in the rate of the metabolic process (here photosynthesis).

To explain it further, say at a given time, only the factor that is most limiting among all will determine the rate of photosynthesis. For example, if CO_2 is available in plenty but light is limiting due to cloudy weather, the rate of photosynthesis under such a situation will be controlled by the light. Furthermore, if both CO_2 and light are limiting, then the factor which is the most limiting of the two, will control the rate of photosynthesis.

Blackman (1905) studied the effect of CO_2 concentration, light intensity and temperature on rate of photosynthesis. All other factors were maintained in optimum concentration. Initially the photosynthetic material was kept at 20°C in an environment having 0.01% CO_2. When no light was provided to photosynthetic material, it did not perform photosynthesis. Instead, it evolved CO_2 and absorbed O_2 from its environment. He provided light of low intensity (say 150 foot candles) and found photosynthesis to occur. When light intensity was increased (say 800 foot candles), the rate of photosynthesis increased initially but soon it leveled off. The rate of photosynthesis could be further enhanced only on the increase in availability of CO_2. Thus, initially light intensity was limiting the rate of photosynthesis.

When sufficient light became available, CO_2 became limiting factor (Fig. 6.17). When both are provided in sufficient quantity, the rate of photosynthesis rose initially but again reached a peak. It could not be increased further. At this time, it was found that increase in temperature could raise the rate of photosynthesis up to 35°C. Further increase was not possible. At this stage, some other factor became limiting. Therefore, at one time only one factor limits the rate of physiological process.

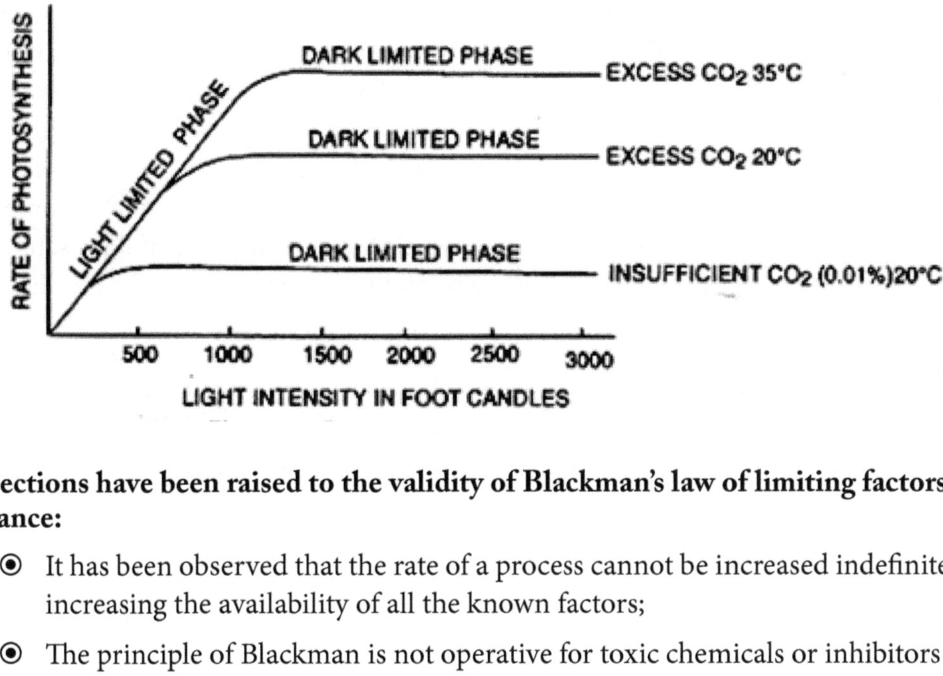

Objections have been raised to the validity of Blackman's law of limiting factors. For instance:

- It has been observed that the rate of a process cannot be increased indefinitely by increasing the availability of all the known factors;
- The principle of Blackman is not operative for toxic chemicals or inhibitors and
- Some workers have shown that the pace of reaction can be controlled simultaneously by two or more factors.

2.22.3 External Factors

The environmental factors which can affect the rate of photosynthesis are carbon dioxide, light, temperature, water, oxygen, minerals, pollutants and inhibitors.

1. Effect of Carbon dioxide

Being one of the raw materials, carbon dioxide concentration has great effect on the rate of photosynthesis. The atmosphere normally contains 0.03 to 0.04 per cent by volume of carbon dioxide. It has been experimentally proved that an increase in carbon dioxide content of the air up to about one per cent will produce a corresponding increase in photosynthesis provided the intensity of light is also increased.

2. Effect of Light

The ultimate source of light for photosynthesis in green plants is solar radiation, which moves in the form of electromagnetic waves. Out of the total solar energy reaching to the earth, about 2% is used in photosynthesis and about 10% is used in other metabolic activities. Light varies in intensity, quality (wavelength) and duration.

The effect of light on photosynthesis can be studied under following three headings:

(i) Intensity of Light

The total light perceived by a plant depends on its general form (viz., height of plant and size of leaves, etc.) and arrangement of leaves. Of the total light falling on a leaf, about 80% is absorbed, 10% is reflected and 10% is transmitted. Intensity of light can be measured by lux meter. Effect of light intensity varies from plant to plant, e.g., more in heliophytes (sun loving plants) and less in sciophytes (shade loving plants). For a complete plant, rate of photosynthesis increases with increase in light intensity, except under very high light intensity where phenomenon of Solarization' occurs, (i.e., photo-oxidation of different cellular components including chlorophyll). It also affects the opening and closing of stomata thereby affecting the gaseous exchange. The value of light saturation at which further increase is not accompanied by an increase in CO2 uptake is called light saturation point.

(ii) Quality of Light

Photosynthetic pigments absorb visible part of the radiation i.e., 380 mμ, to 760 mμ. For example, chlorophyll absorbs blue and red light. Usually plants show high rate of photosynthesis in the blue and red light. Maximum photosynthesis has been observed in red light than in blue light followed by yellow light (monochromatic light). The green light has minimum effect. The rate of photosynthesis is maximum in white light or sunlight (polychromatic light). On the other hand, red algae shows maximum photosynthesis in green light and brown algae in blue light.

(iii) Duration of Light

Longer duration of light period favours photosynthesis. Generally, if the plants get 10 to 12 hrs. of light per day it favours good photosynthesis. Plants can actively exhibit photosynthesis under continuous light without being damaged. Rate of photosynthesis is independent of duration of light.

3. Effect of Temperature

The rate of photosynthesis markedly increases with an increase in temperature provided other factors such as CO_2 and light are not limiting. The temperature affects the velocity of enzyme controlled reactions in the dark stage. When the intensity of light is low, the reaction is limited by the small quantities of reduced coenzymes available so that any increase in temperature has little effect on the overall rate of photosynthesis. At high light intensities, it is the enzyme-controlled dark stage which controls the rate of photosynthesis and there the Q10 = 2. If the temperature is greater than about 30°C, the rate of photosynthesis abruptly falls due to thermal inactivation of enzymes.

4. Effect of Water

Although the amount of water required during photosynthesis is hardly one percent of the total amount of water absorbed by the plant, yet any change in the amount of water absorbed by a plant has significant effect on its rate of photosynthesis. Under normal conditions water rarely seems to be a controlling factor as the chloroplasts normally contain plenty of water. Many experimental observations indicate that in the field the plant is able to withstand a wide range of soil moisture without any significant effect on photosynthesis and it is only when wilting sets in that the photosynthesis is retarded. Some of the effect of drought may be secondary since stomata tend to close when the plant is deprived of water. A more specific effect of drought on photosynthesis results from dehydration of protoplasm.

5. Effect of Oxygen

Excess of O2 may become inhibitory for the process. Enhanced supply of O_2 increases the rate of respiration simultaneously decreasing the rate of photosynthesis by the common intermediate substances. The concentration for oxygen in the atmosphere is about 21% by volume and it seldom fluctuates. O_2 is not a limiting factor of photosynthesis. An increase in oxygen concentration decreases photosynthesis and the phenomenon is called Warburg effect. [Reported by German scientist Warburg (1920) in Chlorella algae]. This is due to competitive inhibition of RuBP-carboxylase at increased O_2 levels, i.e., O_2 competes for active sites of RuBP-carboxylase enzyme with CO_2. The explanation of this problem lies in the phenomenon of photorespiration. If the amount of oxygen in the atmosphere decreases then photosynthesis will increase in C_3 cycle and no change in C_4 cycle.

6. Effect of Minerals

Presence of Mn++ and CI– is essential for smooth operation of light reactions (Photolysis of water/evolution of oxygen) Mg^{++}, Cu^{++} and Fe^{++} ions are important for synthesis of chlorophyll.

7. Effect of Pollutants and Inhibitors

The oxides of nitrogen and hydrocarbons present in smoke react to form peroxyacetyl nitrate (PAN) and ozone. PAN is known to inhibit Hill's reaction. Diquat and Paraquat (commonly called as Viologens) block the transfer of electrons between Q and PQ in PS II. Other inhibitors of photosynthesis are monouron or CMU (Chlorophenyl dimethyl urea), diuron or DCMU (Dichlorophenyl dimethyl urea), bromocil and atrazine etc., which have the same mechanism of action as that of violates. At low light intensities potassium cyanide appears to have no inhibiting effect on photosynthesis.

2.22 4 Internal Factors

The important internal factors that regulate the rate of photosynthesis are:

1. Protoplasmic factors

There is some unknown factor in protoplasm which affects the rate of photosynthesis. This factor affect the dark reactions. The decline in the rate of photosynthesis at temperature above 30°C or at strong light intensities in many plants suggests the enzyme nature of this unknown factor.

2. Chlorophyll content

Chlorophyll is an essential internal factor for photosynthesis. The amount of CO_2 fixed by a gram of chlorophyll in an hour is called photosynthetic number or assimilation number. It is usually constant for a plant species but rarely it varies. The assimilation number of variegated variety of a species was found to be higher than the green leaves variety.

3. Accumulation of end products

Accumulation of food in the chloroplasts reduces the rate of photosynthesis.

4. Structure of leaves

The amount of CO_2 that reaches the chloroplasts depends on structural features of the leaves like the size, position and behaviour of the stomata and the amount of intercellular spaces. Some other characters like thickness of cuticle, epidermis, presence of epidermal hairs, amount of mesophyll tissue, etc., influence the intensity and quality of light reaching the chloroplast.

5. CO_2 Compensation Point

It is that value or point in light intensity and atmospheric CO_2 concentration when the rate of photosynthesis is just equivalent to the rate of respiration in the photosynthetic organs so that there is no net gaseous exchange. The value of light compensation point is 2.5 -100 ft. candles for shade plants and 100-400 ft. candles for sun plants. The value of CO_2 compensation point is very low in C_4 plants (0-5 ppm), where as in C_3 plants it is quite high (25-100 ppm). A plant can not survive for long at compensation point because there is net lose of organic matter due to respiration of non-green organs and dark respiration.

3

ENERGY CONVERSION IN PHOTOSYNTHESIS

Solar energy has the capacity to fulfill global human energy demands in an environmentally and socially responsible manner, provided efficient, low-cost systems can be developed for its capture, conversion, and storage. Toward these ends, a molecular-based understanding of the fundamental principles and mechanistic details of energy conversion in photosynthesis is indispensable. This review addresses aspects of photosynthesis that may prove auspicious to emerging technologies. Conversely, areas in which human ingenuity may offer innovative solutions, resulting in enhanced energy storage efficiencies in artificial photosynthetic constructs, are considered. Emphasis is placed on photoelectrochemical systems that utilize water as a source of electrons for the production of solar fuels.

3.1 PERSPECTIVES AND OVERVIEW

During the winter months of 1855, Benjamin Silliman, Jr., professor of general and applied chemistry at Yale, conducted a series of pioneering experiments on the properties and uses of petroleum gathered from springs near Titusville, Pennsylvania. His detailed and optimistic report, completed April 16, 1855, led to the drilling of the first successful oil well by Edwin L. Drake at Titusville in 1859, and the founding of the modern petroleum industry. -The American Petroleum Institute, 1955

The drilling of the Drake well indeed marked the beginning of the international search for oil and in many ways changed the way we live. Currently, the energy economy of nearly all countries is based on the use of stored energy, primarily fossil fuels (the ancient products of photosynthesis) in the form of coal, oil, and natural gas, as well as nuclear energy in the form of uranium isotopes. Increases in world population and the rise of emerging economies are correlated with a projected increase in the rate of world-marketed energy consumption by 44% over the 2006 to 2030 period (from 15.8 TW in 2006 to 22.7 TW in 2030). In the near term, current global market downturns dampen world demand for energy as manufacturing and consumer demand for goods and services abate. With anticipated economic recovery after 2010, most nations will return to trend growth in energy demand. Although oil supplies will dwindle, coal is abundant! Likewise, estimated fossil energy resources are capable of supporting a 30-TW burn rate for several centuries. The more significant global concern is an alteration of Earth's atmospheric chemistry in the form of increasing anthropogenic CO2 levels stemming from the combustion of fossil fuels. Such issues dictate an imperative for transition to alternative carbon-neutral sources of energy.

Of the available carbon-neutral energy sources, solar is quite promising, as it is abundant (striking the Earth's surface at a rate of ~120,000 TW) but unfortunately diffuse (~100 mW/cm^2 =1.3 horsepower per square meter and diurnal. Thus, devising cost-effective methods for efficiently capturing and storing this energy for human use is among the grand challenges of science. Nature offers inspiration for designing devices capable of such processes and assurance that solar energy conversion can be performed on vast scales. Globally, photosynthesis stores solar energy in reduced carbon at a rate of ~120 TW. As structural and mechanistic details of this process are elucidated, it should be possible to extract the principles involved and incorporate them into synthetic constructs. We must understand the limitations of biological systems imposed by their evolutionary history; such inefficiencies provide opportunities for human ingenuity to develop innovations. In this review, we discuss fundamental principles and parameters of solar energy conversion while drawing on examples from natural and artificial photosynthetic systems. Maximum efficiencies and theoretical limitations of such constructs are addressed as are potential strategies for improving current efficiencies.

Of the available carbon-neutral energy sources, solar is quite promising, as it is abundant (striking the Earth's surface at a rate of ~120,000 TW) but unfortunately diffuse (~100 mW/cm^2 = 1.3 horsepower per square meter and diurnal. Thus, devising cost-effective methods for efficiently capturing and storing this energy for human use is among the grand challenges of science). Nature offers inspiration for designing devices capable of such processes and assurance that solar energy conversion can be performed on vast scales. Globally, photosynthesis stores solar energy in reduced carbon at a rate of ~120 TW. As structural and mechanistic details of this process are elucidated, it should be possible to extract the principles involved and incorporate them into synthetic constructs. We must understand the limitations of biological systems imposed by their evolutionary history; such inefficiencies provide opportunities for human ingenuity to develop innovations. In this review, we discuss fundamental principles and parameters of solar energy conversion while drawing on examples from natural and artificial photosynthetic systems. Maximum efficiencies and theoretical limitations of such constructs are addressed as are potential strategies for improving current efficiencies.

3.2 INSPIRATION FROM NATURE

Photosynthesis is the process by which a wide array of organisms, including archaea, bacteria, algae, and plants capture solar energy and ultimately convert it to chemical energy for storage purposes. In most cases, the process of photosynthesis involves five distinct phases:

- The initial absorption of photonic energy (light),
- The transfer of electronic excitation energy by antenna pigments (conduits) to a reaction center,
- A charge separation event at a reaction center that converts electronic excitation energy to chemical redox energy,
- Stabilization of the charge separation by secondary processes often involving migration of holes and electrons to spatially separate electron donors and acceptors, and
- Synthesis of stable chemical products that store a portion of the initially absorbed photonic energy

Although there are variations in the overall processes of the many photosynthetic organisms, the fundamental principles and physics involved are similar. In this section, we briefly discuss key components involved in the capture, conversion, and storage of solar energy. We then review the photosynthetic process as it occurs in bacterial and oxygenic photosynthesis.

3.3 CAPTURE AND CONVERSION

For light energy to be stored by photosynthetic organisms, it must first be captured and converted to a more readily usable form. The absorption of light, the transfer of its energy, the generation of charge-separated states, and subsequent further spatial separation of electrons and holes (charge stabilization) are fundamental steps of this overall process.

3.2.1 The Absorption of Light by Photoactive Pigments

The sun, as with all stars, is a nearly perfect black-body radiator incandescently emitting radiation that produces a characteristic broad spectrum of light. Figure 1 shows the solar irradiance spectra outside the Earth's atmosphere [Air Mass (AM)0] and at sea level (AM1.5). As solar radiation passes through the atmosphere, incident energy is reduced by various absorption and scattering processes. High-energy ultraviolet (UV) light is effectively screened out by the ozone layer. Consequently, wavelengths less than 400 nm contribute only ~8% to the total solar irradiance. The spectral region from 400 to 700 nm (visible light) makes up ~45% of the solar irradiance whereas wavelengths longer than 700 nm (infrared light) account for ~47%.

The electric field of an incoming photon of light can induce an oscillating electric dipole in one of the chromophores associated with the photosynthetic apparatus, inducing a transition to a higher electronic energy state. Promoting an electron from the highest occupied molecular orbital to the lowest unoccupied molecular orbital gives a simple illustrative description of an electronic configuration constituting the lower energy ground state (S) and the higher energy excited state (S*) of a chromophore following the absorption of light (hn):

$$S + h\nu \rightarrow S^*$$

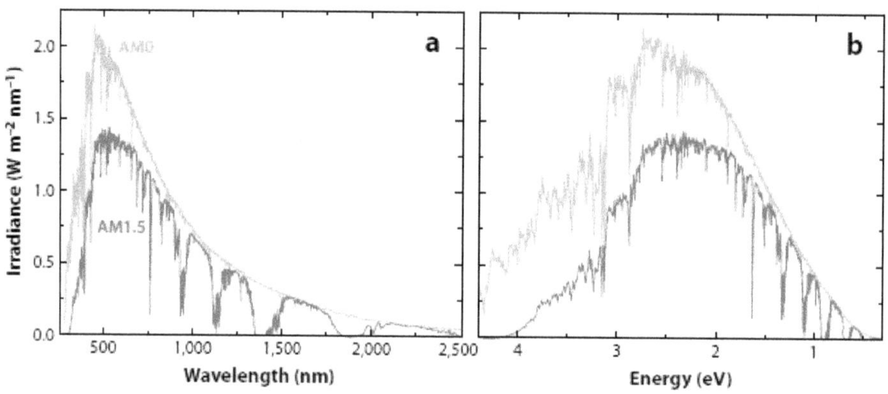

Solar irradiance spectra outside the Earth's atmosphere [denoted by AM0 (green line)] and at sea level [denoted by AM1.5 (blue line)] displayed as a function of wavelength and (b) the same spectra plotted as a function of energy.

Nature makes use of a wide array of pigments, including chlorophylls, bacteriochlorophyll (BChl), carotenoids, and phycobilins, for effectively capturing a wide range of photons. No known organism, however, is capable of utilizing light of wavelengths longer than 1,030 nm for photosynthesis. Although this spectral region accounts for ~30% of the total solar irradiance, the photons that compose it have relatively low energy content.

3.2.2 Energy Transfer via Antenna Systems

The majority of pigments involved in photosynthesis function as antennas, harvesting the relatively diffuse source of photons arriving from the sun and delivering their energy to reaction centers. This function allows reaction centers to turn over at optimal rates and permits organisms to utilize photons over a large spectral range, including wavelengths for which the extinction coefficient of the reaction center pigments may be comparatively low. Most antenna systems incorporate a spatial and energetic organizational scheme, allowing excitation energy to pass from lower wavelength-absorbing pigments, located at the more distal sections of the antenna complex, to longer wavelength-absorbing chromophores that are proximal to reaction centers. Loss of photonic energy as heat (DH), during migration from excited-state donor pigments (D*) to ground-state acceptor pigments (A), provides a degree of irreversibility to the overall process, promoting effective funneling of excitation energy to reaction centers:

$$D^* + A \rightarrow D + A^* + \Delta H \qquad 2$$

The fundamental equation for energy (10, 11), and electron (12), transfer is the Fermi Golden Rule, which accounts for transitions between states due to weak perturbations. The formalism is expressed in Equation 3,

$$T_{i\rightarrow f} = \frac{2\pi}{\hbar}|\langle\psi_i|V|\psi_f\rangle|^2\rho,$$

where Ti→ f is the transition probability; ђ is the Dirac constant (the reduced Plank constant); ci and cf are wave functions for the initial and final states, respectively; V is the operator that couples the initial and final states; and r is the density of states. Conventional Forster Theory uses the Golden Rule expression in the context of a dipole-dipole approximation with the assumption that the donor and acceptor pigments are weakly coupled . This mechanism involves a nonradiative resonant process (neither photon emission nor absorbance is involved) resulting from coulombic coupling of transition dipoles among corresponding energy donors and acceptors. Singlet-singlet energy transfer between chlorophyll pigments is well described by the Forster model. As expressed in Equation 4,

$$k_{D\rightarrow A} = \alpha\frac{\kappa^2 k_D^0}{R_{DA}^6}J(\varepsilon_A),$$

4

THE PHOTOSYNTHETIC PIGMENTS

4.1 INTRODUCTION TO PIGMENT MOLECULES

The most important components of the chloroplast lamellae are colored organic compounds—the photosynthetic pigments. A rule, almost too obvious to deserve the title of the "first law of photochemistry" (which is sometimes bestowed on it), is that only light absorbed in a reacting system can have a chemical effect. However, when light absorption initiates a long chain of exergonic reactions, small, "catalytic" amounts of pigments may suffice. (An example is the photo induction of flowering and other "photomorphological" effects, where the primary light-absorbing component proved very elusive, and was only recently identified as a colored protein, the "phytochrome.") Much larger amounts of pigments must be present to produce a sizable photo chemical reaction when such chains cannot develop. A chain reaction is out of the question in photosynthesis because of the endergonic nature of the overall reaction, and yet photosynthesis must be a very fast process to provide all the organic material needed for the growth and life activity of the plant. This is why photosynthesizing cells have to contain very large amounts of pigments (up to 5% or more of total dry material). Pigments are molecules strongly absorbing visible light. Such absorption is restricted to certain classes of compounds. In painting, we use mostly the cheaper and more stable inorganic pigments (ochre, cobalt blue, cinnabar, chrome yellow, etc.); but good organic pigments have one or two orders of magnitude higher molecular absorption coefficients than the best inorganic ones.

The molecular structure of such organic pigments is characterized by long chains (or closed rings) of so-called "conjugated" double bonds. Carbon atoms, being quadrivalent, can be bound, in organic molecules, by single, double, or triple bonds, of which only single and double bonds are encountered in complex molecular structures. (The four bonds of a carbon atom are directed from the center to the four corners of a tetra hedron, and deflection from these directions needed to form a triple bond causes a stress.) Rings (or linear chains) consisting of regularly alternating ("conjugated") single and double bonds are particularly stable because they are reinforced by "resonance" (for example, that between the structures —C—C—C—C— and =C—C—C—C=)—interaction between several structures with the same arrangement of atoms, but different distribution of electrons. The most common of conjugated rings is that present in benzene, C_6H_6. It contains three single and three double bonds, and two main resonating structures are possible  and  (omit ting the C and H symbols). More complicated rings of the same character are present in other "aromatic" compounds, for example, napthalene, C_{10}HS, 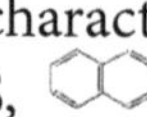 and anthracene, $C_{14}H_{10}$,  (Only one of the resonating structures is represented.) The nitrogen atom, being trivalent, can form a single bond with one carbon atom, and a double bond with another; it can be, therefore, substituted for a carbon atom (or more precisely, for a CH-group) in con jugated systems. Rings containing atoms other than C are called hetero cyclic (in contrast to the homocyclic rings, containing only carbon atoms).

The simplest of them is pyridine, C_5H_5N, or  One heterocyclic ring, common in plant pigments, is the five-membered pyrrol ring: 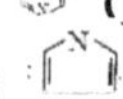

It is typical of molecules containing chains or rings of conjugated bonds to have strong absorption bands at relatively long waves, that is, in the visible or in the near ultraviolet. (Organic molecules containing only single bonds, or isolated double bonds, usually absorb in the far ultraviolet.) As the number of conjugated double bonds increases, absorption shifts towards the longer waves, and becomes stronger. Thus, in benzene, the first absorption band lies below 250 nm, while in anthracene, it lies close to the visible, making this compound yellowish. In still more complex conjugated ring systems, such as those present in *indigo* or in *chlorophyll* (or in long linear conjugated chains, such as those found in *carotenoids*), the absorption bands shift into the visible, making these compounds pigments.

The strong light absorption by conjugated systems has to do with their resonance structures (of which the two above-mentioned energetically equivalent structures of benzene are the prototype). As shown in Fig. 1, the splitting of excited electronic energy levels into several components, due to the existence of these resonating structures, leads to narrowing of the gap between the ground state and the first excited state; the first absorption band thus moves towards longer waves.

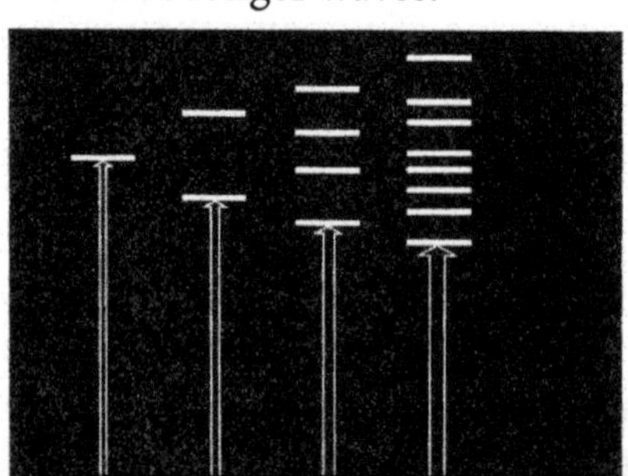

Fig. 1: Splitting of excited energy level by resonance in a chain of identical atoms (or molecules).

4.2 CHLOROPLAST PIGMENTS

Photosynthetic organs of plants always contain an assortment of pig ments. These can be divided into three major classes (see Tables 1–3): (1) chlorophylls and (2) carotenoids—both water insoluble, and (3) phycobilins—water-soluble (because of their attachment to water soluble proteins).

The first general feature of the distribution of chloroplast pigments is the universal presence of chlorophyll a, found in all photosynthesizing cells (except photosynthetic bacteria, where a related pigment bacteriochlorophyll (BChl) is found instead). Chlorophyll a is present also in brown, red, and blue-green algae, where its green color is masked by other pigments. So far, no plant was found capable of true photosynthesis which did not contain chlorophyll a (Chl a). The widespread occurrence, as well as certain chemical properties of chlorophyll a in vivo (and of bacteriochlorophyll in bacteria), suggest that these pigments (or, at least some of their molecules) play an active role in photosynthesis, functioning as photoenzymes. All other pigments (and a large part of Chl a or BChl itself) seem to serve only as physical energy suppliers. Pigments other than Chl a are often called accessory pigments.

An enzyme is a protein catalyst that mediates a chemical reaction by active but reversible participation in it—as a middleman who buys a commodity from the seller to sell it to the buyer. The commodity, in the case of a redox reaction, is hydrogen atoms or electrons. A photo enzyme is a broker who can ply its trade only after having been activated by light absorption. Because of its role as a light-activated enzyme in photosynthesis, chlorophyll has been called "the most important or ganic compound on earth."

The second common, but not universal rule is the presence of a second kind of chlorophyll, in addition to Chl a. Higher plants and green algae contain, in addition to the yellow-green chlorophyll a, the blue-green chlorophyll b (Chl b)—a locally oxidized derivative of Chl a. Several other chlorophyll-type compounds have been described in the literature. "Chlorophyll c" is found in brown algae and diatoms. A "chlorophyll d," identified by H. Strain and others in red algae, has proved elusive in later experiments. A "chlorophyll e" has been reported in golden-yellow algae.

Table 1: The Chlorophylls

Type of chlorophyll	Characteristic absorption peaks		Occurrence
	In organic solvents, nm	In cells, nm	
Chl a	420, 660	435, 670-680 (sev- eral forms)	All photosynthesiz ing plants (except bacteria)
Chl 6	453, 643	480, 650	Higher plants and green algae
Chlc	445, 625	Red band at 645	Diatoms and brown algae
Chld	450, 690	Red band at 740	Reported in some red algae

Chlorobium chlorophyll (also called "bacterio viridin")	Two forms: 425, 650 760 32, 660	Red band at 750 (or760)	Green bacteria
Bacteriochlorophyll a (BChla)	365, 605,770	Red bands at 800, 850, and 890	Purple and green bacteria
Bacteriochlorophyll b (BChlb)	368, 582, 795	Red band at 1,017	Found in a strain of Rhodopseudomonas, (a purple bacte rium)

Green bacteria contain mostly so-called "Chlorobium chlorophyll," or bacterioviridin, in addition to traces of bacteriochlorophyll, which is the main pigment of purple bacteria.

Table 2: The Carotenoids

Types of carotenoids	Characteristic absorption peaks, nm"	Occurrence
I. CAROTENES		
A-Carotene	In hexane, at 420, 440, 470 In hexane, at 425, 450, 480 In hexane, at 440, 460, 495	Many leaves, and certain algae. In red algae and a group of green algae called Siphonales, it is the major carotene Main carotene of all other plants Major carotene of green sulfur bacteria; traces in some plants
II. CAROTENOLS (also called "xanthophylls")		
B-Carotene y-Carotene	Luteol In ethanol, at 425, 445, 475	Major carotenol of green leaves, green algae and red algae
Violaxanthol	In ethanol, at 425, 450, 475	Second major carotenol of leaves
Fucoxanthol	In hexane, at 425, 450, 475	Major carotenol of diatoms and brown algae
Spirilloxanthol	In hexane, at 464, 490, 524	Common in purple bacteria

It has been difficult to establish the exact location of carotenoid bands in vivo (except in the case of purple bacteria) because of their strong overlapping with the blue-violet bands of chlorophylls. The bands in vivo are estimated to be shifted by about 20 nm to the long-wave side from their position in solution. In the case of fucoxanthol, absorption extends in vivo to 550 nm.

Table 3: The Phycobilins

Types of phycobilins	Absorption peaks	Occurrence
Phycoerythrins Phycocyanins Allophycocyanin	In water, and in vivo: 490, 546, and 576 nm At 618 nm, in water and in vivo At 654 nm in phosphate buffer at pH 6.5	Main phycobilin in red algae, also found in some blue-green algae Main phycobilin of blue-green algae; also found in red algae Found in blue-green and red algae

An additional complication—to be discussed later in the chapter arises from the apparent presence in living cells of several forms of one and the same pigment, which become identical upon extraction. This fact was long known in the case of bacteria, where three distinctly separated far-red absorption bands of bacteriochlorophyll are observed in vivo, at approximately 800, 850 and 890 nm, but give rise, upon extraction into an organic solvent, to a single absorption band at 770 nm. It has been surmised that bacteriochlorophyll is present in the living cell in the form of complexes with three different proteins, or in three different aggregation stages. A similar but less conspicuous polymorphism exists also in the case of chlorophyll a in green plants.

The third general rule, from which no exceptions are known, is the presence in all photosynthesizing cells of an assortment of carotenoids relatives of the pigment that accounts for the orange color of carrot roots. Most carotenoids are yellow or orange. Their color is normally masked by chlorophyll, but in chlorophyll-deficient, so-called aurea varieties of plants, or in fall, when chlorophyll disintegrates, the yellow pigments become visible. Some of them become in autumn orange, or even red, through oxidation, thus contributing to the variety of foliage colors. (However, important contributions to the bright red or purple colors of some leaves in autumn are also made by pigments of another class, so-called anthocyanins.)

The specific assortment of carotenoids is different in plants of different classes. In general, there are two major groups: the carotenes (which are hydrocarbons) and the carotenols. The ending "ol" suggests that these compounds are mostly alcohols (although some of them are ketones). The plant carotenols are also designated as xanthophylls (from xanthos = orange-yellow).

Certain algae are brown; examples are large algae such as Fucus, familiar to all visitors on ocean beaches, and the free-swimming micro scopic diatoms (Bacillariophyceae), having rigid silica skeletons. The diatoms are among the most successful plants on earth; they fill the upper layers of the oceans and may account for as much photosyn thesis on earth as the green land plants. The brown color of all these organisms is due to a special carotenol, called fucoxanthol.

The fourth important fact is the presence, in the red marine algae (Rhodophyceae), and the primitive blue-green algae (Cyanophyceae, encountered on land as well as in shallow water), of pigments of a still different class, called phycobilins (from phycos = alga, and bilin = bile pigment), because of their similarity to the pigments of the bile. These pigments may be present in amounts equal to, or greater than those of chlorophyll a. Phycobilin is a generic name for pigments of two kinds: the red phycoerythrins, and the blue phycocyanins. The phycobilins are structurally related to chlorophylls. The phycoerythrins appear to have the definite function of improving the light absorption in the middle of the visible spectrum; the same is probably true of the above-mentioned "brown" carotenoid, fucoxanthol.

Phycobilins can be extracted from the algae simply by placing the latter in distilled water, while chlorophyll extraction requires organic solvents. We have seen that chloroplasts contain hydrophilic proteins and hydrophobic lipoids; the latter, like grease spots, can be extracted only by organic solvents. As mentioned in Chapter 8, chlorophyll mole cules have affinity to both polar and nonpolar molecules; this is why they can be most easily extracted by a mixture of organic solvents, such as acetone or alcohol, with water. The latter disintegrates the protein structure, while the former pulls out the lipoid molecules.

Phycobilins have no phytol chains and are attached to water-soluble proteins; they are thus easily extractable into pure water. Most caro tenoids have no affinity for water and are soluble only in organic solvents.

4.3 CHLOROPHYLLS

4.3.1 Structure

The chlorophylls a and b have a common basic structure. Some fa miliarity with organic chemistry is needed to discuss it. It is a "por phyrin" structure, consisting of four pyrrol rings (Fig. 9.2), joined into a single master ring by CH "bridges." Figure 9.2 represents porphin, the mother substance of all porphyrins. Often a metal atom is found in the center of a porphyrin—iron in heme, magnesium in chlorophyll. In the latter, a long chain of carbon atoms ("phytol chain," from phytos=plant) is attached to the ring system. Other groups attached to the porphin skeleton will be discussed below. These side chains, and the double bonds present in the overall structure, distinguish among themselves the numerous compounds of this class.

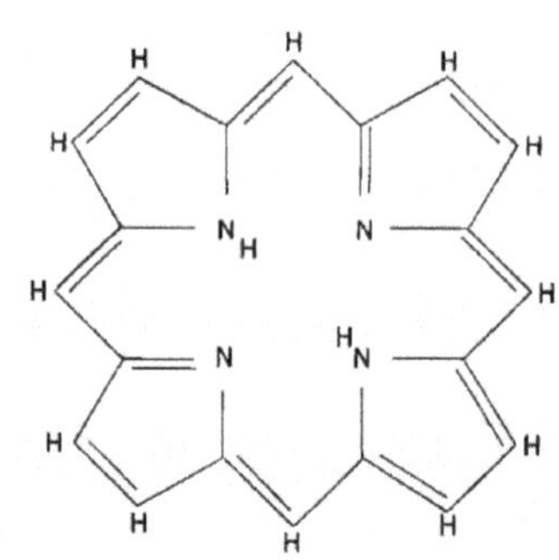

Fig. 2: Porphin structure consisting of four pyrrol rings (all corners except those marked N are occupied by carbon atoms).

The porphin unit plays a very important role in nature. It forms the skeleton of both chlorophyll and heme, the red pigment of blood, and of other compounds of physiological significance (for example, the cytochromes).

The complete molecular structure of chlorophyll a is represented in Fig. 3. (Two Nobel prizes were awarded for the development of this structure, to the German chemists Richard Willstätter, and Hans Fisher, respectively; a third one was awarded to Robert B. Woodward of Har vard for total in vitro synthesis of chlorophyll.) Compared to the porphin structure of Fig. 2, that of chlorophyll a is characterized by a missing double bond in one of the pyrrol rings (ring IV). It is thus derived from a dihydroporphin (a porphin containing two additional hydrogen atoms). Bacteriochlorophyll is derived from tetrahydroporphin, with two fewer double bonds and four more hydrogen atoms than porphin.

A peculiar and apparently important characteristic of chlorophyll (and of the bacteriolchlorophyll, too) is the presence of a fifth, "homocyclic," five-membered ring (ring V in Fig. 3) which carries a carbonyl group, C=0. This additional ring may be the "nerve center" of the chlorophyll molecule, most closely associated with its photocatalytic action in photo synthesis. This conclusion is not based on direct evidence, but rather on general feeling arising from the totality of chemical and photochemical experience with chlorophyll and related compounds.

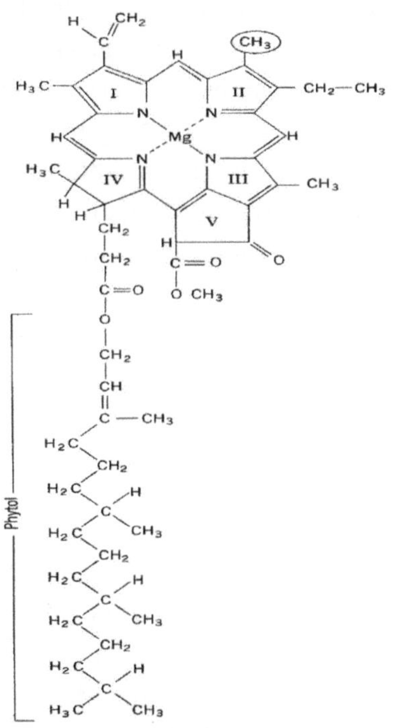

Fig. 3: Chlorophyll a (the circled CH; group is replaced by CHO group in chlorophyll b). All corners except those marked N are occupied by carbon atoms.

Another peculiarity of the chlorophyll molecule is the above-mentioned presence of a magnesium atom in the center of the molecule. We do not know its function. In hemin, the same position is occupied by an atom of iron, and the capacity of iron to exist in two oxidation states, as ferric ion, Fe+3 and ferrous ion, Fe+2, gives a hint as to its possible catalytic function. Magnesium does not have a similar property, and its special appropriateness to serve as center of such a catalytically active molecule is not clear. (Of all commonly available metals, it may perhaps fit best into the available space!) The presence of a metal atom, with its tendency to assume a positive charge, pushes the electrons in the porphyrin ring towards the periphery of the molecule, and thus into ring V. This may be important for its catalytic properties. But, again, this is not a matter of clear-cut evidence, but of chemical "feeling."

Some short side chains in the chlorophyll molecule may be simply residues left over from its synthesis in the organism, which begins with short-chain molecules, such as those of acetic acid, CH_3COOH, and of glycine, NH_2CH_2COOH. When these are combined into rings, methyl groups, CH3, are left hanging outside the ring. However, the presence in ring I of a nonsaturated (vinyl) side chain, $-CH=CH_2$, may be of some significance because its double bond is conjugated on one side with the ring system, and thus affects its overall properties. Also, impor tant is the presence, in ring IV, of the long and almost saturated phytol chain. (Phytol contains only one double bond, and fourteen single bonds.) This chain causes chlorophyll molecules to attach themselves to other saturated long-chain molecules in the chloroplasts, that is, it gives it lipoid solubility (lipoids being a generic name for "fatlike" compounds). The general rule in organic chemistry is that polar molecules associate with other polar molecules or groups (and are, therefore, attracted to water, which is the most polar of all simple solvents), while nonpolar chains associate selectively with other nonpolar chains. (In inorganic chemistry, on the other hand, the most important bonds involve positive ions associating themselves with negative ions, as in the formation of acids, alkalis, and salts.)

Chlorophyll b—the second variety of chlorophyll present in green plants and green algae-differs from chlorophyll a only by local oxidation of one side chain in ring II (marked by a ring in Fig. 3). This side chain is $--CH_3$ in Chl a and —C—H in Chl b.

4.3.2 Absorption Spectra

As a pigment, chlorophyll a is characterized by two strong absorption bands, located in the blue-violet and in the red region of the spectrum (Fig. 9.4a). The first of them, often called the Soret band, is common to all porphin derivatives; but the second one is peculiar of chlorophyll and other compounds derived from dihydroporphine. In diethyl ether, the first has its maximum at 430 nm and the second at 660 nm. In the living cell, these bands lie at about 440 and 675 nm, respectively, and are complex. The absorption is weak between the two bands, that is, in the green part of the spectrum, causing the green color of vegetation.

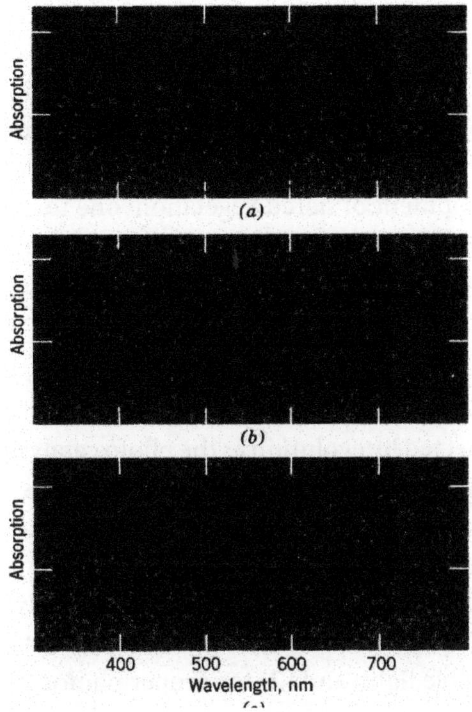

Fig. 4: Absorption spectra of three types of chloroplast pigments. (a) chloro phylls; (b) carotenoids; (c) phycoerythrins and phycocyanins.

Chlorophyll b spectrum differs from that of Chl a by the two bands being closer together: the Soret band of Chl b is located in ether) at 453 nm, and the red band at 643 nm. The ratio of intensities of the two bands is shifted in Chl b strongly in favor of the Soret band compared to chlorophyll a.

Bacteriochlorophyll a (BChl a) also has two main bands, but they lie farther apart; the Soret band in the near ultraviolet (at 365 nm) and the long-wave band, at the limit of the visible, at 770 nm (in methanol). In the living cell, the same bands lie at about 367 and 800-900 nm. The second one is usually split into three components, at about 800, 850, and 890 nm. A bacteriochlorophyll 6 has been recently dis covered in some bacteria; it has an infrared absorption band at 1014 nm. Another chlorophyll derivative (mentioned before) is found in green bacteria; it is the "Chlorobium chlorophyll" or "bacterioviridin." In the living cell, the long-wave bands of this chlorophyll are found at 750 or 760 nm. Chlorobium chlorophyll-carrying green bacteria always contain also small quantities of bacteriochlorophyll a.

4.3.4 Chromatic Adaptation

The presence of a strong, long-wave absorption band, covering much of the red and orange region, in which sunlight is most abundant, may be one of the reasons why chlorophyll has been selected by nature as the main pigment in photosynthesis. Without this absorption band, the parts of the solar spectrum that contain the greatest amounts of useful energy would not be efficiently absorbed by cells. On the other hand, the specific form of the absorption spectrum of chlorophyll, with its absorption gap in the green region, which accounts for the pleasant green color of vegetation, implies a low capacity of these plants to utilize light in the middle of the visible spectrum. This deficiency is made up, at least to some extent, by the development of tiered vegetation: one tier of leaves lets enough light through for a second and a third layer to survive. In deciduous forests, bushes and grass can live on the ground. (Coniferous forests, on the other hand, often have such dense crowns of needles that the shadow under them is too deep for any vegetation to develop on the ground.)

The position of the absorption bands in the red cannot be the only reason why chlorophyll a had been selected by evolution as the photo catalyst in photosynthesis. Other pigments exist that would provide a better coverage of the visible spectrum. Additional reasons for this selec tion must be sought in specific physicochemical properties that bind chlorophyll in appropriate locations in the cell; and even more, in specific catalytic properties, which permit it to serve as an effective photocatalyst in photosynthesis. The selection of chlorophyll may be the best solution nature found to satisfy two needs: (1) for effective absorption of visible light, and (2) for proper photocatalytic properties.

As mentioned above, green land plants and green algae contain chloro phyll b in addition to chlorophyll a. Because of the shift of its Soret band, chlorophyll b in solution is blue-green, while chlorophyll a is yel lowish green. The presence of chlorophyll b, therefore, narrows somewhat the "green gap" in the absorption spectrum of leaves; this may be the rationale for its presence, in relatively larger quantities, in shade-adapted plants.

We have also encountered another method of plant adaptation to life under water. The brown algae, as well as the diatoms, narrow the green gap by the provision of two other pigments—chlorophyll c, (whose chem ical nature is not yet well known; it may be a porphin rather than a chlorin derivative), and of a specific carotenoid pigment, fucoxanthol (to be discussed below). Even more effective in filling the "green gap" is phycoerythrin, the main pigment of red algae (cf. below).

4.4 PHYCOBILINS (PHYCOERYTHRINS AND PHYCOCYANINS)

4.4.1 Structure

The chemical structure of the phycobilins is related to that of chloro phylls. If one snips the ring system of chlorophyll and permits the four pyrrol rings to straighten out, the magnesium atom slips out and what remains is an open conjugated system of four pyrrol rings. This is the fundamental structure of the *bilin* pigments, which can be considered as derived from a common parent, the hydrocarbon *bilan* (see Fig. 5*a* and b). The bilins are so called because they were first discovered in bile, to which they give its dark color. They may be, in fact, products of metabolic transformation of hemoglobin and chlorophyll, ingested with animal and plant food. The name phycobilin means algal bilin. Recently, it has been shown that the pigment phytochrome, responsible for the light control of seed germination and of flowering, is chemically related to the phycobilins.

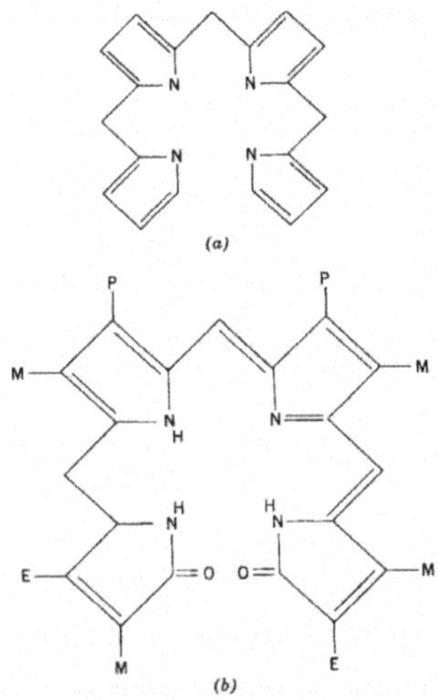

Fig. 5: Structure of bilin pigments. (a) bilan; (6) phycoerythrobilin. (All un marked corners are occupied by carbon atoms.) (E, ethyl; M, methyl; P, propionyl groups.)

4.4.2 Absorption Spectra

In algae living deep under the sea, the deficiencies of chlorophyll as a light absorber become critical because light reaching these organisms is filtered through thick greenish-blue layers of water. (Dissolved salts, and organisms living in the surface layers, contribute to the absorption of red and blue-violet light.) Plants living under those conditions have evolved auxiliary pigments that absorb green light. Red algae, in particular, have solved the difficulty by means of a phycobilin called phycoerythrin, with absorption bands in the middle of the visible spectrum. For example, one type of phycoerythrin has bands at about 500, 545 and 570 nm (Fig. .4c). This permits these algae to perform photosynthesis (and thus to grow) in dim blue-green light prevailing in the depths of the ocean. Together, chlorophyll and phycoerythrin let through only a band in the far-red. This is why the algae which contain phycoerythrin appear red.

The deeper under the sea a red alga lives, the more phycoerythrin it contains in relation to chlorophyll.

However, complications make the overall picture more complex. For example, some algae encountered in great depth are green. They manage to survive by slowing down their life processes sufficiently to permit the small amount of light that their green pigment is able to catch to cover their needs.

Some surface, or even terrestrial, algae contain a pigment related to phycoerythrin, called phycocyanin. This is blue rather than red because its main absorption band is located at about 630 nm (Fig. 4c). It does not fill up the gap in the chlorophyll a spectrum, but it narrows it, similarly to Chl b. The algae that contain this pigment are bluish green (Cyanophyceae). They are primitive organisms, intermediates between true plants and bacteria. Like bacteria, the blue-green algae do not have photosynthetic organelles bounded by a membrane (chloroplasts); like green bacteria, they contain a pigment that absorbs at 750 nm. It can be surmised that primitive plants of the type of blue-green algae have later developed in two directions—one in which the phycobilins have disappeared, and one in which one particular form of phycobilin, the phycoerythrin, has accumulated. Plants of the first type have taken over land and the surface layers of the ocean, while plants of the second kind have concentrated in the depths of the ocean. However, the division is not sharp, with some species of one kind invading the domain of the other.

4.5 CAROTENOIDS

4.5.1 Structure

The third type of pigments, present in all photosynthesizing cells, the carotenoids, contain an open-chain conjugated double bond system of so-called polyene type (Fig. 9.6), ending with so-called "ionone" rings. Carotenoids are either hydrocarbons, in which case they are called carotenes, or oxygen-containing compounds, in which case they are called carotenols (or xanthophylls). The three common carotenes are *a, b,* and y carotene. They are "stereoisomers" having the same molecular formula C40H56 (Fig. 9.6), and distinguished only by the arrangement of their molecules in space. The carotenols have their oxygen in the form of hydroxyl, carbonyl, or carboxyl groups attached to the "ionone" rings.

9.5.2 Absorption Spectra

The carotenoids generally are yellow or orange; they have a group of two or three absorption bands located in the blue-violet part of the spectrum (Fig. 4b). For example, ß-carotene, dissolved in hexane, has absorption bands at 430, 450 and at 480 nm. Certain algae diatoms (Bacillariophyceae) and brown algae (Phaeophyceae, from phaeos = brown) —contain abundantly a special carotenoid, called fucoxanthol. In the form in which it is present in the cells, fucoxanthol has a much broader absorption band than other carotenoids; it absorbs not only in the blue but also in the green part of the spectrum. Instead of falling off sharply to zero at about 500 nm, as in Fig. 4b, the blue absorption band of fucoxanthol declines slowly, covering much of the gap left by chlorophyll in the green, and destroying the pure green color of the plant.

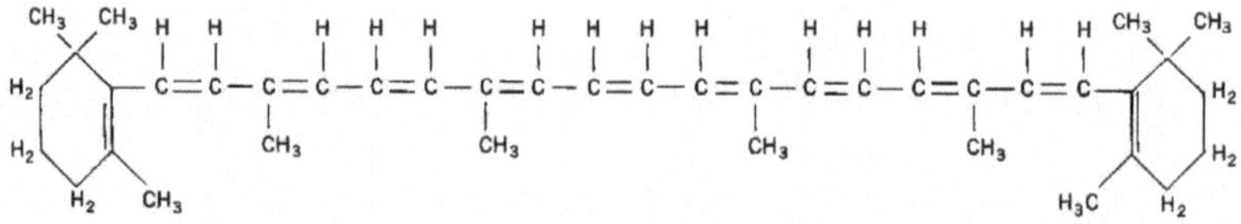

Fig. 6: Structure of B carotene (a and y forms are stereoisomers). (All corners on the rings at the two ends are occupied by carbon atoms.)

To sum up, plants have developed, for the needs of photosynthesis, not a single pigment, but various pigment assortments. These change from one phylum of plants to another. The relative amounts of the several components change not only from phylum to phylum, or species to species, but also from specimen to specimen—from a shade-adapted to a sun-adapted leaf on the same tree, from a young to an old algal cell, from a specimen grown in green light to a specimen grown in red light, etc. The varying ratios of different pigments in the living cell reflect a dynamic balance of continuous synthesis and decomposition of the pigments. It has been long a subject of guessing why chloroplast pigments occur in pairs, such as chlorophyll a and chlorophyll b, phycoerythrin and phycocyanin (a little phycocyanin is always found in red algae), and carotenes and carotenols. One member of the pair is slightly more oxidized than the other. So far, no explanation exists.

4.6 MULTIPLICITY OF CHLOROPHYLL A FORMS IN VIVO

Looking at the red absorption band of chlorophyll a in the living cell, one notes its greater width compared to the corresponding band in solution. (The "half-width" of the band—that is, its width at the level corresponding to half its maximum intensity—is 30 nm in vivo, as against about 18 nm in organic solvents.) Recently, careful observations suggested that the width of the red band in vivo is due to the presence of at least two and probably three (or even four) components. The analysis of the absorption bands of chlorophyll a in suspensions of chloroplasts, or in whole algal cells, encounters several difficulties.

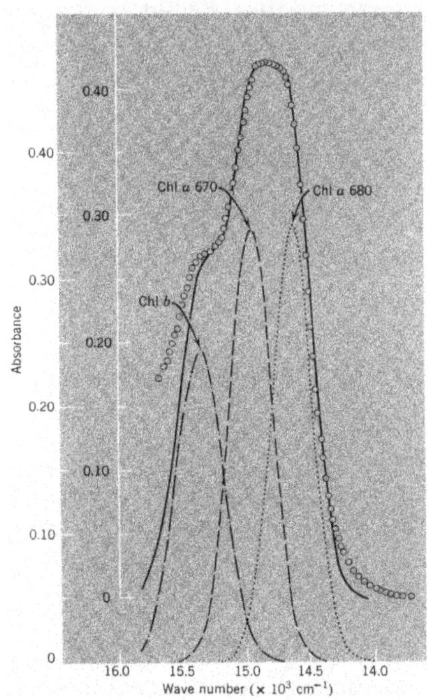

Fig. 7: Analysis of red chlorophyll absorption band in Chlorella pyrenoidosa, into two major Chl a components (Chl a 670 and Chl a 680), and one Chl b com ponent (Chl b 650).

Minor components of Chl a are supposed to exist at 690–700 nm. (C. Cederstrand, E. Rabinowitch, and Govindjee, 1966.) This figure is plotted on a wavenumber (reciprocal wavelength) scale.

- Other pigments besides chlorophyll a are present; this makes, in particular, the analysis of the Soret band almost impossible.
- Light is strongly scattered by the suspension. There is general scattering and, more importantly, there is "selective scattering" around the absorption band, found in 1956 by P. Latimer in our laboratory. Scattering increases the apparent absorption, if measured in an ordinary spectrophotometer, and can totally distort the absorption spectra.
- In a scattering suspension, light that is less strongly absorbed has a larger effective path length than the more strongly absorbed light (this is known as the *detour effect).*
- A mutual shading of the pigment molecules takes place in densely colored particles, such as the chloroplasts, while some light passes the suspension without hitting any particle at all (this has been called the *sieve effect).*
- Fluorescence contributes to the transmitted light, particularly significantly when the absorption is strong and the fluorescence yield is high.

Two methods have been used to remedy some of these difficulties: (a) An opal glass can be placed between the sample and the detector, so that most of the forward-scattered light is collected (much of the scat tered light in a suspension of large particles leaves the suspension in the forward direction). (b) An integrating sphere (Ulbricht sphere) can be used for collecting all (or, at least, almost all) the scattered energy. However, neither of the two methods eliminates the sieve or the detour effect.

C. S. French at the Carnegie Institution of Washington at Stanford, California, has constructed a "derivative" spectrophotometer in which the first derivative of the absorbance is plotted automatically as function of wavelength. This procedure accentuates the complex structure of a band, since every inflection in the band envelope appears either as crossing of the abscissa or as a peak. The "derivative absorption spectra" of various algae, obtained in this way, were analyzed by French and co-workers into three components (to which a certain standard shape, that of a so-called Gaussian error curve, has been ascribed for the pur pose of analysis). They were identified as Chl a 672, Chl a 683, and Chl a 694. In our laboratory, Carl Cederstrand has measured the absorption spectra of algae in an integrating dodecahedron with one photocell on each of its 12 faces, and analyzed the results with the help of a computer. The red band of chlorophyll a was found to be essentially double with its envelope covering two neighboring bands (Fig..7), like a couple huddling under a single raincoat (a couple with a child, if the chlorophyll b band is included). One band was located at 668 nm (French's "Chl a 672'), the other at 683 nm (French's "Chl a 683") - again assuming a Gaussian shape of the component bands. This assump tion is not completely arbitrary, since a Gaussian curve was found to match very closely the red band of Chl a in solution. (Deviations from the Gaussian curve are, however, visible on both ends of the band in Fig. 7.) Later, we found that elimination of the "sieve effect" by breaking Chlorella cells with ultrasonic waves (making the particles small enough to reduce shading to insignificance), does not affect the qualitative results of the analysis; but, as expected, it moves the peaks of the two chlorophyll a components closer together—to 670 and 680 nm, respectively. The half band-width of the Chl a 670 and Chl a 680 bands was found to be about 17 nm, that is very similar to that of the Chl a band in ether.

Cederstrand's analysis gave no evidence for the existence of a third long-wave component (such as French's Chl a 694). The computer, asked to explain in the simplest possible way the shape of the red band envelope in terms of Gaussian components, answered that three components (Chl b and two forms of chlorophyll a) provide as good an approximation as can be reasonably expected. Because of deviations from Gaussian shape in the wings in the Chl a band in vitro, it seemed unjustifiable to postulate additional components in order to obtain a better fit in the long-wave wing of the band in vivo.

However, quite apart from this analysis, there is considerable evidence for the existence of one (or two) minor absorption bands of Chl a in the 690–700 nm region. One is the position and the shape of the "red drop" in the action spectra of Chl a fluorescence and of photosynthesis. It indicates a minor Chl a form, not contributing to fluorescence, and contributing only partially to photosynthesis, with an absorption band in the region of 695 nm. Another is the "difference spectrum" obtained by subtraction of the spectrum of Chlorella sonicates obtained at neutral pH and under anaerobic conditions, from that of sonicates obtained at acid pH under aerobic conditions. This spectrum, too, shows a band at 693 nm, suggesting that Chl a 693 is present in a small amount in living cells, and is particularly easily de stroyed by ultrasonic irradiation in an aerobic acid medium.

If one introduces a 693 nm band into Cederstrand's analysis, the result is a slight shift of the main long-wave component, Chl a 680, towards the shorter waves, and a change in the ratios of the two peaks in favor of Chl a 670. Additional minor Chl a component (0.3% or less of total chlorophyll), with maxima at 700 nm (P700) and at 682–690 (P690) have been sug gested on the basis of difference spectroscopy. Their con centrations are too small for them to be noticeable in the analysis of the absorption spectrum.

Spectroscopic evidence for the complexity of chlorophyll a in vivo is supported by chemical evidence, presented particularly by T. M. Godnev and A. A. Shlyk in Minsk (Byelorussia). They noted that during gradual extraction of chlorophyll from leaves, the absorption spectrum changed. J. H. C. Smith and co-workers, at the Carnegie Institution in Stanford, have similarly demonstrated that during the greening of leaves, the red absorption band undergoes shifts. In these experiments, leaves were permitted to form in the dark. They are then colorless or "etiolated." When exposed to light, they rapidly become green. Altogether, it seems as if two or three spectroscopically different forms of chlorophyll a are formed at different times, and are extracted with different ease.

Fractionation of these two (or three) Chl a forms by breaking the cells mechanically and solubilizing the pigment complexes selectively by means of detergents such as desoxycholate or digitonin, or by extraction with solvents of varying polarity, has been attempted in several laboratories. These observations are relevant to the question of whether some components of Chl a are associated with one or the other of the two photochemical reactions postulated in photosynthesis and with the two "pigment systems" supposed to sensitize them.

5

PHOTOSYNTHETIC ENERGY CONVERSION

5.1 AN INTRODUCTION: AN OVERVIEW

Photosynthesis is the most important natural process on earth, which transformed the once lifeless planet into a living world. While primitive photosynthetic bacteria such as purple sulfur bacteria and green sulfur bacteria carry out anoxygenic photosynthesis, producing elemental sulfur from hydrogen sulfide with the help of sunlight, cyanobacteria, algae and plants carry out oxygenic photosynthesis to convert water and carbon dioxide to sugars with the help of sunlight and release oxygen as a byproduct. The conversion of solar energy to chemical energy via photosynthesis with the release of oxygen has an evolutionary significance on life as we know it today. In fact, photosynthesis is the only natural process known on earth to form oxygen from water. Further, fossil fuels such as coal, petroleum and natural gas are formed from the remains of the dead plants by exposure to heat and pressure in the earth's crust over millions of years. With increasing energy crisis and environmental issues lately, now is the time to revisit photosynthesis in order to address these issues. In this context, a great deal of ongoing research is focused on utilizing photosynthetic energy conversion as a renewable, self-sustainable and environment friendly source of energy. When compared to the finite reserve of fossil fuels, sunlight, the energy source for photosynthesis, is abundant around the planet and is inexhaustible.

The earth receives solar energy at the rate of about 120,000 TW, which far exceeds our current global demand of ~16 TW.1 However the only major technology available for solar energy conversion is photovoltaics (PV). PV devices such as solar panels generate electrical power by converting solar radiation into direct current electricity using semiconductors. Solar cells include first generation conventional wafer-based cells made up of crystalline silicon (Fig. 1a), second generation thin film solar cells (Fig. 1b) made up of amorphous silicon, cadmium telluride, and copper indium gallium selenide, and third generation emerging solar cells such as organic solar cells (Fig. 1c), dye sensitized solar cells (DSSC) (Fig. 1d), and quantum dots solar cells, each with their own efficiency limits, advantages and disadvantages. Compared to the single junction solar cells (maximum efficiency of ~27% for crystalline Si cell with concentrator), gallium arsenide based multiple junction solar cells can reach efficiencies as high as 44%. However, the PV-based technologies (some of which remain expensive today) are not suited for pushing the efficiency limits higher. This is main reason why alternate solar capture technologies based on natural photosynthesis are being explored. The internal quantum efficiency of the charge separation step in natural photosynthesis is ~100%. Further the charge carrier recombination time is much lower for photosynthesis-based light capture (>10-1 for PSII stacked membrane converter), compared to PV technologies (>10-3 for Si PV and >10-6 for organic PV) as shown in Table 1. Another remarkable feature that separates photosynthesis from traditional PV technologies is that photosynthesis contributes to global CO_2 sequestration (via the dark reaction), which can never be accomplished by any of the PV systems designed so far.

Table 1. Performance comparison of natural photosynthesis versus different photovoltaic technologies (DSSC: dye sensitized solar cells; PV: phtotovoltaics; adopted from Pace 200553).

Solar Energy Technology	Internal Quantum Efficency	Theoretical Maximum Effciency	Present Efficiency Range	Charge Recombination Duration
Silicon PV	~55 %	~ 25%	18-24 %	< 10-3 s
Orgenic PV	10-30%	-	2-6 %	< 10-6 s
DSSC	~ 90%	~ 33%	10-12 %	~ 10-1 s
Natural photosynthesis	~ 100%	>40 %	-	> 10-1 s

5.2 MECHANISM OF PHOTOSYNTHESIS

In higher plants, photosynthesis takes place in compartmentalized organelles called chloroplasts (Fig. 2). The inner matrix of chloroplast is called the stroma, in which thylakoids are organized into a structure similar to a stack of coins called granum (plural: grana) (Fig. 2). Photosynthesis consists of two reactions: (1) a light dependent reaction called the light reaction that takes place inside thylakoids (lumen); and (2) a light independent reaction called the dark reaction (Calvin cycle) that takes place in the stroma. During the light reaction, with the help of sunlight, water is oxidized to protons, electrons, and oxygen. The sequential steps of transferring the electrons from water to NADP+ to generate NADPH are collectively called the photosynthetic electron transport chain (P-ETC) as shown in Fig. 3. P-ETC encompasses various protein complexes such as photosystem II (PSII), cytochrome b6f, photosystem I (PSI) and ATP synthase, soluble proteins such as plastocyanin, ferredoxin and NADP reductase and plastoquinone that are embedded in the thylakoid membrane (Fig. 2). Both the photosystem complexes, namely PSI and PSII, contain chlorophylls at their core reaction center, where the actual charge separation happens upon light absorption thereby generating the excited electrons. When these photo-excited high energy electrons transfer down the energy gradient, the energy drop is harnessed to pump protons outside the thylakoid membrane. This generates the proton motive force necessary to generate ATP by ATP synthase. The NADPH and ATP generated in the light reactions are used in the dark reactions, during which the absorbed CO_2 is reduced to metabolites that subsequently form sugar (Fig. 4). The net reaction of photosynthesis is the synthesis of sugar using sunlight, water, and CO_2 while oxygen is released as a byproduct.

Light energy can be harnessed in many ways using photosynthesis for either direct generation of electricity or the production of energy rich fuels like ethanol, propanol, butanol or hydrogen (Fig. 4).2 For electricity and hydrogen generation, the light reaction is of the most concern; however, for the production of energy-rich carbon-based fuels, the dark reaction is exploited. Figure 5 shows different schemes of electricity generation in photobioelectrochemical cells (PBEC). The PBEC is made up of an anode containing a biocatalyst such as photosystem II (Fig. 5a), thylakoid membrane (Fig. 5b), or whole cell microorganism such as cyanobacteria (Fig. 5c) coupled with an enzymatic cathode. With the help of light, the biocatalyst on the anode transfers the electron generated from oxidation of water. On the cathode, enzymes such as laccase or bilirubin oxidase are employed that catalyze the reduction of oxygen to water. When such anodes are coupled with cathodes containing enzymes such as hydrogenase or nitrogenase, the system generates hydrogen (Fig. 5d). In either case, water and light are the only raw materials used to generate electricity and hydrogen, thereby making the system sustainable, economic, and environment friendly. On the other hand, the pathways in the Calvin cycle and the central carbon metabolism are manipulated to produce carbon based biofuels like ethanol, propanol, and butanol.

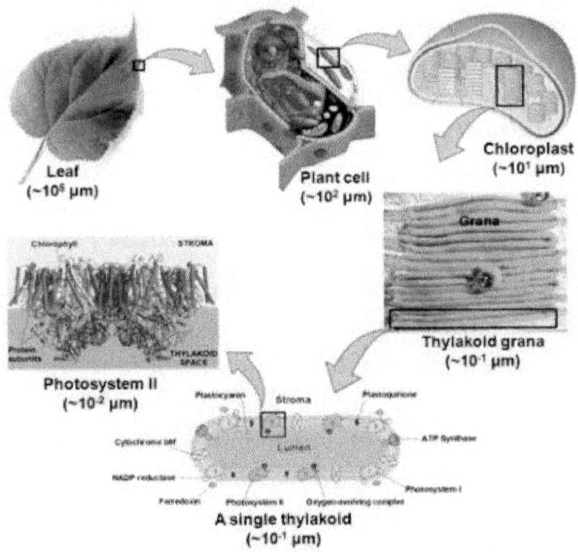

Fig. 2: Hierarchical organization of photosynthetic machinery in thylakoids inside the chloroplast of plant cell (with relative dimensions mentioned in micrometers)

5.3 ELECTRICITY GENERATION

Direct conversion of light to electricity can be achieved in a photo-bioelectrochemical cell (PBEC) using natural photosynthetic machines as biocatalysts. The photosynthetic machines include bacterial photosynthetic reaction center (RC), photosystem I and II (PSI and PSII) complexes, organelles such as thylakoids that are isolated from algae and plants, and whole cell photosynthetic microorganisms such as cyanobacteria and microalgae. When the biocatalyst used is a photosynthetic microorganism, the system is also called as photosynthetic microbial fuel cell (PMFC). The PBEC systems based on the various biocatalysts (Fig. 6a) are discussed below.

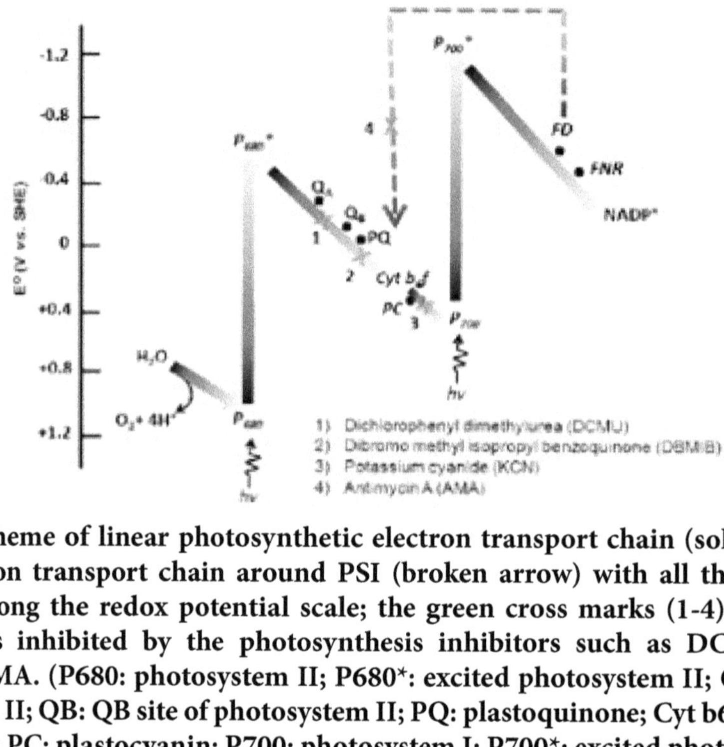

Fig. 3: Z- scheme of linear photosynthetic electron transport chain (solid arrow) and cyclic electron transport chain around PSI (broken arrow) with all the components expressed along the redox potential scale; the green cross marks (1-4) represent the specific sites inhibited by the photosynthesis inhibitors such as DCMU, DBMIB, KCN and AMA. (P680: photosystem II; P680*: excited photosystem II; QA: QA site of photosystem II; QB: QB site of photosystem II; PQ: plastoquinone; Cyt b6f: cytochrome b6f complex; PC: plastocyanin; P700: photosystem I; P700*: excited photosystem I; FD: ferredoxin; FNR: ferredoxin NADPH reductase; NADP+: oxidised form of NADPH; Adopted from Sekar et al. 20152)

Various attempts have been made to immobilize the isolated thylakoids onto different support matrices such as albumin glutaraldehyde cross-linked matrix,4 multi-walled carbon nanotubes, encapsulating the thylakoid membrane vesicles onto conductive nanofibers by electrospinning, and vapor deposition of thin layer of silica onto the thylakoid layer. PSI complexes have been immobilized using several different strategies such as gold nanoparticle (GNP) modified electrode,8 functionalized nano-porous gold leaf electrodes,9 PSI-GNP hybrid electrode modified with 3-mercapto1-propanesulfonic acid,10 and via self-assembly onto zinc oxide nanomaterials. The immobilization strategy is instrumental in dictating the orientation of biocatalyst on the surface as well as its proximity to the electrode. These two factors are very important for electrochemical reactions because the electron transport pathway inside the huge PSI and PSII complexes is strictly vectorial. A poly-histidine (His) tag and Ni(II)- nitrilotriacetic acid (Ni-NTA) system has been found to be profoundly useful in immobilizing PSII for its efficient photo-electrochemistry. Other significant improvements are witnessed by immobilizing PSII in osmium-containing redox polymer based on poly(1- vinylimidazole) and a matrix of 2-mercapto-1,4- benzoquinone (MBQ), electro-polymerized on the gold surface.18 Further, a lot of effort has been taken to maintain the activity of the isolated thylakoids and photosystems by mimicking the natural environment on the electrodeor by preventing the loss of activity through catalytic quenching of reactive oxygen species which otherwise reduces the activity of the photosynthetic machines.

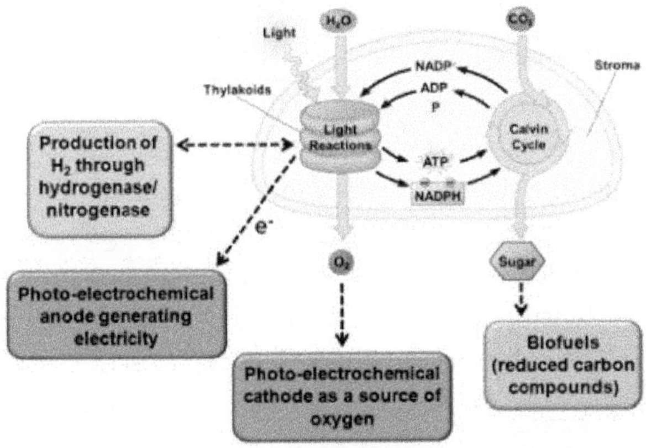

Fig. 4: Chloroplast showing light reaction in thylakoid and dark reaction in stroma and utilization of these reactions for various applications such as generation of electricity, hydrogen and carbon based biofuels.

5.4 PHOTOSYNTHETIC MICROORGANISM

While relatively higher power densities were achieved using isolated photosynthetic machines such as RC, PSI, PSII or thylakoids, they are not practical for energy conversion systems due to: requirement of laborious, skillful isolation procedures; requirement of specific environmental conditions (pH, temperature, ionic concentration of surrounding media etc.,); instability caused by photo damage; and inability to self-repair upon photo damage, since they are present in an artificial environment devoid of their natural counterparts. All the above caveats can be overcome by employing the whole cell photosynthetic microorganisms in PBEC/PMFC. The whole cells retain all their native biological functions and therefore possess superior stability upon immobilization on electrode surfaces. Cyanobacteria such as Synechococcus elongatus, Synechocystis sp., Nostoc sp., Anabaena variabilis, and Spirulina platensis and green algae such as Chlamydomonas reinhardtii, Chlorella vulgaris, and Ulva lactucaare employed in the PMFC for light induced electric current generation. Compared to growing the culture of cyanobacteria in PMFCs with bare untreated electrodes, growing cyanobacterial biofilm or immobilizing the cyanobacterial cells onto electrodes modified with nanostructure-based support matrix such as polyaniline, polypyrrole, multi-walled carbon nanotubes, and electrodes modified with osmium redox polymer, indium tin oxide have been shown to significantly improve the power density. When such an anode is combined with an oxygen reducing cathode, the oxygen evolved in photosynthesis would be subsequently reduced at the cathode and the entire system would be completely sustainable, environment friendly and requires only water and light for the generation of electricity. However, the extracellular electron transport flux from P-ETC of the photosynthetic microorganisms to the electrode is fairly low compared to the extracellular electron transport flux of dissimilatory metal reducing bacteria (DMRB) such as Geobacter and Shewanella in microbial fuel cells (MFC). One way to address this is through the use of redox mediators such as 1,4-benzoquinone, 2,6-dimethyl-1,4-benzoquinone, 2-hydroxy-1,4-napthaquinone, or phenazine methosulfate36 that are permeable through the outer membrane of the living cell and greatly enhance the power density. However, from an energy conversion standpoint, the use of redox mediators decreases the overall cell voltage and therefore is not attractive. Moreover, mediators must be regenerated constantly, which introduces additional complexity to the system and hence is not attractive from a design standpoint. High power density can also be achieved by engineering a novel miniaturized systems such as microfluidic bio-photovoltaic devices and micro-sized bio-solar cells, however these systems are only suited for ultra-low power applications.

5.5 GENETIC ENGINEERING

Regardless of the mode of extracellular electron transport (direct or mediated), the PMFC generally suffers from low electron flux because the electrons must be diverted from their native routes to alternate pathways to reach the electrode. These alternate electron-harvesting pathway are very difficult to achieve. A robust approach to efficiently collect more electrons from the P-ETC without using the redox mediators would be a welcome strategy to improve the performance of PMFC on par with that of MFC. The photosynthetic microorganisms such as cyanobacteria have been performing photosynthesis for over 3.5 billion years. However, they have not evolved for extracellular electron transport. Nonetheless, with the advancement of genetic engineering and molecular biology, these smart microorganisms can be made smarter to benefit our needs to generate more electricity. Analyses of electron transfer pathways from the P-ETC to the electrode leading to photocurrent generation on the PMFC anode would greatly help in understanding the mechanism of extracellular electron transfer as well as other bottlenecks which facilitates further optimization for enhancing photocurrent. This can be accomplished with the help of site-specific photosynthesis inhibitors that block a specific pathway in P-ETC as shown in Fig. 3. Using these inhibitors both individually as well in certain combinations, more precise source of photocurrent can be ascertained and is very useful to engineer appropriate strategy to improve the photocurrent. For example, diverting electrons from earlier steps in P-ETC, say from PSII complex, contributes to more photocurrent and the conversion efficiency can be further increased (Table 1). Accordingly various efforts have been undertaken to collect more electrons from PSII.

Certain genetic engineering approaches have been used to redirect the electrons from the photosystem complexes, thereby manipulating P-ETC for generating higher photocurrent. Electrons from QA - in PSII were redirected to engineered collection sites approximately 13 Å away on the stromal side of the thylakoid membrane. The positively charged amino acid lysine (Lys, K) at position 238 of D1 protein in PSII is important for the insulation of the PSII electron flow from external oxidation by soluble species and is also highly conserved in PSII of higher plants and algae. Modification of this lysine to glutamate (Glu, E), i.e., K238E resulted in alternative electron transfer pathway to soluble electron acceptor protein (cytochrome c) near the QA - site. This redirection along with the addition of an herbicide that blocks the electron flow at QB site resulted in decreased oxidative damage. In another attempt, cationic redox-active metal complexes like CoIII complexes, small enough to fit in a 5 Å diameter negatively charged patch along the stromal side of the membrane were also used to redirect electrons from QA - site.

Cyanobacteria, algae and green plants contain chlorophyll-a (Chl-a) as the primary photosynthetic pigment in their PSI and PSII complexes. Chl-a strongly absorbs only blue (430 nm) and red (670-680 nm) lights. Light harvesting complexes in algae and green plants and phycobilins in cyanobacteria contain accessory pigments that absorb visible light in the range of wavelengths neglected by the Chl-a, thereby increasing the action spectrum of photosynthesis. In any case, they cannot absorb light beyond the red end of visible spectrum. However, another class of photosynthetic bacteria such as green bacteria and purple bacteria contain bacteriochlorophylls (BChl) that absorb strongly in near infrared region spectrum (705– 1040 nm range). Further, Chl-a present in both PSI and PSII compete for the same light which theoretically reduces the efficiency by half. Replacing one of these Chl-a in the photosystems with BChl could effectively enhance the existing efficiency by a factor of two, similar to the multi-junction solar cell with different bandgaps and greatly broadens the action spectrum of photosynthesis to the near infra-red region. Besides the recommendation and discussion of the prospects this genetic modification could bring, no attempts have been made yet in this pursuit.

An interesting feature unique to cyanobacteria is that both P-ETC and R-ETC are functional in their thylakoid membrane. The photosynthesis and respiration reactions share certain components such as plastoquinone pool, cytochrome b6f complex and cytochrome c6 or plastocyanin. This peculiar organization in cyanobacteria is the primary reason for its capability to generate electricity under both light and dark conditions unlike PBEC based on isolated thylakoids, PSI and PSII, which can generate electricity only in light. The light induced generation of electrons in P-ETC can be used up by either quinol oxidase or cytochrome oxidase in R-ETC and vice versa. Indeed the overlapping of photosynthesis and respiration is a protective mechanism evolved to handle excess electrons. The photocurrent generated by live cyanobacteria can be attributed to the overflow of the excess electrons from the P-ETC on excessive light absorption. From a PMFC perspective, the R-ETC can be considered as a competitive pathway for electron transport that decreases the electron flux to the electrode thereby decreasing the photocurrent generation. Some or all of the competing oxidases in R-ETC were knocked out of cyanobacteria through genetic manipulations that resulted in more electrons being available for photocurrent generation. Further, the superior electron transfer ability of DMRB such as Geobacter and Shewanella can be predominantly attributed to the cytochromes present in their outer membrane collectively called as outer membrane cytochromes (OMC). Imparting these efficient extracellular electron transfer ability of DMRB to cyanobacteria through genetic manipulations can be a ground breaking and remarkable milestone in cyanobacterial photosynthetic energy conversion. Recently, one such novel and innovative manipulation has been successfully carried out by our group.

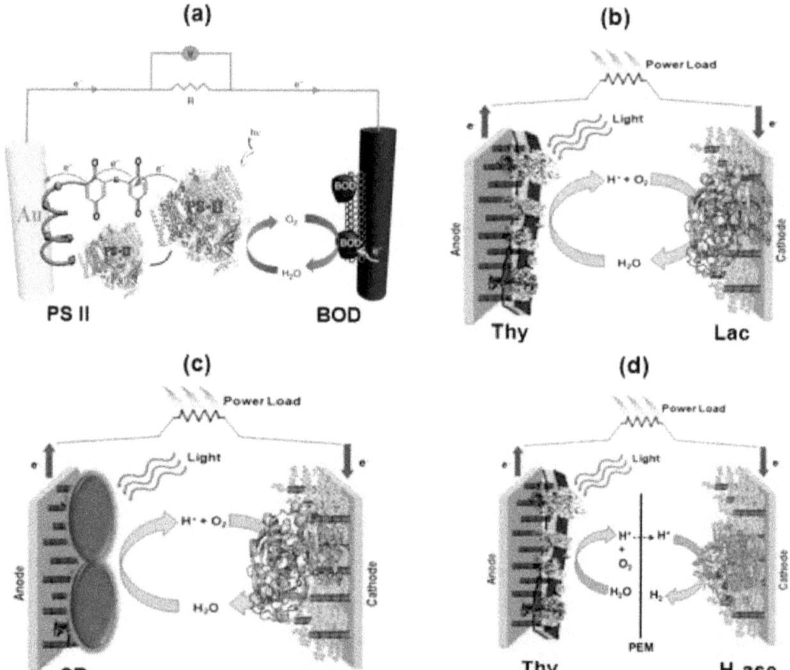

Fig. 5: Schematic showing different photo-bioelectrochemical cell architectures for photosynthesis based electricity generation using biocatalysts such as (a) photosystem II (PSII);18 (b) thylakoid (Thy);5 (c) whole cell photosynthetic microorganisms like cyanobacteria (CB)$_2$ on the anode and enzymatic cathode employing laccase (Lac) and bilirubin oxidase (BOD); (d) hydrogen generation using thylakoid (Thy) on the anode and hydrogenase (H_2ase) on the cathode (PEM: proton exchange membrane).

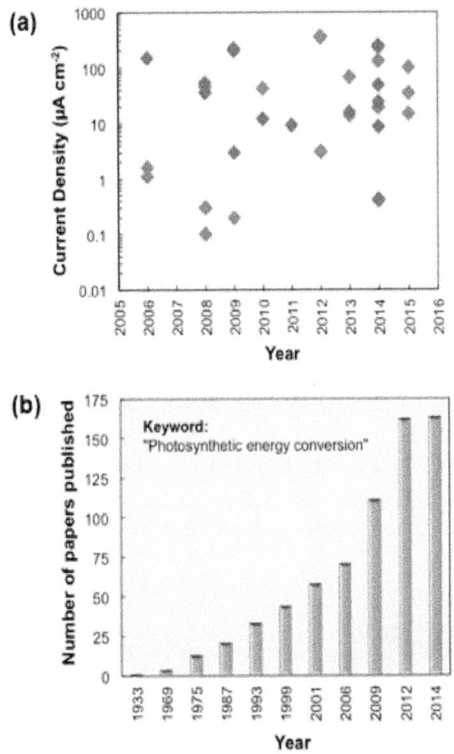

Fig. 6: (a) Current density achieved in photo-bioelectrochemical cells in the last decade using thylakoids or photosystems (red filled marker) and whole cell microorganisms such as cyanobacteria or microalgae (green filled marker) as a biocatalyst on the anode; (b) Number of papers published in the research area of photosynthetic energy conversion (data collected from Web of Science database using the keyword "photosynthetic energy conversion").

5.6 HYDROGEN AND OTHER BIOFUELS

Hydrogen is one of the most attractive alternate energy carriers that are expected to replace fossil fuels for transportation applications. Currently, the production of hydrogen is carried out through conventional steam reforming, electrolysis and thermolysis. All these three methods are expensive and operate at high temperatures and pressures. Steam reforming uses hydrocarbons, which are nonsustainable, and generates CO_2 as byproduct. On the contrary, photosynthesis based hydrogen production is carried out at ambient temperature using photosynthetic microorganisms which generates hydrogen as part of their metabolism by simultaneously sequestering CO_2 in a photobioreactor.48 Further, cyanobacteria such as Synechocystis are employed in bio-photoelectrolysis cells49 for generating hydrogen. In vivo, hydrogen is produced with the help of metallo-enzyme complex called hydrogenases, which catalyze the reversible oxidation of hydrogen to protons and electrons. Hydrogen is also produced as a by-product during nitrogen fixation by other enzyme complexes called nitrogenases in nitrogen fixing cyanobacteria such as Cyanothece and Anabaena. The ferredoxin in the P-ETC is the electron donor for both nitrogenases and hydrogenases. In vitro, the photosynthetic route for hydrogen production would involve an electrochemical full cell with photosynthetic machinery such as PSII as biocatalyst on the anode and enzyme that can catalyze the reduction of protons such as hydrogenase on the cathode as shown in Fig. 7. To date, the highest production rate was achieved by Melnicki et al. (2012), who witnessed sustained H2 production by a unicellular cyanobacterium Cyanothece through photosynthetic process in an electrochemical cell. Under continuous illumination and a nitrogen-deprived environment, cyanobacteria cells in the photo-bioreactor generated H2 at the rate of 400 µmol/ mgChl/h. Though the production of H_2 through photosynthetic means is quite inspiring and challenging, a huge leap in productivity is required for this technology to supersede the conventional production of H_2. With the advancement of metabolic engineering and synthetic

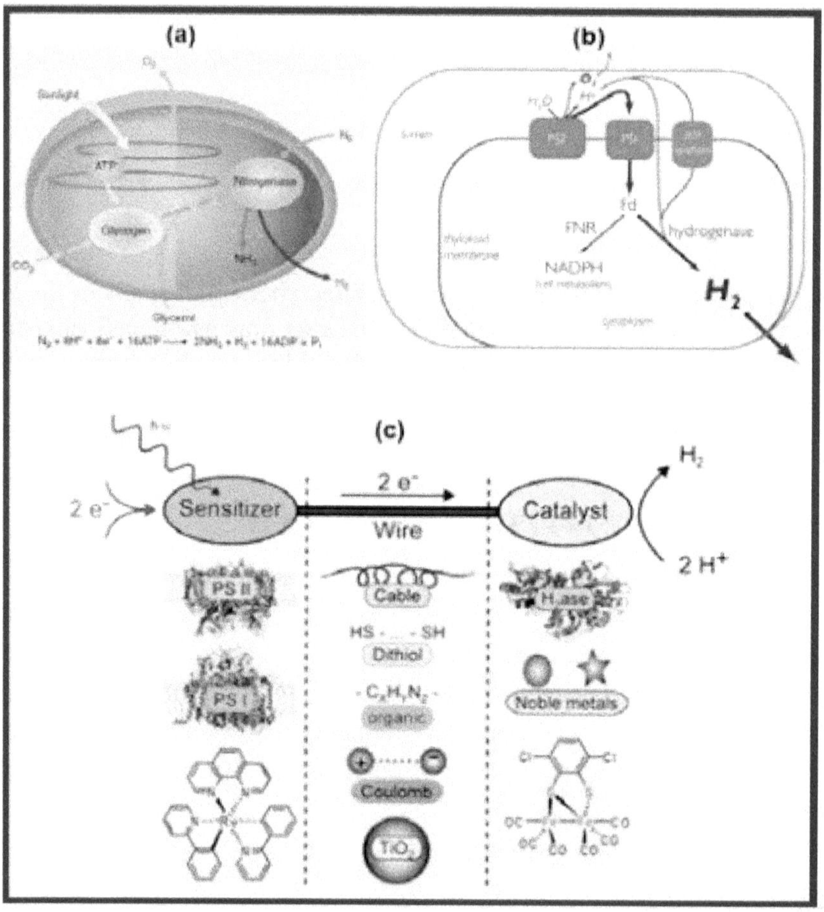

Fig. 7: Photosynthesis based hydrogen generation in vivo using (a) nitrogenase; (b) hydrogenase; and (c) in vitro by combining photosystem (PSII/PSI) and hydrogenase (H2ase)

5.7 AT A GLANCE

Photosynthetic energy conversion is a sustainable, renewable, clean and environment-friendly process. It has an enormous potential to be an alternative energy technology that could effectively replace the finite fossil energy. The field of research has already witnessed major breakthroughs towards generating electricity, hydrogen and other chemical fuels (Fig. 6b). However the technology is still primitive and the performance parameters such as power density, biocatalyst stability, and output require huge improvements before the technology can be considered for any practical application. Modern genetic engineering tools offer prospects for engineering the biology towards enhanced energy conversion. There have been a few endeavors towards this outlook, but more interdisciplinary research work encompassing molecular biology, protein engineering and metabolic engineering along with electrochemistry is required to reach a paradigm shift towards realizing a true biological solar cell based on natural photosynthesis as an alternative power source.

Photosynthesis is the most important natural process on earth, which transformed the once lifeless planet into a living world. While primitive photosynthetic bacteria such as purple sulfur bacteria and green sulfur bacteria carry out anoxygenic photosynthesis, producing elemental sulfur from hydrogen sulfide with the help of sunlight, cyanobacteria, algae and plants carry out oxygenic photosynthesis to convert water and carbon dioxide to sugars with the help of sunlight and release oxygen as a byproduct. The conversion of solar energy to chemical energy via photosynthesis with the release of oxygen has an evolutionary significance on life as we know it today. In fact, photosynthesis is the only natural process known on earth to form oxygen from water. Further, fossil fuels such as coal, petroleum and natural gas are formed from the remains of the dead plants by exposure to heat and pressure in the earth's crust over millions of years. With increasing energy crisis and environmental issues lately, now is the time to revisit photosynthesis in order to address these issues. In this context, a great deal of ongoing research is focused on utilizing photosynthetic energy conversion as a renewable, self-sustainable and environment friendly source of energy. When compared to the finite reserve of fossil fuels, sunlight, the energy source for photosynthesis, is abundant around the planet and is inexhaustible.

6

UTILIZATION OF BIOLOGICAL SOLAR ENERGY AND DEVELOPMENT OF ARTIFICIAL PHOTOSYNTHESIS IN COMBATING CLIMATE CHANGE

6.1. INTRODUCTION

Fossil fuel provide us with almost all the energy needed to power our technological devices, heat our homes and produce different chemicals and raw material that support everyday life, obviously the quantities of fossil fuel available to us will dwindle at no distant time, and now we are faced with the challenges of increasing level of carbon dioxide in the atmosphere and the consequences of global warming. It is worthy of note that fossil fuel reserves are derived from the process of photosynthesis which is the process by which plants, algae and certain types of bacteria capture sunlight efficiently and convert into organic molecule that is the building blocks of all living organisms. Photosynthesis produces more than 100 billion tons of dry biomass annually, which would be equivalent to a hundred times the weight of the total human population on our planet at the present time and equal to about 100 TJ of stored energy.

Biological solar energy is the energy of the sun that is utilized by living organisms such as plants and animals through the process of photosynthesis. This process is sub-divided into light and dark reaction which enable the earth to maintain its carbon cycles. This chapter discusses solar energy efficiencies and how artificial photosynthesis can support the natural one by creating either artificial sun or artificial plant and artificial photosynthesis can be one of the better options in combating climate charge.

6.2. PRINCIPLES OF PHOTOSYNTHESIS

The raw materials of photosynthesis which are Sunlight, water and carbon dioxide are available in almost unlimited amounts. At the heart of the reaction is the splitting of water by sunlight into oxygen and Hydrogen. Oxygen which is the by-product of photosynthesis is released into the atmosphere where it is available for us to breathe and to use for burning our fuels. The hydrogen is not release into the atmosphere as H2, it is combined with carbon dioxide to make organic molecules of various types. When we burn fossil fuels, we combine the carbon stored hydrogen at these organic molecules with oxygen, releasing water and carbon dioxide and effectively reversing the chemical reactions of photosynthesis. Energy is also released from organic molecules when they are metabolized within our bodies by the process of respiration. It is worthy of note that, all energy derived from the products of photosynthesis originates from solar energy as illustrated below:

6.2.1. Energy Flow in Nature

The light reaction of photosynthesis include light absorption, charge separation, water splitting, and electron/proton transfer and provides the reducing equivalents or hydrogen electrons (e) and protons (Hc) to convert carbon dioxide (CO_2) to sugars and other organic molecules that make up living organisms (biomass) including those that lie at the bottom of food chains and provide food. Photosynthesis reactions gave rise to the fossil fuel formed millions of years ago.

The reverse of photosynthesis such as oxidation of organic molecules by respiration (controlled oxidation within our bodies) or burning fossil fuels releases CO_2 combining the hydrogen back with oxygen to form water. The energy release in this process originated from the sunlight. This reaction is subdivided into light and dark reaction, the light reaction occurs in the presence of sunlight and water at the heart of the process the of water take place within the red region at the visible spectrum which have the largest wave length, the visible spectrum has a wavelength of 7x10-7m comprises of red, orange, yellow, green, blue, indigo and violet rays and the equation of light reaction.

Oxygen is obtained as a by-product and a co-enzyme Nicotinamide Ademine Denucleotide (NADP) and Adenosine Triphosphate (ATP) is formed whereas, the dark reaction take place in the absence of light. The energy provided by the ATP with the reduced compound NADP lead to the assimilation of carbon dioxide under a controlled enzyme to form simple sugar and represented in the equation below.

$$\underset{\text{Hydroxyl Ion}}{4H^+} + CO_2 \longrightarrow \underset{\text{Sugar}}{CH_2O} + \underset{\text{Water}}{H_2O}$$

The beauty of this reversible reaction is that the dark reaction returns back the water and CH_2O is the carbon structure from which simple sugar, proton, fat and oil are form.

$$\underset{\text{Water}}{4H_2O} \xrightarrow{\text{Chloroplast}} \underset{\text{Hydrogen Ion}}{4H^+} + \underset{\text{Hydroxyl Ion}}{4(OH)}$$

Then the hydroxyl (OH) is converted to water

$$\underset{\text{Hydroxyl Ion}}{4(OH)} \longrightarrow \underset{\text{Water}}{2H_2O} + O_2$$

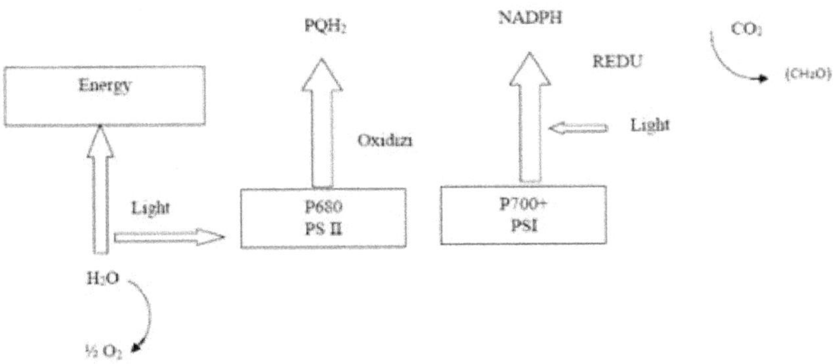

Fig. 1: Light Reactions or Photosynthesis

6.2.2 Efficiency of Photosynthesis

It is worthy of note that photosynthetic organisms can efficiency trap light energy at all wavelengths of visible solar radiation, however, the energy used for splitting water and converting carbon dioxide to carbohydrate is only equivalent to the red region of the spectrum. Higher energy photons (shorter wavelength light, towards the violet region) are degraded to heat by internal conversion within the light harvest pigment to the energy level of red photons. For each electron/proton contracted from water and used to reduce carbon dioxide (CO_2), the energy of two red photons is required which can be accomplished by linking together in series two photo systems namely photo system 1 (PSI) and photo system II (PSH). The Photo system II (PSII) uses light to extract electrons/protons from water while the photo system I (PSI) uses light to give additional conversion process to carbohydrate.

(PSII) which generates a strong oxidizing species (P6800), that help to drive the water Each electron extracted from water and transferred to carbon dioxide enquires the energy of two photons of light, one is absorbed by photo system II splitting reaction and a reductant, plastoquinol (PQH_2). The other generates a strong reducing species that donate hydrogen to carbon dioxide CO_2 to produce sugars and other organic molecules, symbolized as CH_2O and a weak Oxidant P7000 . Electron and proton flow fromPQH_2 to P7000 results in the release energy to convert ADP to ATP. The ATP produced is a store of energy needed along with $NADPH_2$ to convert carbon dioxide to carbohydrate. The production of Oxygen O_2 involves the splitting of two water molecules; the overall process requires eight photons of light (Bridgewater, 2004). A typical product of carbon

Fixation is glucose ($C_6 H1_2O_6$) whose energy content is 2800KJ per mole; this implies that to make a glucose molecule the energy of 48 red photons is required the free energy G stored per Photon as:

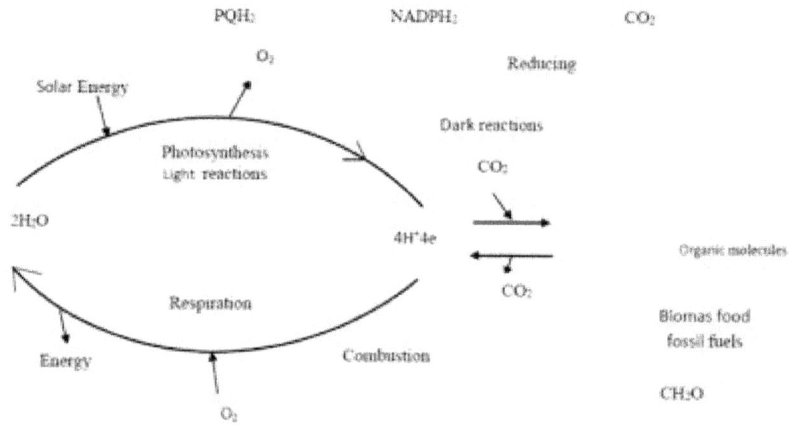

Fig. 2: Energy flow in Nature

$$G = \Omega T\ \Omega R\ \Omega S\ \Omega L\ \Omega_0\ Hv_0 \text{ (Thorndike, 1996)}$$

Where ΩT, Correction for saturation effects; ΩR, Entropy Loss; ΩS, Special distribution of light; ΩL, factor accounting for irreversible energy losses in photo-chemical red biochemical pathways; Ω0, thermodynamic efficiency; Hv0, energy of a photon at the optimum frequency for conversion at red photon.

The process of oxygenic photosynthesis principally linked to the light driven water splitting reaction that occurs in enzymes found in plants, algae and cynobacterial known as photo system (PSII) where solar energy is absorbed by chlorophyII and other pigment and is transferred efficiently to the PSII reaction centre in the process called charge separation take place. The initial conversion of light into electrical energy occur at maximumthermodynamic efficiency of greater than 700c and generate a radical perstate, (P680 pheo) wh + ere p680 is the chlorophyII. A molecule and pheo is a pheophytin , a molecule (a chlorophyll molecule without mg ion ligated into its tetrapyrole head group. The essential point is that the energy of the photon has been used to remove an electron from one site and place it on another(pheo).

The splitting of water into O_2 and the equivalent of $2H_2$ (two NADPH) requires four electrons and PS11 must absorb four photons(4hr) to drive this reaction;

$$2H_2O \xrightarrow[4hr]{} O2 + 4H^{+\ \text{and}}\ 4e$$

6.2.3 Artificial Photosynthesis Model

Artificial photosynthesis model explained that photosynthesis can be successfully mimicked in two different ways, namely Design and production of artificial sun that can function using the basic operation principles of the light driven charge separation that occurs in photosynthesis reaction such prototype sun will have water as it primary source of power as:

$$4H2O \quad + \quad 4H \quad + \quad 4(OH)$$

Chloroplast Light   Hydroxyl ion

Then $2H_2 \quad + \quad O_2 \quad 4H_2O \quad + \quad \text{Heat} \quad + \quad \text{electricity}$

This device can be effectively deployed to extremely cold part of the globe to support natural plants for photosynthesized man for heat and electricity. The second option is the design and production of massive prototypes artificial plants that can trap light at all wavelength but most importantly at the heart of its reaction be able to split water within the red region of the spectrum.

6.2.5 Discussion

The first model attempt to aid photosynthesis by creating artificial prototype solar system that work with natural plants to achieve and balanced the system using the equation.

$$2H2 \quad + \quad O_2 \rightleftharpoons 2H2O \quad + \quad \text{Heat} \quad + \quad \text{electricity}$$

The second model utilizes artificial prototype plant that mimic natural plant that will be able tosplit water at the pick of the reaction, that is light driven reaction, using the equation,

$$4H^{2}O \xrightarrow[\text{Catalyst}]{\text{Light Emzyme of}} 4H^{+} \quad + \quad 4(OH)$$

Hydrogen ion   Hydroxyl ion

Then the hydroxyl (OH) is further converted to water as:

$$4(OH) \quad 4H2O \quad + \quad 02$$

Hydrogen ion   Water   Oxygen

6.3 COMBATING CLIMATE CHANGE

Emission from fossil fuel burning is virtually certain to drive the upward trend in atmosphere carbon dioxide level as a result of greenhouse gases such as carbon, nitrous oxide, perfluorocarbons and sulphur hexafluoride that have a lasting effect on atmosphere composition, radiation forcing and climate. It is currently difficult to know the correct greenhouse gases in the atmosphere that is safe, developing countries are experiencing a dilemma when energy consumption, economic growth, standards of living and transition from civilian to civilian and that affect other researches and innovations, while developed countries are beginning to create energy choices that include low carbon technologies.

No single technology or concept will achieve the global goal of tackling climate change as well as no single policy approach will yield success. The options for combating climate change are decarbonizes energy economics from an established fossil fuel base, development of cleaner and low-carbon technologies. Artificial photosynthesis that is light driven can be achieved as a way of decarbonizing the atmosphere.

6.3.1 Policies and Implementations

Photosynthesis is the most successful solar energy converter on Earth. It provides energy for all life on our planet and is the source of the fossil fuels that pilot our technologies. The solar energy cell can be massively achieved using a good knowledge of molecular processes of photosynthesis and plant molecular biology to attach the challenge of providing non-polluting renewable energy for the benefit of mankind.

The following two avenues can be useful if vigorously explore:

Artificial photosynthesis: The need to establish or venture into the construction of artificial system capable of exploiting solar energy in order to split water with the aim of producing fuels like methane and simple sugar at an efficiency of at least 10% using the available knowledge of photovoltaic's and nanotechnology.

Biomass: Invest massively in the technology of biomass by world government with the intension of reducing CO_2 emission that will mimic man-made global climate change.

6.3.2 Future Energy Option

The most sustainable and reliant sources of future energy is the renewable and water, so it is also imperative to consider a quote from Jules Verne: "I believe that water will one day be used as fuel, because the hydrogen and Oxygen which constitute it, used separately or together, will furnish an inexhaustible source of heat and light. I therefore believe that, when coal deposits are oxidized, we will heat ourselves by means of water. Water is the coal of the future".

6.4 CONCLUSIONS

Having assessed future energy policy options using artificial photosynthesis as tool for decarbonizing the atmosphere, the most sustainable energy source is renewable energy and water.

7

PHOTOSYNTHESIS AND RESPIRATION

7.1 ATP IS THE ENERGY CURRENCY OF THE CELL

- Cells need to carry out many reactions that are energetically unfavorable. You have seen some examples of these non-spontaneous reactions in earlier chapters: the synthesis of nucleic acids and proteins from their corresponding nucleotide and amino acid building blocks and the transport of certain ions against concentration gradients across a membrane.
- In many cases, unfavorable reactions like these are coupled to the hydrolysis of ATP in order to make them energetically favorable under cellular conditions; we have learned that for these reactions the free energy released in breaking the phosphodiester bonds in ATP exceeds the energy consumed by the uphill reaction such that the sum of the free energy of the two reactions is negative ($\Delta G < 0$). To perform these reactions, cells must then have a way of generating ATP efficiently so that a sufficient supply is always available. The amount of ATP used by a mammalian cell has been estimated to be on the order of 109 molecules per second. In other words, ATP is the principal energy currency of the cell.
- How does the cell produce enough ATP to sustain life and what is the source of the energy required to drive the uphill reactions needed to form ATP? A biochemist by the name of Peter Mitchell considered these questions in the 1950s and 1960s and made the surprising discovery that membranes play a critical role. Specifically, this work showed, and as we shall see, that the energy released by transporting protons across the membranes of certain specialized organelles—chloroplasts and mitochondria—drives the synthesis of ATP from ADP and inorganic phosphate (Pi). Figure 1 depicts the structure of these organelles and the following sections explain how these organelles capture energy from sunlight or energy from specific chemicals (e.g. carbohydrates) and convert it to the cell's energy currency.

7.2 PETER MITCHELL AND THE CHEMIOSMOTIC THEORY

- In the 1950s, it was a mystery how the phosphorylation of ADP to ATP was mechanistically connected to the oxidation of glucose. Two main theories emerged—promoted by two men who were educated in the Biochemistry department at Cambridge in the years after World War II. Edward Charles Slater developed methods during the war to measure vitamin concentrations in food and later studied small molecule inhibitors of oxidative phosphorylation. These studies led him to suggest that the oxidation of glucose in some way led to the production of a high energy intermediate metabolite.
- This theory suggested that that intermediate was then used to convert ADP to ATP. Slater spent years trying to identify and isolate this proposed high energy intermediate without success. Around the same time, Peter Mitchell began working on the same problem. During the war, Mitchell helped develop a treatment for skin lesions caused by a chemical warfare agent, Lewisite. The molecule they developed would not cross biological membranes and would scavenge the chemical agent from blood, allowing it to be excreted. After the war, Mitchell studied the mechanism of action of penicillin and the bilayer structure of bacterial membranes. His work on membranes and the transport of metabolites across them led him to propose the chemiosmotic theory of oxidative phosphorylation.
- This theory suggests that the energy in glucose is in some way stored in an electrochemical gradient across a biological membrane. The energy released upon movement of an ion down that gradient is then used to convert ADP and Pi to ATP. This was a radical idea at the time and Slater and Mitchell debated their theories in print for many years until Mitchell's gained widespread acceptance. He was awarded the Nobel Prize in Chemistry in 1978 for his work on the chemiosmotic theory. Mitchell and his colleague, Jennifer Moyle, performed many of the definitive experiments on chemiosmosis at a research facility he built in a neglected mansion, Glynn House, in Cornwall, England. After stepping down from his position at Edinburgh University for health reasons, he spent several years restoring the mansion and building the laboratory, which was named the Glynn Research Institute. Mitchell continued to work there with a small group of research associates until 1987.

7.3 CHLOROPLASTS HARVEST LIGHT ENERGY TO GENERATE ELECTRICAL ENERGY

- The ultimate source of chemical energy in the biosphere is the sun. Photosynthetic organisms capture the sun's energy and use it to generate chemical fuels (simple sugars), which are directly or indirectly consumed through the food chain. This conversion of sunlight into chemical energy takes place in three steps. First, photosynthetic organisms use light energy to generate an electrochemical gradient across a membrane, converting energy from light into electrical potential energy. A remarkable protein machine then converts this potential energy into chemical energy in the form of ATP. Finally, the ATP is used to drive the formation of carbohydrates by fixation of carbon dioxide from the atmosphere. In what follows, we will consider these three phases of photosynthesis: the generation of an electrochemical gradient, the synthesis of ATP, and the conversion of carbon dioxide to carbohydrate.

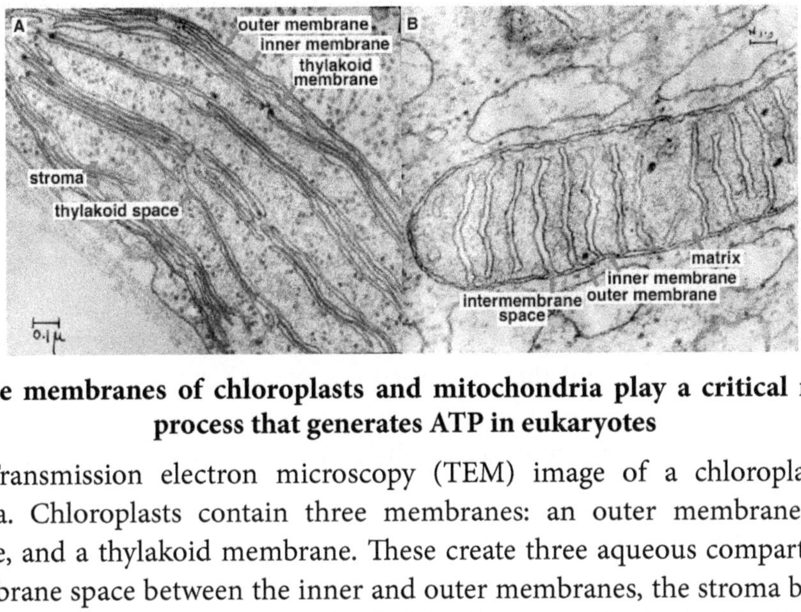

Fig. 1: The membranes of chloroplasts and mitochondria play a critical role in the process that generates ATP in eukaryotes

(A) Transmission electron microscopy (TEM) image of a chloroplast from a green alga. Chloroplasts contain three membranes: an outer membrane, an inner membrane, and a thylakoid membrane. These create three aqueous compartments: the intermembrane space between the inner and outer membranes, the stroma between the inner and thylakoid membranes, and the thylakoid space inside the thylakoid membrane. The thylakoids pack against each other into stacks called grana.

(B) TEM image of a mitochondrion from a guinea pig pancreas. The organelle contains two membranes: an inner and an outer membrane. The inner membrane is folded into extended internal structures called cristae, which give the inner membrane a larger surface area than the outer membrane. The membranes create two internal aqueous compartments: the intermembrane space (IMS) between the two membranes and the matrix inside the inner membrane. Both TEM images were collected by George E. Palade (cellimagelibrary.org).

- The molecules that absorb light in plants are found in the thylakoid membranes of their chloroplasts. Chloroplasts have three membrane bilayers: outer and inner membranes, which surround the organelle, and a thylakoid membrane, which folds back and forth upon itself into stacks of membranes called grana (Figure 1). Light-absorbing protein complexes called photosystems span the thylakoid bilayer and consist of several integral membrane proteins and hundreds of small pigment molecules. The most important of these pigments is chlorophyll—the molecule that gives plants their green color. Most of the chlorophyll molecules are bound to the proteins at the periphery of the photosystem in what are called antenna complexes, but two special chlorophyll molecules are also found in the middle of the photosystem in the reaction center (Figure 2). Photosystems convert light energy into electrical energy at this reaction center.

- Chlorophyll molecules contain a chemical functional group, a porphyrin ring, that is capable of absorbing sunlight in the red region of the visible spectrum (Figure 2). When one of the chlorophyll molecules in the antenna complexes absorbs light, the electrons in its porphyrin ring move to a higher energy state. As the electrons return to their normal, lower energy state, they transfer their absorbed energy to a neighboring chlorophyll molecule. This second chlorophyll molecule, in turn, transfers its absorbed energy to yet another neighbor in a chain reaction that continues until the energy reaches the reaction center of the photosystem complex. The electrons in the pair of chlorophyll molecules in the reaction center also move to a higher state when they absorb the energy from the neighboring antenna chlorophyll, but they then transfer those high energy electrons to an acceptor molecule. In other words, the reaction center chlorophyll molecules transfer electrons with their absorbed energy rather than just transferring the energy (as the antenna chlorophyll molecules do). This electron transfer is the step in photosynthesis that converts light energy into electrical energy.

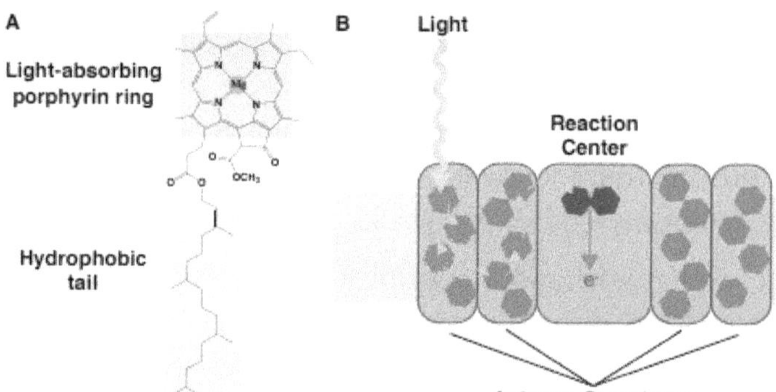

Fig. 2: Light-harvesting photosystem complexes in the chloroplast thylakoid membrane contain many bound chlorophyll molecules, which absorb light energy and funnel it to the reaction center

(A) The chemical structure of the chlorophyll molecule contains a light-absorbing porphyrin ring and a hydrophobic tail. A single photosystem can contain hundreds of chlorophyll molecules.

(B) The trajectory of energy transfers between chlorophyll molecules in the photosystem is shown. Any one of the antenna chlorophyll molecules (green hexagons) can absorb light and then transfer it to its neighbors. The individual chlorophyll molecules are bound by the proteins in slightly different ways that allow the light energy to be transferred towards the reaction center. When the pair of chlorophyll molecules at the reaction center (dark green hexagons) receive the energy, they eject electrons, initiating an electron transport chain.

- The transfer of electrons from the reaction center creates a charge separation in which the chlorophyll molecules now carry positive charges. This leaves us with two questions: How do the special chlorophyll molecules regain their electrons and what is the fate the electrons that had been ejected from the reaction center? We shall address the fate of the ejected electrons after we explain the answer to the first question. The chlorophyll molecules regain their proper number of electrons by extracting them from water molecules in the thylakoid space. Reactions of this type, in which electrons are transferred from one molecule to another, are called redox reactions. The molecule that loses electrons is said to be "oxidized," and the molecule that gains electrons is said to be "reduced." As shown below, we can write the oxidation and reduction steps of a reaction as two separate "half reactions," in which we explicitly show the lost or gained electrons as products or reactants, respectively.

Half Reactions:

Oxidation: $A \rightarrow A^+ + e^-$

Reduction: $B^+ + e^- \rightarrow B$

Complete Redox Reaction:

$A + B^+ \rightarrow A^+ + B$

In photosynthesis, the oxygen atoms in water molecules undergo oxidation, transferring some of their electrons to reduce the special chlorophyll molecules in the reaction center and in the process the water breaks down to form O2 and H+ ions (Figure 3). The oxidation and reduction half reactions for this process are:

Half Reactions:

Oxidation: $2H_2O \rightarrow 4H^+ + O_2 + 4\,e^-$

Reduction: $2\text{ chlorophyll}^{2+} + 4\,e^- \rightarrow 2\text{ chlorophyll}$

Complete Redox Reaction:

$2H_2O + 2\text{ chlorophyll}^{2+} \rightarrow 4H^+ + O_2 + 2\text{ chlorophyll}$

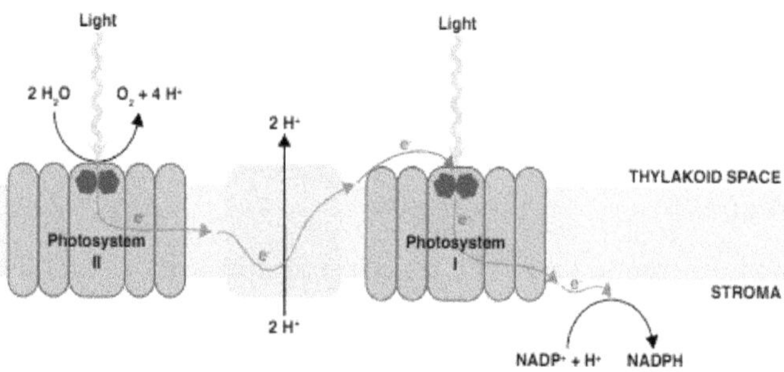

Fig. 3: The electron transport chain in the chloroplast thylakoid membrane.

Two photosystems harvest light energy and eject high energy electrons, which are then transferred to other electron carrier molecules in the chain. As the electrons move from water to NADPH, six protons are released into the thylakoid space, creating a proton gradient across the membrane. For a more detailed description of each of the electron transfers in this chain.

- This oxidative water splitting reaction starts the process of generating an electrochemical gradient across the thylakoid membrane; as the water molecules react, they produce more protons on the inside of the membrane (in the thylakoid space) than there are on the outside (in the stroma). The electrons that are released from the special reaction center chlorophyll molecules initiate a series of electron transfer reactions that are energetically favorable—that is, the electrons become more stable with each transfer from one molecule to the next in the chain (Figure 3). The electrons pass from the first photosystem (which is counterintuitively named photosystem II) through a set of intermediate electron carriers. Some of the energy that is released by these downhill electron transfers is used to transport two protons from the stroma into the thylakoid space, increasing the proton gradient across the membrane. The electrons then arrive at a second photosystem complex (named photosystem I) where they receive another input of light energy. The antenna chlorophyll molecules of this complex again absorb light and funnel its energy to the special reaction center chlorophyll molecules, which then lose electrons. These high energy electrons are transferred through another set of intermediates on the stromal side of the membrane to the ultimate acceptor in the chain, NADP+, generating NADPH. NADPH is a small molecule that, by carrying high energy electrons, functions as another energy currency in the cell.
- Overall, electrons are transferred through the chloroplast electron transport chain from H_2O to NADP+, generating O_2, protons, and NADPH in the process. The "Z-scheme" summarizes the energetic changes associated with these electron transfers (Figure 4). The two photosystems each increase the energy of the electrons in the pathway; all of the other transfers are energetically downhill. Two inputs

of light energy are required because photosystem II cannot generate an electron with high enough energy to be transferred to NADP+ and photosystem I cannot directly extract electrons from water. In tandem, however, the two photosystems are capable of transferring electrons uphill from H_2O to NADPH. The electrons on NADPH are then used to make a carbohydrate from carbon dioxide in the last phase of photosynthesis; however, this carbohydrate synthesis also requires ATP, which is made using the other product of the electron transport chain, the proton gradient across the thylakoid membrane.

- For each NADPH molecule produced by the electron transport chain, six protons are released into the thylakoid space inside the membrane. Four of these come from the water splitting reaction at photosystem II and two are transported across the membrane using the energy of the subsequent downhill electron transfers. The resulting pH difference between the thylakoid space and the stroma is approximately 3 units (i.e., a 1,000-fold difference in [H+]). This proton gradient is a high energy, low entropy state, which effectively conserves some of the energy absorbed from the sun. Because protons are positively charged, this gradient has both chemical and electrical components (Figure 5). The chemical gradient is created by the difference in the concentration of the protons (a pH difference), and the electrical gradient is created by the charge separation (a potential difference, or voltage). Both of these gradients contribute to the proton motive force— i.e., a force that drives the movement of protons down these gradients from the thylakoid space into the stroma.

7.4 COMPLEXITIES AND SIMILARITIES OF ELECTRON TRANSPORT CHAINS

- The electron transport chains of chloroplasts and mitochondria share several similar components despite the fact that they derive their high energy electrons from very different sources. A more detailed description of the electron transfers in these organelles illustrates these similarities. From the first photosystem in the chloroplast electron transport chain (photosystem II), high energy electrons are transferred to a small molecule, plastoquinone (pQ). Plastoquinone is soluble in the lipid bilayer and carries the electrons to the next protein complex in the thylakoid membrane, the cytochrome b6f complex (yellow region in Figure 3). The cytochrome b6 f complex transfers the electrons from plastoquinone to plastocyanin (pC), a small water-soluble protein in the thylakoid space. As the cytochrome b6 f complex performs these electron transfers, it transports two protons across the thylakoid membrane. Plastocyanin then carries the electrons from the cytochrome b6 f complex to the second photosystem in the chain (Photosystem I). The electrons on plastocyanin are used to replace those that are ejected from the reaction center of Photosystem I. The high energy electrons from Photosystem I are transferred to ferredoxin (Fd), a protein on the stromal side of the membrane. The final protein complex in the chain is the ferredoxin-NADP+ reductase, which transfers the electrons from ferredoxin to NADP+, generating NADPH.

- The mitochondrial electron transport chain starts at Complex I, which accepts electrons from NADH and transfers them to a hydrophobic small molecule, quinone, in the membrane bilayer. Quinone has similar properties to plastoquinone in chloroplasts; it is soluble in the membrane lipids and carries electrons through the membrane from one protein complex to the next. As complex I transfers electrons to quinone, it transports four protons across the membrane. Complex II also transfers electrons to quinone but it accepts them from a different substrate, succinate instead of NADH, and does not transport any protons. Quinone then carries the electrons it has accepted from Complex I or II to Complex III. Complex III pumps another four protons across the membrane as it transfers the electrons from quinone to cytochrome c. Cytochrome c is a soluble protein on the outer surface of the membrane (like plastocyanin in chloroplasts) that can move between complexes III and IV. When cytochrome c binds to complex IV, it transfers its electrons through complex IV to oxygen on the opposite side of the membrane. Oxygen is the ultimate electron acceptor in the electron transport chain. Complex IV reduces oxygen to water and transports another two protons across the membrane in this process.
- There are thus electron transfers in both chains that involve a small lipid-soluble molecule, plastoquinone or quinone, and a soluble protein outside the membrane, plastocyanin or cytochrome c. These carrier molecules move the electrons from one membrane protein complex to the next in the chain. Both electron transport chains also clearly generate the proton gradients across their respective membranes that are critical for synthesizing ATP.

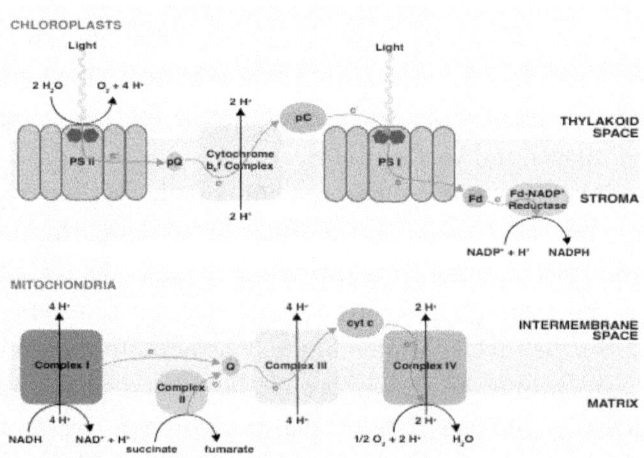

7.5 OTHER CELLULAR ENGERY CURRENCIES: NADPH, NADH, AND $FADH_2$

- Photosynthesis and aerobic glucose metabolism are redox processes in which electrons are transferred between carbon and oxygen. In photosynthesis, the carbon in carbon dioxide is reduced and the oxygen in water is oxidized; in aerobic glucose metabolism, the opposite occurs as the carbon in glucose is oxidized and oxygen is reduced to form water. The electrons in these processes, however, are not transferred directly between the carbon and oxygen atoms. The electrons are instead first transferred to or "captured" by carrier molecules, NADP+, NAD+, and FAD, which become NADPH, NADH, and $FADH_2$ upon reduction. These molecules are water-soluble dinucleotides that readily undergo reversible reduction and oxidation. The structures on the left show both the oxidized (left) and reduced (right) forms of these molecules and how the equilibrium mixture between these states is governed by whether they are receiving or donating electrons.
- The nicotinamide group (highlighted in yellow) of the nicotinamide adenine dinucleotides (NADP+ and NAD+) readily accepts and donates two electrons and a proton to other molecules in the cell. These two molecules differ only by the presence of a 2'-phosphate on NADP+, but they are used in different processes in the cell. NADPH is used in biosynthetic processes like photosynthesis. By contrast, NAD+ is used to oxidize substrates like glucose to make ATP. A similar molecule, FAD, is also used to metabolize substrates. The flavin group (also highlighted in yellow) carries two electrons, but it can transfer them one at a time to other molecules in the cell.
- Note that because the oxidation of NADPH, NADH, and $FADH_2$ is energetically favorable, these molecules can be thought of as another energy currency in the cell. Their ability to donate electrons to other molecules can be harnessed to make other forms of chemical energy or an electrochemical gradient across a membrane.

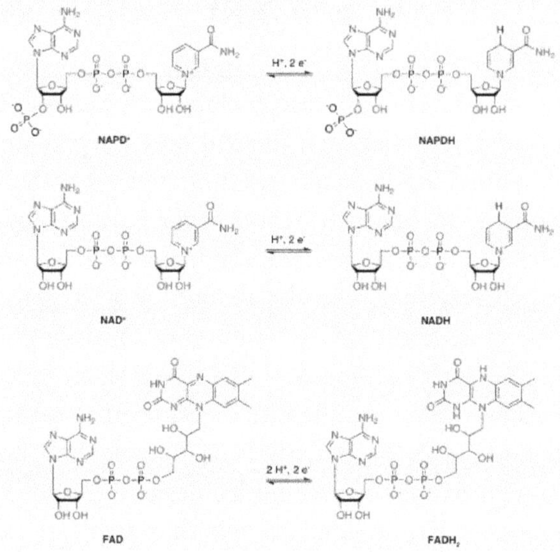

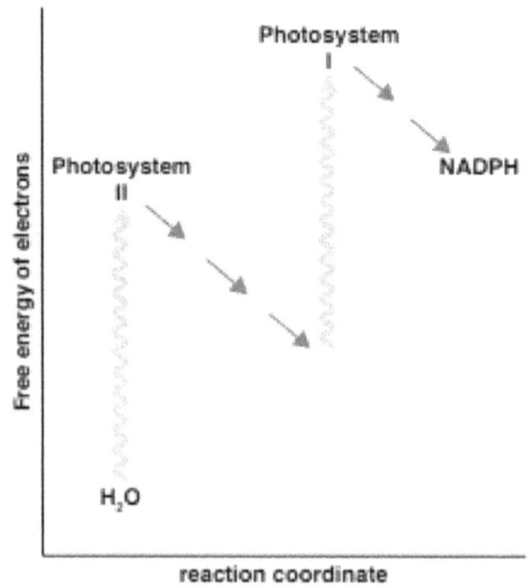

Fig. 4: The Z – Scheme

The "Z-scheme" describes how light energy is used in two steps in the chloroplast electron transport chain to increase the energy of electrons from water so that they are capable of ultimately reducing NADP+ to NADPH. The two photosystems increase the energy of the electrons, whereas all the other transfers between chemical groups in the chain stabilize the electrons.

7.6 ATP SYNTHASE IS A MOLECULAR MACHINE THAT COUPLES PROTON TRANSPORT TO ATP SYNTHESIS THROUGH MECHANICAL ROTATION

- When protons flow down their electrochemical gradient across the thylakoid membrane, energy is released that can be used to produce ATP from ADP and inorganic phosphate (Pi). The ATP Synthase is a molecular machine that directly couples the downhill transport of protons across the membrane to the uphill chemical reaction that synthesizes ATP. This remarkable machine is found in the membranes of chloroplasts, mitochondria, and bacteria where it performs the same, fundamental energy-converting reaction. The potential energy stored in the proton gradient is first converted to mechanical energy and then into chemical energy by a series of conformational changes that alter the catalytic sites of the ATP synthase as we will explain. The concept that an electrochemical gradient can be harnessed to generate chemical energy in the form of ATP came as a huge surprise and was the basis for a Nobel Prize to its advocate Peter Mitchell.

◉ The structure of the ATP Synthase reveals how the enzyme accomplishes these energy interconversions. The synthase has a lollipop structure with its base embedded in the membrane and its head protruding into the chloroplast stroma (or mitochondrial matrix or bacterial cytoplasm) (Figure 6). The membrane portion of the synthase contains the proton transporting activity; protons passively flow down their gradient through a channel formed by the transmembrane protein subunits. The movement of protons across the membrane induces the transmembrane proteins to rotate in the plane of the membrane, which in turn makes the axle of the ATP synthase rotate as well. The axle consists of a set of extended proteins which link the transmembrane proteins to the enzymatic, ATP-generating part of the ATP synthase. These axle proteins sit inside the central cavity of the ATP generator (the head of the lollipop). The ATP generator contains 6 subunits (3 α and 3 β), which are arranged like segments of an orange around the axle. Each β subunit contains a catalytic site for synthesizing ATP. The ATP generator does not rotate with the axle; it is held in place by the stator protein, which is anchored in the membrane. When the axle rotates inside the ATP generator, it interacts differently with the three individual β subunits; these changing interactions drive conformational changes in the β subunits that facilitate ATP synthesis.

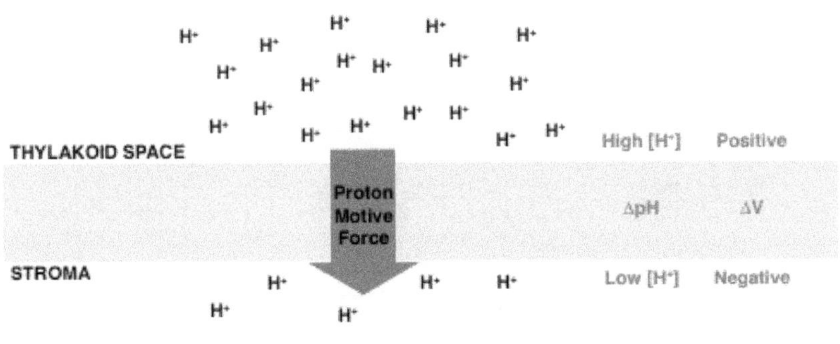

Fig. 5: The electron transport chain generates an electrochemical gradient across the thylakoid membrane

Both the chemical (pH) and electrical (voltage) components of the gradient favor the movement of protons into the stroma from the thylakoid space.

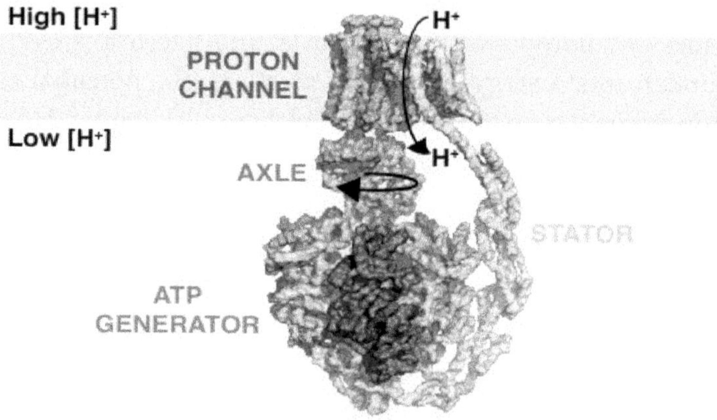

Fig. 6: The ATP synthase couples proton transport to ATP synthesis through mechanical rotation.

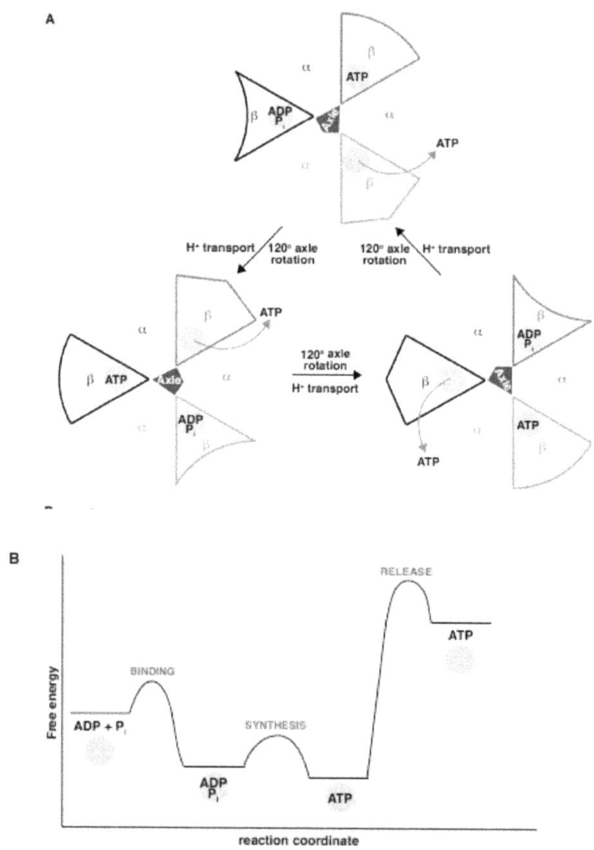

Fig. 7: ATP synthesis is catalyzed by a binding change mechanism that dramatically stabilizes ATP in the enzyme active site

(A) The subunits of the ATP generator exist in different conformations (indicated by the different shapes) depending on their interactions with the axle proteins at their center. The subunits of the ATP generator do not rotate, but each 120° rotation of the axle induces their conformations to interconvert such that the affinity of the β subunits for ADP and ATP changes. The text describes the reaction cycle from the perspective of the black β subunit, but notice that all three of the β subunits are engaged in synthesizing ATP.

(B) The energetic changes associated with each step of the ATP synthesis reaction are shown. Each step corresponds to a 120° rotation of the axle proteins. The initial association of the substrates, ADP and Pi, with the enzyme is spontaneous. Upon rotation of the axle, the active site converts to a conformation that binds ATP tightly and the new bond between ADP and Pi accordingly forms. The final rotation of the axle provides the energy to release the newly formed ATP from the active site.

- The rotation of the axle proteins induce conformational changes in the structure of the β subunit active sites (Figure 7A). These changes cause the affinities for ATP and ADP and Pi to vary as the axel rotates (Figure 7B. Paul Boyer described this "binding change mechanism" using biochemical experiments, and John Walker verified it by determining the structure of the ATP synthase; they received the Nobel Prize in Chemistry in 1997 for their work. This mechanism dictates that at any given time, the three β subunits are each in a different state: one has ADP and Pi bound, the second has ATP bound tightly; and the third is empty. The rotation of the axle induces the subunit states to sequentially interconvert (Figure 7A). If we consider the reaction cycle from the perspective of a single β subunit that starts in the conformation that binds ADP and Pi tightly, it will spontaneously pick up these substrates from the chloroplast stroma as the binding equilibrium is favorable ($\Delta G < 0$). The first rotation of the axle then induces this subunit to adopt the conformation that binds ATP tightly. This conformation binds ATP so tightly that the free energies of the ADP- and ATP-bound states are nearly equal even though ATP has a much higher energy than ADP and Pi when the nucleotides are not bound to the enzyme. The activation barrier to forming ATP in the enzyme active site is also therefore relatively small compared to that in the uncatalyzed reaction (Figure 7B). The ATP generator thus makes a reaction that is nonspontaneous ($\Delta G > 0$) in solution readily reversible ($\Delta G \sim 0$) in the active site.
- The final rotation of the axle then induces the ATP-bound β subunit to convert to the state that has low affinity for ATP. The newly synthesized ATP accordingly dissociates from the enzyme and is released back into the stroma. This release step is the slowest step in the reaction on the ATP generator; by binding ATP so tightly, the enzyme increases its barrier to dissociation. The rotation of the axle provides the energy required to overcome this barrier and shift the β subunit to

the low-affinity state, thereby converting mechanical energy into chemical energy and driving the ATP synthesis reaction in the forward direction.

The dark reactions of photosynthesis generate a sugar from carbon dioxide

- Carbon dioxide is fixed, that is, removed from the atmosphere and incorporated into a small, water-soluble carbohydrate in the last phase of photosynthesis. (Carbohydrates are molecules that have the general formula Cn (H_2 O)n in which each carbon is hydrated with water. They can consist of a single sugar such as glucose [C_6 $H1_2O_6$] or ribulose [see below] or a polymer of multiple sugars as in starch and cellulose.) This last phase consists of a cycle of reactions that are collectively termed the dark reactions of photosynthesis or, alternatively, the Calvin cycle. These reactions utilize the NADPH and ATP produced by the electron transport chain and the ATP synthase. In addition to producing carbohydrates, the dark reactions also regenerate NADP+ and ADP so that photosynthesis can continue (Figure 8).
- The critical carbon-fixing reaction of the Calvin cycle is carried out by an enzyme named ribulose 1,5-bisphosphate carboxylase/oxygenase, or rubisco for short. Rubisco attaches CO_2 to a 5-carbon molecule, ribulose 1,5-bisphosphate. [This molecule is a derivative of the sugar ribulose in which two hydroxyls (OH groups) have been replaced by phosphates.] This generates an unstable 6-carbon intermediate, which breaks down into two 3-carbon molecules of 3-phosphoglycerate (Figure 9). (A note on nomenclature: bisphosphate molecules contain two phosphate groups that are not directly attached to one another. In diphosphates, like ADP, the phosphates are directly linked.) Rubisco is an unusual enzyme in that it is exceptionally inefficient; it converts only 3 molecules of CO_2 to 3-phosphoglycerate each second at 25°C. Rubisco's inefficiency results from its need to discriminate between CO_2 and O_2 . O_2 is much more abundant in the atmosphere than CO_2 , and as O_2 is smaller than CO_2 , it also fits into rubisco's active site. When O_2 binds to rubisco, it reacts with ribulose 1,5-bisphosphate, thereby destroying this critical substrate for carbon fixation. The active site of rubisco has accordingly evolved to bind CO_2 more tightly than O_2 , but as the enzyme binds CO_2 tighter, the energetic barrier to its reaction increases and the reaction rate decreases.

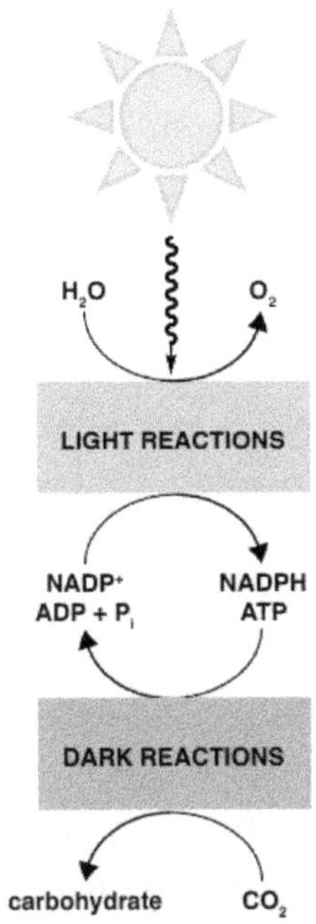

Fig. 8: Overview of photosynthesis

The oxygen in water is oxidized and the carbon in carbon dioxide is reduced in two separate steps of photosynthesis: the light and dark reactions, respectively. NADPH acts an intermediary electron carrier between these two reactions. The NADPH and ATP produced by the light-dependent oxidation of water is used to fix carbon dioxide into a water-soluble carbohydrate building block.

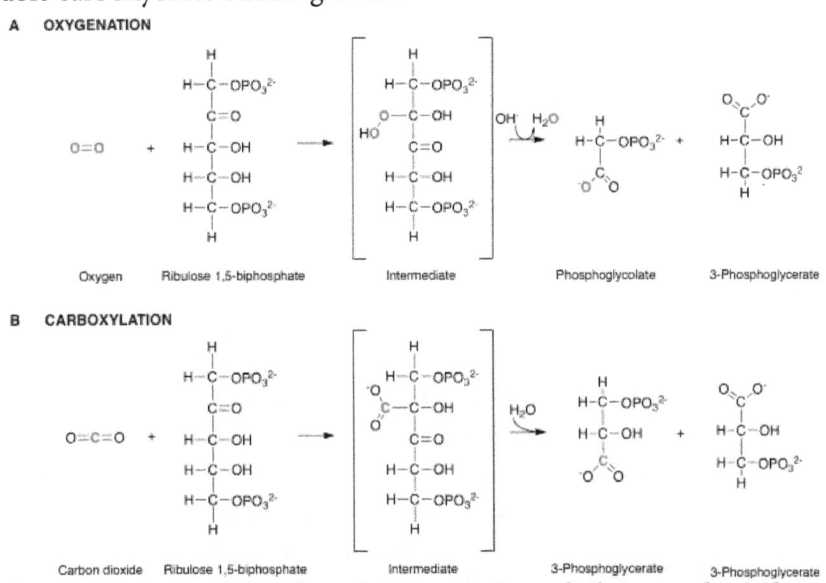

Fig. 9: Rubisco catalyzes the carbon-fixing reaction of photosynthesis by attaching a carbon dioxide molecule to ribulose 1,5-bisphosphate, but the enzyme must contend with a competing oxygenation reaction

- This oxygenation reaction (A) creates a product, phosphoglycolate, which cannot be easily recycled to make ribulose 1,5-bisphosphate. The enzyme has consequently evolved to suppress the oxygenation reaction but with detrimental effects on the efficiency of carboxylation reaction (B).
- Rubisco thus illustrates the tradeoff between substrate specificity and catalytic activity. Enzymes must bind their substrates with enough affinity to allow those substrates to outcompete other similar non-substrate molecules that might fit in the active site; however, enzymes cannot bind their substrates too tightly or they will make them too stable to react. The balance of these two competing factors forces rubisco to a low catalytic efficiency, but plants make lots of this protein to compensate. Rubisco accounts for 50% of the protein weight of chloroplasts and is consequently one of the most abundant enzymes on the planet.
- The remaining reactions in the Calvin cycle convert the 3-phosphoglycerate made by rubisco to glyceraldehyde 3-phosphate and regenerate the rubisco substrate, ribulose 1,5-bisphosphate (see Box 4 for details). The glyceraldehyde 3-phosphate is used as a substrate to generate the sugar glucose and, in turn, the polymer of glucose, starch. Plants (and animals) generally store chemical energy that is not immediately required to perform cellular reactions in the form of sugar or starch rather than as ATP because sugars are more soluble in water than ATP. More sugar can be dissolved in the cell and, therefore, more energy can be stored in the cell in this form.
- In summary, through the light and dark reactions of photosynthesis, plants convert light energy first to potential energy in the form an electrochemical gradient, then to chemical energy in the form of ATP and the reducing potential of NADPH, and finally to chemical energy in the form of a simple sugar. Glyceraldehyde 3-phosphate molecules can be combined to make sugar molecules like glucose and then assembled into larger starch molecules for storage. Starch can then be broken down to make ATP according to the cell's needs.

7.7 CARBON FIXATION BY THE CALVIN CYCLE

- Melvin Calvin received the Nobel Prize in Chemistry in 1961 for describing how plants assimilate carbon dioxide. This process is accordingly named the Calvin cycle and it consists of three steps. The first step is catalyzed by rubisco and results in the incorporation of carbon dioxide molecules into ribulose 1,5-bisphosphate to produce 3-phosphoglycerate. In the second step, 3-phosphoglycerate is phosphorylated and reduced to make glyceraldehyde 3-phosphate using the ATP and NADPH produced in the light reactions of photosynthesis. This molecule, glyceraldehyde 3-phosphate, is the product of photosynthesis; it is a simple sugar that can be used as a biosynthetic building block for larger sugars, fatty acids, and amino acids.

- In step three of the Calvin cycle, however, some of the glyceraldehyde 3-phosphate molecules produced by step two are recombined and phosphorylated to regenerate ribulose 1,5-bisphosphate. This step also requires ATP and is necessary to keep the Calvin cycle running in the forward direction. If the rubisco substrate, ribulose 1,5-bisphosphate, was depleted, CO_2 could no longer be fixed.

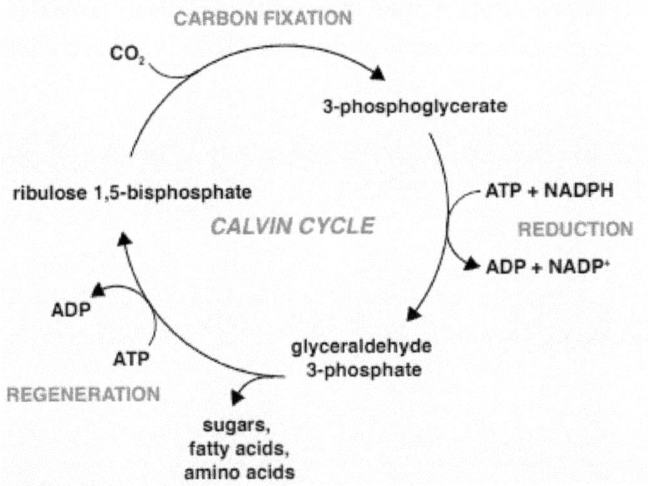

7.8 NON-PHOTOSYNTHETIC ORGANISMS GENERATE ATP BY METABOLIZING GLUCOSE AND OTHER CARBON-CONTAINING MOLECULES

Many organisms are not photosynthetic and hence are unable to capture the sun's energy to make ATP. Instead, they consume carbon-containing compounds, such as carbohydrates, lipids and proteins produced by other organisms and metabolize them to generate ATP. Ultimately, this food chain originates from glucose generated by photosynthetic organisms. (Even photosynthetic cells in plants rely on glucose generated by their own chloroplasts to generate ATP and many plant cells, such as those of the roots and stem, are not photosynthetic.) Here we will focus on generating ATP from glucose. This can be achieved in two ways: one requires oxygen (O_2) and the other does not. The anaerobic (oxygen-independent) process is called fermentation and is much simpler than the aerobic (oxygen-dependent) process.

Fermentation begins with a series of ten reactions that are collectively called glycolysis. Glucose, a sugar ("glyco") molecule containing 6-carbons, is split ("lysis") into two 3-carbon molecules. The first five reactions are the preparatory phase of glycolysis, in which two ATP molecules are expended to phosphorylate the sugar substrate and enable its cleavage into two 3-carbon molecules. The second five reactions constitute the payoff phase, as molecules containing high energy phosphates are generated and then used to directly phosphorylate ADP. In total, these ten reactions produce two molecules of ATP and two molecules of NADH.

- The sixth and seventh steps of glycolysis are the key reactions that generate the first molecule containing a high energy phosphate, 1,3-bisphosphoglycerate (Figure 10B). Importantly, the phosphate bond in 1,3-bisphosphoglycerate (highlighted in yellow) is at a higher energy state than ATP. This molecule can, therefore, spontaneously transfer a phosphate to ADP to make ATP. The remaining steps in glycolysis generate another molecule with a similarly unstable phosphate bond, phosphoenolpyruvate, which also directly phosphorylates ADP.

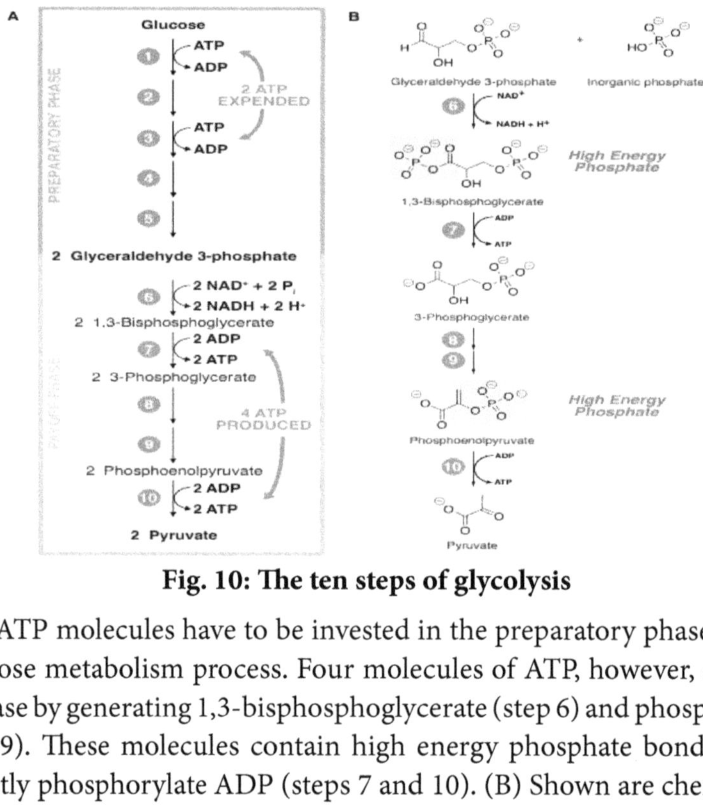

Fig. 10: The ten steps of glycolysis

(A) Two ATP molecules have to be invested in the preparatory phase of glycolysis to start the glucose metabolism process. Four molecules of ATP, however, are produced in the payoff phase by generating 1,3-bisphosphoglycerate (step 6) and phosphoenolpyruvate (steps 8 and 9). These molecules contain high energy phosphate bonds, which enable them to directly phosphorylate ADP (steps 7 and 10). (B) Shown are chemical structures depicting the molecules involved in steps 6 through 10. (Note that steps 6 and 7 reverse the reactions of the second step of the Calvin cycle. Regions of molecules highlighted in yellow depict unstable phosphate bonds. ADP reacts with these reactive phosphate groups to make ATP.

Glycolysis produces a net yield of 2 molecules of ATP, which can be used to carry out cellular reactions; however, and as we have seen, glycolysis also consumes 2 molecules of NAD+ (Figure 10, step 6). These NAD+ molecules must be regenerated so that glycolysis can continue breaking down more molecules of glucose. In cells that are oxygen-deprived, the NAD+ is regenerated by reducing the pyruvate from glycolysis to either lactate or ethanol using NADH. This process is ancient; early cellular organisms evolved in an atmosphere that contained very little oxygen and needed to generate energy anaerobically, but many contemporary organisms have retained the ability to ferment glucose in this way. For example, when muscle cells are being strenuously exerted, they become depleted of oxygen and start to convert pyruvate to lactate (Figure 11). The build-up of the acidic lactate molecule in these cells causes muscle soreness. Many microorganisms, including brewer's and baker's yeast, perform a different type of fermentation in which pyruvate is converted into ethanol and carbon dioxide. The carbon dioxide produced by these organisms makes bread rise and champagne bubble, but the critical product for sustaining life is NAD+. The entire fermentation process thus produces a net yield of 2 molecules of ATP without consuming oxygen or changing the concentrations of NAD+ and NADH in the cell.

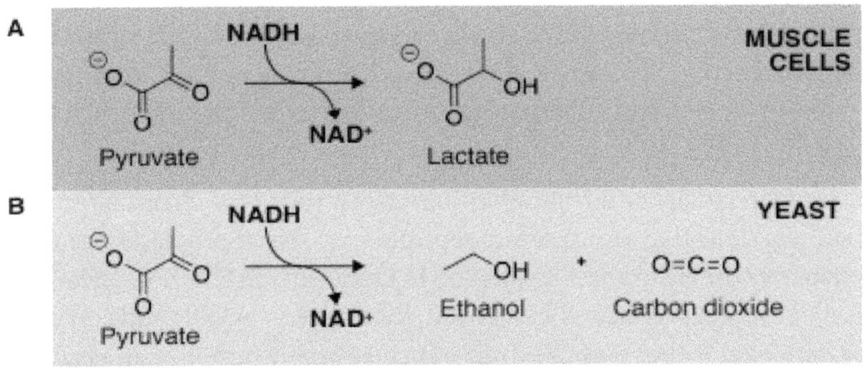

Fig. 11: Cells that live in oxygen-poor environments or that are transiently exposed to such conditions regenerate the NAD+ required for glycolysis by fermenting pyruvate.

The NADH produced by glycolysis is used to reduce pyruvate to (A) lactate or to (B) ethanol and carbon dioxide. Another glucose molecule can then be broken down using the recycled NAD+ to make more ATP for the cell.

7.9 RESPIRATION EXTRACTS MUCH MORE ADDITIONAL ENERGY FROM GLUCOSE

- Fermentation does not extract all of the chemical energy stored in glucose. Much more ATP can be produced if pyruvate is further metabolized and NADH is re-oxidized using oxygen. Instead of fermenting pyruvate to lactate or ethanol, pyruvate can be completely converted into carbon dioxide through the action of the enzyme pyruvate dehydrogenase and a series of reactions called the citric acid cycle (Figure 12). These phases of the glucose metabolism process produce two more molecules of ATP and several more molecules of NADH and $FADH_2$. These NADH molecules carry much of the energy that was originally contained in the chemical bonds of glucose in the form of high energy electrons.
- NADH then acts as the starting material for a chain of electron transfer reactions that end with the electrons being transferred to molecular oxygen (O_2) to produce water (H_2O). These reactions are catalyzed by a set of protein complexes

in the inner membrane of the mitochondria. The electrons are progressively stabilized as they are transferred down the chain, and the released energy is used to transport protons from the most internal aqueous compartment in the mitochondria, the matrix, to the space between the two membranes, the intermembrane space (Figure 13). The protein complexes of the mitochondrial electron transport chain thus have a similar function to those in the chloroplasts of photosynthetic organisms; these protein complexes create a proton gradient that is harnessed by an ATP synthase to make ATP. The production of ATP in the mitochondria is, however, coupled to the consumption of oxygen in what is referred to as respiration. Interestingly, in certain cells the proton gradient across the mitochondrial membrane also serves as an energy source for generating heat (thermal energy) instead of ATP (chemical energy). This alternative process is used by babies and hibernating animals to stay warm, and it illustrates the generality of the electrochemical gradient as an energy reservoir.

- The mitochondrial and chloroplastic electron transport chains affect the energy of the electrons they transport in opposite ways. In chloroplasts, the net reaction destabilizes the electrons; they start on H_2O and end on NADPH. In mitochondria, the net reaction stabilizes the electrons; they start on NADH and end on H_2O. The key, however, is that both electron transport chains produce an electrochemical gradient across their respective membranes. These gradient conserves some of the energy derived from sunlight or the chemical bonds of glucose. The source of the high energy electrons that are used to make the gradient then becomes immaterial and ATP is made by a common machine that is found in all organisms.

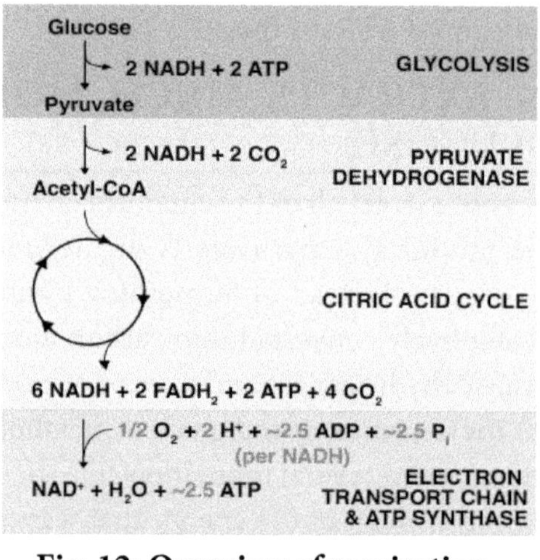

Fig. 12: Overview of respiration

Glucose is oxidized to carbon dioxide in the first three stages of the pathway. The electrons that are lost in this oxidation process are transferred to NAD+ and FAD to produce NADH and $FADH_2$. In the last stage of the pathway, the electrons are transferred from these molecules to oxygen, reducing O_2 to water.

(Note that the stoichiometry of ATP production by the electron transport chain and the ATP synthase is shown per NADH molecule. FADH2 is used by the electron transport chain in a similar manner.)

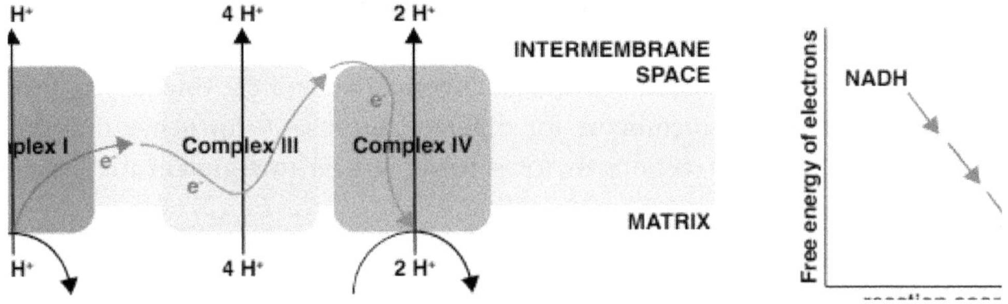

Fig. 13: The mitochondrial electron transport chain uses the energy from the electrons in NADH to transport 10 protons across the mitochondrial inner membrane from the matrix into the intermembrane space.

(A) Electrons are transferred from NADH to H_2O through a series of protein complexes in the membrane and other carrier molecules. (The carrier molecules and one additional protein complex are not shown in this simplified depiction of the electron transport chain; see Box 2 for a more detailed description.) Each of the transmembrane protein complexes transports protons across the membrane as it transfers electrons. Complex IV, the last complex in the chain, carries out the critical oxygen-consuming reaction; this one complex consumes approximately 90% of the oxygen taken up by the cell.

(B) All of the electron transfers in the mitochondrial electron transport chain are energetically favorable.

7.10 BROWN FAT, HIBERNATION, AND THE UNCOUPLING PROTEINS

- The efficiency of the ATP synthase depends on the tight coupling of proton transport to ATP synthesis; in other words, protons do not cross (or "leak" through) the membrane without passing through the ATP synthase and generating ATP. In certain circumstances, however, it is advantageous to decouple proton transport across the inner mitochondrial membrane from ATP synthesis. Some fat cells produce a protein that creates a pore in the inner mitochondrial membrane. This protein is called an uncoupling protein because it allows protons to flow freely down the gradient created by the electron transport chain without producing ATP.
- The uncoupling protein effectively dissipates the electrochemical gradient and releases the potential energy stored in the membrane as heat. The fat cells that produce this protein have a higher-than-average number of mitochondria to enable this heat production, and they are named brown fat cells because the many

iron containing proteins in their mitochondria give them a brown appearance. Newborn babies and hibernating animals have more brown fat than (alert) adults because they are more susceptible to temperature changes and less able to regulate their temperature via other means (e.g., shivering, moving to a different location, putting on clothes). Different cell types can thus harness the energy stored in the electrochemical gradient. Brown Fat, Hibernation, and the Uncoupling Proteins the mitochondrial membrane for different purposes - for generating ATP to perform chemical reactions or for generating heat to maintain the organism's homeostasis.

7.11 AT A GLANCE

- Living systems must maintain conditions that are far from equilibrium. This requires a constant input of energy. The energy that drives living systems ultimately derives from the sun. Energy from sunlight is captured by photosynthetic organisms, which convert light energy into electrical energy using special complexes in the membranes of their chloroplasts. These photosystem complexes produce high energy electrons that are then used to pump protons across the thylakoid membrane. The electrons are derived from water, which is split (oxidized) to generate oxygen, protons and electrons, which become a source of reducing power in the form of NADPH.
- Next, a rotary motor, ATP synthase, converts the potential energy stored in the proton gradient to chemical energy in the form of ATP through a mechanical process that changes the nucleotide binding affinities of its active sites. The ATP and NADPH from these first two phases of photosynthesis are then used to reduce ("fix") carbon dioxide to generate chemical energy in the form of three-carbon sugar phosphate molecules by a series of chemical reactions known as the Calvin cycle or the dark reactions of photosynthesis, which are catalyzed by the enzyme rubisco. The three-carbon sugar phosphates can be used to create more complex biological molecules or stored in the form of carbohydrates and later metabolized to generate ATP again.
- Non-photosynthetic organisms metabolize the carbohydrates produced by photosynthetic organisms to generate ATP. They can do so anaerobically in a process called fermentation. Fermentation produces ATP by extracting some of the chemical energy stored in glucose. The carbons in glucose are partially oxidized to produce molecules containing high energy phosphates that can be transferred to ADP. The carbons are subsequently re-reduced so that there is no net oxidation in fermentation.
- Aerobic metabolism or respiration, on the other hand, extracts much more energy from each glucose molecule by fully oxidizing each carbon atom to generate carbon dioxide. The energy in the chemical bonds of glucose is captured

as high energy electrons in the form of NADH and related molecules that can be fed into an electron transport chain in the mitochondrial membrane. As in photosynthesis, the energy carried by those electrons is used to generate a proton gradient across the membrane, which is, in turn, converted into ATP by the same kind of ATP synthase machine as found in chloroplasts. Thus, life is powered by the excitation of electrons, which drive oxidation and reduction reactions that, in turn, generate chemical energy in the form of ATP and sugars.

8

PHOTOSYNTHESIS IN BACTERIA

8.1. INTRODUCTION: AN OVERVIEW

Although the process of photosynthesis is most commonly associated with plants and algae, much of our understanding of the molecular basis for light energy capture and photochemical energy transduction has come from studies of photosynthetic bacteria. Space precludes a detailed description of the different types of photosynthetic bacteria, and so this article will focus in the main on the most heavily studied group of anoxygenic phototrophs, with only brief mention of their less well-known cousins. The cyanobacteria, which carry out a process of oxygenic photosynthesis common to that found in green plants and algae, are not covered.

Even with a limited focus the subject of bacterial photosynthesis is vast, and so the following description is necessarily selective, focusing on the most heavily-studied systems and key issues, and inevitably reflecting the author's personal bias. Accordingly, the bibliography at the end of the article includes some general works that provide more detailed and comprehensive information, and in some instances, in-the-text citations are to recent review articles rather than original research papers. Regarding general works, in particular, the reader is guided to a book by Blankenship (2002), which provides a detailed and accessible account of the light reactions of photosynthesis, and the underlying physical chemistry, and also to a recent book edited by Hunter et al. titled "The Purple Photosynthetic Bacteria" (2008), which contains authoritative reviews on many of the aspects of bacterial photosynthesis outlined below.

8.2 ENERGY TRANSDUCTION IN THE (ANOXYGENIC) PHOTOSYNTHETIC BACTERIA

The bacteria that use chlorophyll-type molecules to exploit sunlight as an energy source comprise the purple phototrophic bacteria, green sulphur bacteria, green filamentous bacteria, the heliobacteria and the cyanobacteria. All but the last of these transduce light energy into a biologically-useful form without the generation of oxygen from the oxidation of water, and are classed as anoxygenic photosynthetic bacteria. The most heavily-studied group are the purple phototrophic bacteria, for a variety of reasons including metabolic flexibility, genetic accessibility, and the relatively simple and modular nature of their photosynthetic apparatus.

In general terms, the strategy for solar energy utilization in all organisms that contain chlorophyll (Chl) or bacteriochlorophyll (BChl) is the same.

- Light energy is captured by pigment molecules in the light harvesting or "antenna" region of the photosystem, and is stored temporarily as an excited electronic state of the pigment.
- Excited state energy is channeled to the reaction centre region of the photosystem, a pigment-protein complex embedded in a chargeimpermeable lipid bilayer membrane.
- Arrival of the excited state energy at a particular bacteriochorophyll (BChl), or pair of BChls in the reaction centre triggers a photochemical reaction that separates a positive and negative charge across the width of the membrane.
- Charge separation initiates a series of electron transfer reactions that are coupled to the translocation of protons across the membrane, generating an electrochemical proton gradient [protonmotive force (pmf)] that can be used to power reactions such as the synthesis of ATP.

The minimal photosynthetic unit in purple phototrophic bacteria comprises a reaction centre (RC) surrounded by a light harvesting complex called LH1. Together these form the so-called RC-LH1 complex, and this is capable of converting light energy into a pmf in partnership with a second membrane-embedded electron transfer protein, the cytochrome (cyt) bc1 complex (termed bc1 below). In some species, light harvesting capacity is augmented by one or more types of peripheral antenna complex, termed LH2, LH3, and so on.

8.3 COMPONENT STRUCTURE

Pigments. The main light harvesting pigment in purple photosynthetic bacteria is not chlorophyll but bacteriochlorophyll (BChl), a closely related magnesium porphyrin that has a more saturated tetrapyrrole ring (Figure 1A). This causes BChl to absorb at significantly longer wavelengths than chlorophyll in the near infrared, the absorbance spectrum being dictated by the details of the conjugated electron system of the macrocycle (Blankenship, 2002) (Figure 1A). Light harvesting is also carried out by a variety of carotenoids that provide the main pigmentation in the visible region of the spectrum, and so make purple bacteria purple (or a variety of other colours). Figure 1B shows the absorbance spectrum of photosynthetic membranes prepared from Rhodobacter (Rba.)

sphaeroides; the strong absorbance in the near infrared is attributable to BChl (Qy bands) as is the band around 600 nm (Qx bands), whilst absorbance between 450 and 550 nm arises mainly from carotenoid. The so-called Soret bands of BChl are located in the near-UV between 300 and 400 nm (not shown)

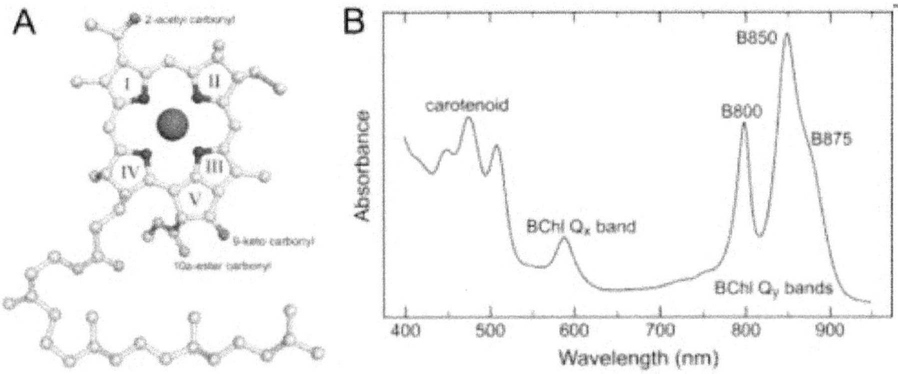

Fig. 1: Pigments and absorption. (A) The structure of bacteriochlorophyll a. Carbon and oxygen atoms that form part of the electron system are shown in yellow and red, respectively, with other carbon and oxygen atoms in white and orange. The central Mg (magenta sphere) is coordinated in-plane by the four pyrrole nitrogens (blue), plus a fifth out-of-plane ligand donated by the protein (not shown). The five rings are numbered as shown, and the labelled carbonyl groups provide additional potential points of attachment to the protein. (B) Visible/near-infrared absorbance spectrum of photosynthetic membranes from Rba. sphaeroides.

Peripheral Antenna. X-ray crystallography and other structural techniques have revealed that the light harvesting pigment-proteins of purple bacteria have a cylindrical architecture. In Rhodopseudomonas (Rps.) acidophila the peripheral LH2 complex is formed from nine copies of each of two short polypeptides, termed and ß, that each have a single membrane-spanning -helix (Figure 2). These and ß polypeptides form two concentric protein cylinders in the membrane, between which are sandwiched the light harvesting BChls and carotenoids.

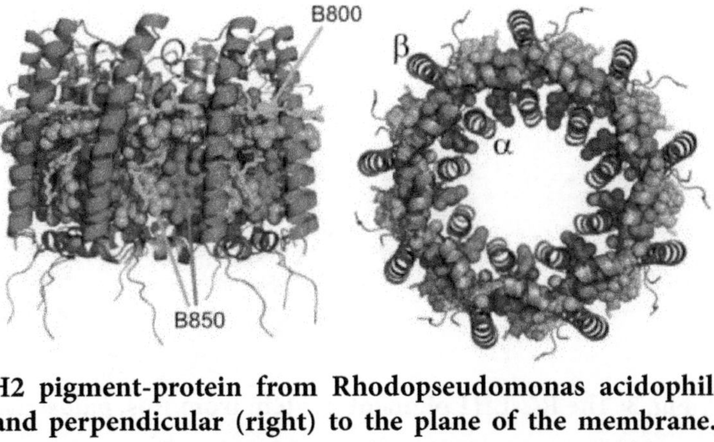

Fig. 2: The LH2 pigment-protein from Rhodopseudomonas acidophila. Views are parallel (left) and perpendicular (right) to the plane of the membrane. The protein

scaffold comprises concentric cylinders of nine (inner), and nine ß (outer) polypeptides (teal and purple ribbons, respectively). The 18 B850 BChls (which absorb strongly at 850 nm) are shown as spheres, in alternating red and orange, and the nine B800 BChls (which absorb strongly at 800 nm) are shown as green sticks (with the central Mg shown as a sphere). The macrocycles of the B850 and B800 BChls are arranged perpendicular and parallel, respectively, to the plane of the membrane.

The BChls of LH_2 form two rings that are arranged approximately in the plane of the membrane. The first of these comprises 18 BChls, one for each and ß polypeptide, oriented with the macrocyle of each BChl approximately perpendicular to the plane of the membrane (Figure 2). These "B850" BChls have a prominent absorbance band at 850 nm. The second ring comprises nine "B800" BChls, which are arranged with the macrocycle of each BChl parallel to the plane of the membrane, and gives rise to a prominent absorbance band at 800 nm. The LH2 complex also contains light harvesting carotenoids, although for clarity these are not shown in Figure 2. Peripheral LH complexes from other species show variations on this theme, e.g., the LH_2 from Rhodospirillum molischianum has eight pairs of and ß polypeptides. The LH_2 from Rba. sphaeroides is very similar to that from Rps. acidophila, and its 850 and 800 nm absorbance bands are apparent in the spectrum shown in Figure 1B.

Core Antenna. To date a high resolution X-ray crystal structure for a RC-LH1 complex has not been reported, and the most detailed information available is from a 4.8 Å structure for the RC-LH1 complex from Rps. palustris (Roszak et al., 2003) (Figure 3). This shows a central RC surrounded by a cylindrical LH1 with a roughly oval cross section in the membrane plane. The LH1 pigment-protein has a similar overall architecture to LH2, being formed from concentric cylinders of two polypeptides also called and ß. Each polypeptide has a single membrane-spanning -helix, with a ring of BChls (one per or ß polypeptide) sandwiched between the protein cylinders. LH1 is larger than LH2 to accommodate the RC in the central vestibule, with 15 pairs of and ß polypeptides in Rps. palustris. LH1 contains carotenoids and single type of "B875" BChl, arranged in a ring with each BChl macrocycle perpendicular to the membrane. The 875 nm absorbance band of the Rba. sphaeroides LH1 complex can be seen as a shoulder in the spectrum in Figure 1B.

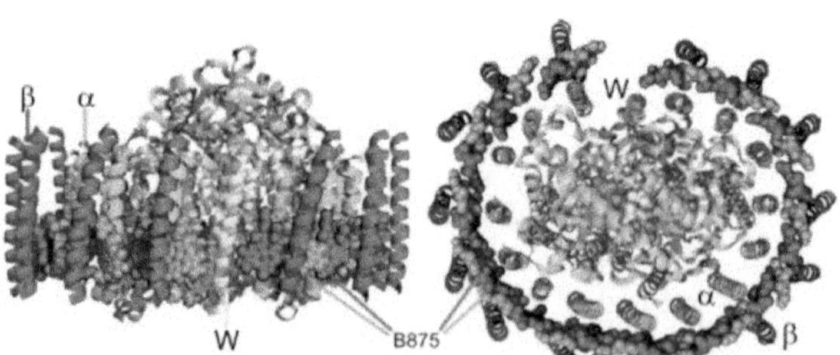

Fig. 3: The RC-LH1 pigment-protein from Rps. palustris.

Views are parallel (left) and perpendicular (right) to the plane of the membrane. The protein scaffold comprises concentric cylinders of 15 (inner) and 15 ß (outer) polypeptides (teal and purple ribbons, respectively), plus a single W polypeptide (yellow ribbon). The 30 B875 BChls are shown as spheres, in alternating red and orange, their macrocycles being arranged perpendicular to the plane of the membrane. The central RC is shown as ribbons and spheres for polypeptides and cofactors, respectively (see Figure 4). The figure was constructed using PDB file 1PYH.

In Rps. palustris, the ring of LH1 pigment-protein does not completely surround the central RC, and an additional membrane-spanning polypeptide is present in the gap in the LH1 ring. The function of this component is unclear, but it has been speculated that it is related to the PufX polypeptide, which is a minor component of the RC-LH1 complex in Rhodobacter (Rba.) sphaeroides, and is important for efficient quinone exchange between the RC and the bc1.

Reaction Centre. The function of the carotenoids and BChls of the antenna is to feed the RC with excited state energy. The related PSO module by Yocum on RCs gives a detailed account of the structure, and mechanism of the purple bacterial RC. The following is a brief summary of the heavily-studied Rba. sphaeroides complex, included here for completeness. The module by Yocum describes the closely-related RC from Blastochloris (Blc.) viridis.

The Rba. sphaeroides RC comprises three polypeptides and ten cofactors, four BChls, two bacteriopheophytins (BPhe), which is BChl lacking the central Mg, two ubiquinones, a carotenoid and a non-heme iron (Figure 4A). The BChl, BPhe and quinone cofactors are arranged within a protein scaffold formed by the L- and M-polypeptides in two approximately symmetrical membrane-spanning branches (Figure 4B). The arrival of excitation energy at a "special pair" of 870 nm-absorbing BChls (denoted P870) at the periplasmic end of the RC triggers a membrane-spanning four-step electron transfer along the A-branch of cofactors, which results in the reduction of a quinone at the so-called QB site near the cytoplasmic side of the membrane.

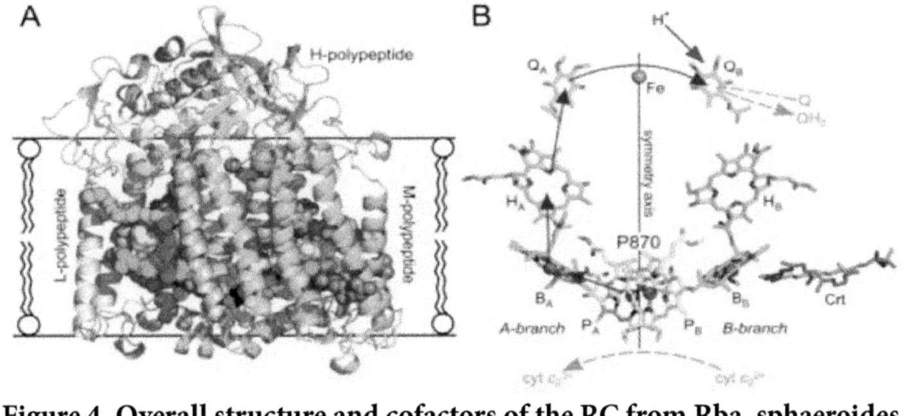

Figure 4. Overall structure and cofactors of the RC from Rba. sphaeroides.

(A) The cofactors (spheres) are encased by the largely intramembrane L and M polypeptides (beige and light-green ribbons, respectively). The H polypeptide has an extra-membrane domain, and a single anchoring -helix (salmon ribbons).

(B) The cofactors are shown as sticks, with Mg and Fe atoms shown as purple or brown spheres, respectively. Hydrocarbon side chains of the BChl, BPhe and quinone cofactors are truncated for clarity. Carbon atoms are coloured thus: P870 BChl dimer (PA and PB) are yellow, monomeric BChls (BA and BB) are green, BPhes (HA and HB) are pink, quinones (QA and QB) are cyan, and carotenoid (Crt) is teal. Oxygens and nitrogens are shown in red and blue, respectively. The BChl, BPhe, and quinone cofactors are arranged in two membrane-spanning branches around an axis of two-fold pseudo-symmetry that connects P870 with the Fe atom. Black arrows show the route of electron transfer, blue arrow shows the site of H+ uptake, and dashed green arrows show Q/QH2 association/disassociation and delivery of electrons to P870+ by cyt c2 2+. The figure was constructed using PDB file 2BOZ.

As outlined below, the RC operates as a light-powered cytochrome c2:ubiquinone oxidoreductase, and participates in a cyclic protonmotive electron transfer system with a partner ubiquinol:cytochrome c2 oxidoreductase, the bc1 (Figure 5). This protein, which has similarities to the bc1 of mitochondria and the b6f complex of oxygenic phototrophs, catalyses a bifurcated electron transfer in which one electron removed from ubiquinol is used to reduce cyt c2 3+, and the second one is used to reduce quinone on the opposite side of the membrane (Hunte et al., 2003; Crofts, 2004). The X-ray crystal structure of the Rba. sphaeroides bc1 has been determined (Esser et al., 2008) and, in common with other bc1 complexes, this protein is dimeric (Figure 5A). At the core of the bc1 is a cyt b, comprising a membrane-spanning helix bundle encasing two hemes termed bL and bH. These form a membrane-spanning electron transfer chain that links a ubiquinol oxidase site near the periplasmic side of the membrane (Qo) with a quinone reductase site near the periplasmic side (Qi). The bc1 also contains a Rieske iron sulphur (Fe-S) protein containing a 2Fe-2S centre and a cyt c1, both of which have domains on the periplasmic side of the membrane, and a single membrane-spanning -helix. Each Rieske protein connects across the dimer, the membranespanning -helix associating with one monomer, and the extra-membrane domain associating with the second monomer.

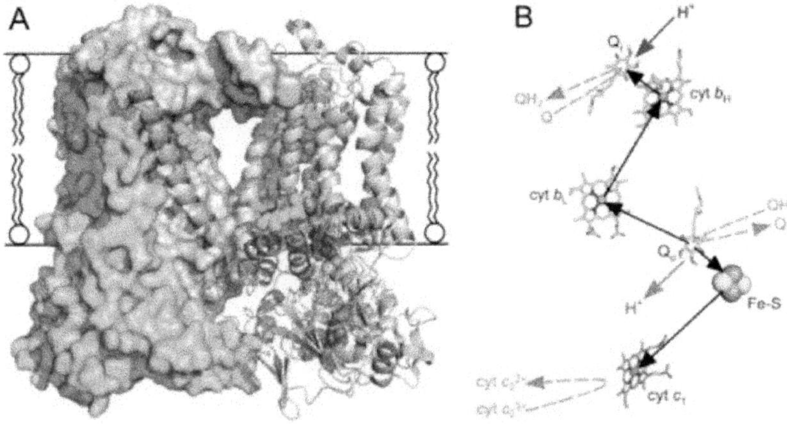

Figure 5. Overall structure and cofactors of the bc1 dimer from Rba. sphaeroides.

(A) In the view presented, parallel to the plane of the membrane, one half is shown as a solid object, and the second half as ribbons and spheres for the protein and cofactors, respectively. Each monomer comprises a cyt b (beige), cyt c1 (green) and Rieske Fe-S protein (pink).

(B) View in the plane of the membrane of a stick model of the cofactors of one half of the bc1 dimer. Heme Fe atoms are shown as small spheres, and the FeS centre as large spheres. Carbon atoms are coloured thus: b hemes are green, c heme is orange, and quinones are cyan. Oxygens, nitrogens, irons and sulphurs are shown in red, blue, brown and yellow, respectively. Black arrows show the routes of electron transfer, blue/red arrows show sites of proton uptake/release and dashed green arrows show association and disassociation of Q/QH_2, and the reduction of cyt c2 3+ by cyt c1. The Figure was constructed using PDB file 2QJY , which has the inhibitor stigmatellin at the Qo site; this has been digitally replaced by quinone to construct the Figure.

8.4 PHOTOSYSTEM MECHANISM

The purple bacterial photosystem is capable of harvesting light energy over a range of wavelengths (see above). Photon absorption produces a singlet excited electronic state of the absorbing pigment, and energy harvested by a carotenoid pigment of an antenna complex is transferred to an adjacent BChl. As outlined above, the BChls have three principal absorbance bands, the Soret band in the blue/near-UV region, the Qx band in the red region, and the Qy band in the near-infrared (the latter two are shown in Figure 1B). Energy absorbed in the Soret or Qx regions is converted to the lowestenergy Qy excited state through internal conversion.

The Qy excited state energies of the LH2 and LH1 BChls are arranged to ensure the funneling of energy into the RC. Within an individual LH2 protein, energy absorbed by the B800 BChls is passed to the ring of lower energy B850 BChls. This energy is then passed either to a neighbouring LH2 or into the ring of lower-energy LH1 B875 BChls surrounding the RC) (Figure 6). The final step involves the transfer of excited state energy to the P870 dimer of BChls in the RC, triggering photochemical charge separation. Funneling of excited state energy into the RC is therefore achieved by having red-shifted (low-energy) BChls closest to the RC.

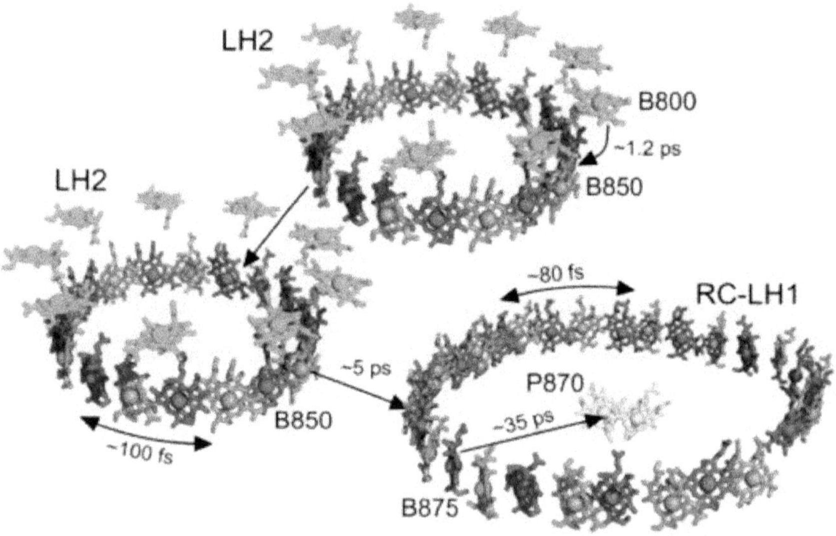

Fig. 6: Energy transfer in the purple bacterial photosystem. B800 BChls of LH1 are shown in green, B850 BChls of LH2 and B875 BChls of LH1 are shown in alternating red and orange, and P870 BChls of the RC are shown in yellow. Lifetimes in picoseconds (ps) are given for the energy transfer events indicated by the arrows. Energy absorbed by antenna carotenoids is passed to neighbouring BChls. Double-headed arrows indicate femtosecond (fs) energy migration within the B850 or B875 pigment rings.

As the rate of energy transfer decreases with the sixth power of the distance between donor and acceptor (Sundström et al., 1999; Blankenship, 2002), light harvesting complexes need to be tightly-packed within the membrane to ensure that excited state energy is passed efficiently to the RC with minimal losses from emission. The excited state lifetime of BChl is of the order of a few nanoseconds, which means that energy absorbed by antenna pigments has to be funneled to the RC on a time scale of a few tens to hundreds of ps. Rates of particular steps in this process are summarised in Figure 6 (reviewed in Sundström et al., 1999). Energy transfer between weakly coupled BChls in adjacent LH complexes takes place on time scales of a few ps and, along with B800 B850 energy transfer, can be described by the Förster mechanism (non-radiative resonance energy transfer through coupling between two transition dipoles) (Sundström et al., 1999; Blankenship, 2002). Within the B850 and B875 rings, where the BChls are strongly coupled, the excitation is delocalized over the whole ring with extremely fast fs "hopping times" between adjacent BChls, and this process is better described using an exciton approach involving delocalized electronic transitions (Sundström et al., 1999; Blankenship, 2002).

The slowest step is the transfer of the excited state from the B875 BChls to the P870 BChl dimer (referred to as 'trapping'), which, due to the relatively large distance, takes place in around 30-50 ps. This arrangement, where the charge-separating Chls or BChls of the RC are separated from the light-harvesting Chls or BChls of the antenna by an exclusion zone formed by the protein scaffold, is also a feature of the structure of both PS1 and PS2 of oxygenic photosynthesis. This architecture appears to have two principal functions, to ensure that unproductive back transfer of excitation energy (detrapping) is slow compared to productive charge-separation (which takes place in a few ps), and to ensure that the efficiency of membrane-spanning charge separation is not interfered with by unwanted electron transfer reactions between RC and antenna BChls.

The process of photochemical charge separation is described in detail in the related module by Yocum on RCs. A summary of the reactions catalysed by the Rba. sphaeroides RC is shown in Figure 4B. Acquisition of excited state energy converts P870 into a sufficiently powerful reductant to be able to donate an electron to the adjacent BA BChl. This electron is subsequently passed in three steps to the quinone at the QB site near the cytoplasmic side of the membrane, the photooxidised P870 (P870+) located near the periplasmic face of the membrane being re-reduced (in Rba. sphaeroides) by a water-soluble c-type cytochrome (cyt c2). A second light-induced charge separation results in double reduction of the QB quinone, delivery of the second electron being accompanied by the uptake of two protons from the cytoplasm to form ubiquinol [dihydroquinone (QH2)]. The immediate products of light absorption, energy funnelling and charge separation are therefore one QH2, which dissociates into the intra-membrane phase, and two oxidised cyt c2 (cyt c2 3+) which can diffuse in the periplasmic space. These mobile products are then used as substrates by the bc1 (Figure 5) (Hunte et al., 2003; Crofts, 2004).

Oxidation of QH2 takes place at a site (Qo) near the periplasmic side of the membrane and is a bifurcated reaction, one electron being used to reduce cyt c2 by passage through the so-called high-potential chain formed by the Rieske Fe-S protein and cyt c1 (Figure 5), accompanied by the release of a proton. A notable feature of this reaction is a large scale change in the conformation of the Rieske protein, that moves the Fe-S centre away from the Qo site and towards the cyt c1 heme. This movement prevents the second electron from being passed to the Rieske Fe-S centre. Instead, accompanied by release of the second proton, the second electron is passed back across the membrane via two b-type hemes to a second quinone reductase site (Qi). Oxidation of a second QH2 at the Qo site results in the reduction of a second cyt c2 via the high potential chain, and double reduction and protonation of the quinone at the Qi site. As the sites for reduction/ protonation of quinone in both the RC and bc1 are located on the cytoplasmic side of the membrane, and the site for quinol oxidation in the latter is located on the periplasmic side, light-powered cyclic electron flow in this system is coupled to the translocation of protons from the cytoplasmic compartment into the periplasm. Figure 7 summarises the movement of quinone and cyt c2 between binding sites on the RC and bc1, and the proton translocation that is powered by sunlight.

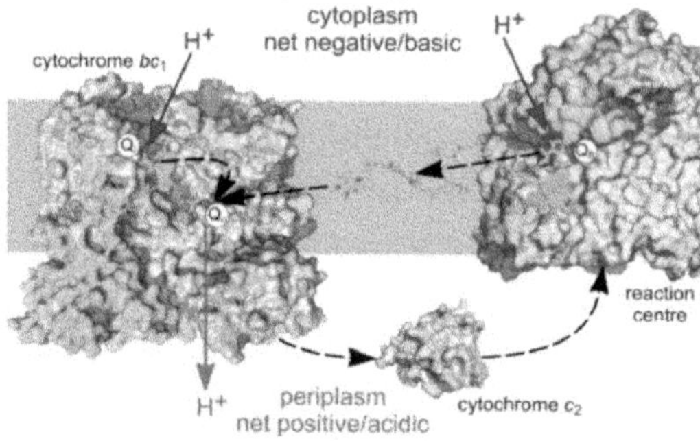

Fig. 7: Light driven cyclic electron transfer and proton pumping. The RC acts in partnership with the bc1 to translocate protons across the photosynthetic membrane, transducing sunlight into the energy of the protonmotive force. Dashed black arrows show the movement of reducing equivalents between the RC and bc1, or between the cytoplasmic and periplasmic side sides of the bc1, by the mobile carriers ubiquinol and cyt c2. Blue arrows show sites for the uptake of protons from the cytoplasm, and red arrows the site of proton release into the periplasm. The migration of electrons internal to the RC and bc1 is not shown. The view of the bc1 dimer shows the Qi site of one monomer, and the Qo site of the second monomer.

Summaries of this process from the points of view of redox potential and free energy are shown in Figure 8. Transformation of the redox potential of P870 through light absorption (Figure 8A) triggers a cascade of redox reactions, as electrons flow via the QB site into the intra-membrane Q-pool, through the high potential chain of the bc1 , and through the cyt c2 pool to re-reduce P870+. Figure 8B shows operation of the RC from the standpoint of free energy. The additional energy of the initially-formed P870* excited state (~1.45 eV for an 870 nm photon) is progressively lost as charge is separated through the protein. However, part of this energy is preserved in the form of the energy of the pmf formed through the proton translocation that is coupled to electron flow. This pmf is then used to power a variety of energy-requiring reactions, including ATP synthesis, active transport, motion of the bacterial flagellum, and so on.

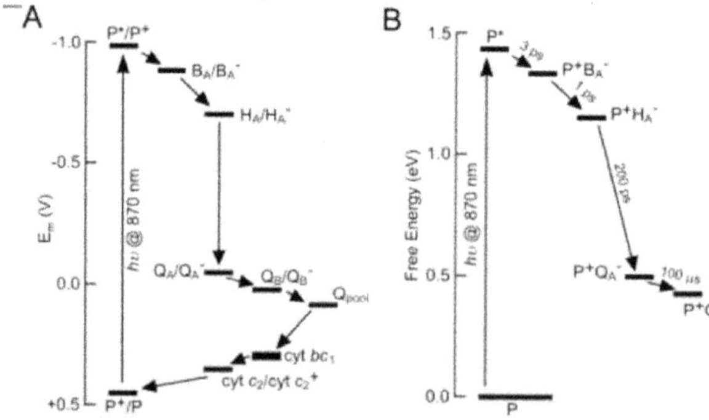

Fig. 8: Redox potential and energetics of electron transfer in the Rba. sphaeroides photosystem. (A) Formation of the excited state through light absorption transforms

the redox potential of P870 (abbreviated to P in this diagram), triggering reduction of BA and subsequent electron transfer. "cyt bc1" refers to the high-potential chain in the bc1 complex. Redox couples are written with the notation reactant/product. (B) Free energy of charge separation in the RC.

8.5 PHOTOSYSTEM DIVERSITY

The outline description given in the last section is based on the photosystem from Rba. sphaeroides, which has been the main workhorse for structural and functional studies. The photosystems from other species of purple bacteria operate on the same principles, but the details vary from organism to organism. So, for example, the first RC to yield a high resolution X-ray crystal structure, that from Blc. viridis, has an additional subunit comprising an extra-membrane tetra-heme cytochrome that is attached to the periplasmic side of the complex (Deisenhofer et al., 1995). One of the hemes of this cytochrome subunit acts as the initial reductant of the photo-oxidised primary electron donor in this RC, termed P960+, the water-soluble c-type cytochrome delivering electrons from the bc1 to this tetra-heme cytochrome subunit, and so only indirectly reducing P960+. This RC contains BChl b rather than BChl a, and has menaquinone in the QA quinone site.

Perhaps the greatest variety comes in the type and organization of lightharvesting complexes present in purple bacteria. It has been known for many years that many species contain one or more types of peripheral light harvesting complex, the levels of which are affected by environmental factors such as light intensity or oxygen levels. In other species, including Blc. viridis and Rsp. rubrum, this peripheral antenna is absent, the light harvesting function being the sole responsibility of the LH1 antenna that surrounds each RC.

A full understanding of the process of photosynthetic energy transduction in purple bacteria not only requires detailed information on the structure and mechanism of individual components, but also an appreciation of how these are put together to form a fully functioning photosynthetic membrane. In recent years the technique of atomic force microscopy (AFM) has provided detailed information on this aspect of the photosystem, the distinctive ring-like topography of light harvesting complexes proving particularly amenable to imaging and interpretation (for reviews see Scheuring, 2006; Sturgis et al., 2008). In Blc. viridis, which lacks peripheral antenna complexes, the RC-LH1 complexes form very regular arrays, each unit in the array comprising a central RC surrounded by a complete ring of LH1 pigment-protein (comprised of 16 pairs of and ß polypeptides, see above).

In Rsp. photometricum, the RC-LH1 complex also has a closed ring of 16 /ß pairs, but the presence of LH2 means that the membrane is less ordered, with individual RC-LH1 complexes separated by variable amounts of smaller rings attributable to the peripheral LH2 complex. In Rps. palustris, the photosynthetic membrane is also disordered in this way, and in agreement with the low resolution X-ray data described above, the ring of

LH1 surrounding each Rps. palustris RC is not complete, comprising only 15 /ß pairs. In Rba. blasticus, two forms of the RC-LH1 complex are apparent, a minor monomeric form comprising an RC surrounded by an incomplete ring of LH1 pigment-protein, and a major dimeric form in which two RCs are surrounded by a S-shaped antenna, both with 13 LH1 /ß pairs per RC. This dimeric arrangement was first reported for the Rba. sphaeroides RC-LH1 complex, and in this species RC-LH1 dimers are arranged in linear arrays several complexes in length, and surrounded by LH2 complexes (Bahatyrova et al., 2004). Although AFM has been successfully used to characterize the organization of RCs and LH complexes in photosynthetic membranes from a variety of purple bacteria, the location of the bc1 and ATP synthase in these membranes has not been established (Scheuring, 2006; Sturgis et al., 2008). However, the information from AFM and kinetic spectroscopy has been used to build spectacular models of the photosynthetic membrane (e.g., see Sener et al., 2007), which include speculations on the location of bc1 and ATP synthase.

Metabolic Diversity The mechanistic or physiological significance of observed variations in the organization of the photosynthetic membrane is not understood, but it may be linked, at least in part, to the metabolic flexibility displayed by this group of bacteria. The species displaying the greatest flexibility, including Rba. sphaeroides and the closely related Rba. capsulatus, are capable of both photoautotrophy (carry out photosynthesis to acquire energy and fix carbon dioxide), and photoheterotrophy (photosynthesis requiring a source of carbon other than carbon dioxide), and can also grow in the dark through aerobic respiration, or anaerobic respiration with a variety of terminal electron acceptors. Fermentative growth is also well documented, as is chemolithotrophy (deriving energy from the oxidation of inorganic compounds) with hydrogen or thiosulphate as electron donors. Most purple photosynthetic bacteria can fix N2 to form NH3 and H2, and some can carry out denitrification of nitrate to N2. Some species are obligate anaerobes, but others can tolerate high levels of oxygen, although in species such as Rba. sphaeroides expression of the photosynthetic apparatus is down-regulated by oxygen. In another group, termed aerobic anoxygenic phototrophs, expression of the photosynthetic apparatus takes place only in the presence of oxygen. These bacteria are widespread with a number of intriguing characteristics, including low levels of RCs and LH complexes, repression of BChl synthesis by light, and high levels of carotenoid.

A major factor in the prominent use of Rba. sphaeroides as a model organism for studying purple bacterial photosynthesis is its metabolic flexibility. In addition to growing in the light, this species will grow in the dark in the presence of a suitable source of carbon and electrons, such as succinate or malate for example, and a suitable electron acceptor, such as oxygen. Although the expression of the photosynthetic apparatus is repressed at very high levels of oxygen, a fully-functional photosystem is assembled at moderate-to-low oxygen tensions, although it is not required for growth. As a result, it is possible to introduce potentially lethal mutations into the RC and LH complexes of the photosystem, and study the structural and functional consequences of these in proteins or membranes prepared from cells grown in the absence of light. Such experiments are not possible in obligate phototrophs, where impairment of the photosynthetic apparatus can lead to cell death, and incubation under illuminated conditions can give rise to reversion or suppression mutations.

8.6 PHOTOSYNTHESIS IN OTHER ANOXYGENIC PHOTOSYNTHETIC BACTERIA

Although by far the most heavily studied, the purple phototrophs are only one of several groups of bacteria that use BChl proteins to exploit sunlight as an energy source. The green non-sulphur bacteria (also called green filamentous bacteria), such as Chloroflexus (Cf.) aurantiacus, have a photosystem similar to that of purple photosynthetic bacteria (Frigaard and Bryant, 2004). The RC of Cf. aurantiacus is similar to that of Rba. sphaeroides, with an intramembrane heterodimer of L and M polypeptides that scaffold the electron transfer cofactors. However, this RC lacks the H-polypeptide, both quinones are menaquinone, and the bacteriochlorin cofactors comprise three BChls and three BPhes. The RC is associated with a B808-866 antenna complex that has spectroscopic similarities to the purple bacterial LH2 antenna, but shows sequence similarities to the polypeptides of the purple bacterial LH1. In addition to possessing an intra-membrane light harvesting system, green non-sulphur bacteria possess an extensive light harvesting system called a chlorosome, which is attached to the cytoplasmic face of the photosynthetic membrane (Blankenship, 2002). This comprises an aggregate of up to 10,000 molecules of BChl c, with smaller amounts of protein, lipid and carotenoids, and the chlorosome transfers energy to the intra-membrane antenna via a set of baseplate proteins, and thence to the RC.

The RCs from purple bacteria and green filamentous bacteria are both classed as quinone-pheophytin type, or type-II RCs, along with the PS2 RC from oxygenic photosynthesis. The remaining anoxygenic photosynthetic bacteria, the green sulphur bacteria and heliobacteria, contain an Fe-S type or type-I RC similar to the PS1 RC from oxygenic photosynthesis (Heathcote et al., 2002). Again the reader is referred to the module by Yocum on RCs for a description of the PS1 RC; the key differences with the type-II RCs discussed thus far, is that the membranespanning electron transfer chain terminates in Fe-S centres rather than dissociable quinones, and within the RC pigment-protein the central electron transfer domain is flanked by two additional symmetrical domains that house antenna BChls or Chls and carotenoids. In green sulphur bacteria, such as Chlorobium limicola, the light harvesting function carried out by the intra-membrane antenna regions of the RC (Hauska et al., 2001) is again augmented by an extra-membrane chlorosome antenna comprising largely of BChls c, d, and e (Blankenship, 2002). This is also connected to the intramembrane antenna via baseplate proteins, but the green sulphur bacteria also contain the Fenna-Matthews-Olson (FMO) protein that participates in energy flow from the chlorosome to the intramembrane components. The FMO protein is water-soluble, and was the first chlorophyll protein to be structurally characterized to a high resolution by X-ray crystallography (Olson, 2004).

The final group of anoxygenic photosynthetic bacteria are the heliobacteria. These also have a type-I RC, with a central electron transfer domain and two antenna domains, and unlike other photosynthetic bacteria this RC contains BChl g as a minor component (Hirozo, 2007). Heliobacteria do not have additional intra- or extra-membrane antenna complexes. A feature of the heliobacterial RC, shared with the type-I RC from green sulphur bacteria, is that it is formed from a homodimer of a single polypeptide, rather than the L/M heterodimer seen in purple bacteria (or the D1/D2 and PsaA/PsaB heterodimers seen in PS1 and PS2). Another feature established in recent is years is that, in common with PS1, both branches of membrane-spanning cofactors are used to transfer electrons across the membrane during photochemical charge separation. In the type-II RCs described above, the two quinones have discrete functional roles, the QA quinone acting as a one electron relay but the QB quinone being specialized for accumulation of two electrons and undergoing double protonation in order to generate a quinol. However, in the type-I RCs the quinone cofactors are both non-dissociable, and on reduction pass the electron on to an Fe-S centre located on the symmetry axis. As a result, in principle, either branch of cofactors could be used to catalyse membrane-spanning electron transfer, and findings in recent years have shown that this is indeed the case (see Heathcote et al., 2002, for a review).

8.7 SUMMARY

Anoxygenic photosynthetic bacteria are a widespread and metabolicallydiverse group of organisms that make a significant contribution to energy input into the biosphere from sunlight. In addition to their native contribution, these bacteria have been at the forefront of laboratory research into the mechanism of photosynthetic energy transduction by natural systems. A small number of species of purple phototrophs, in particular, have provided major insights into a number of key biochemical and biophysical processes, including the molecular details of solar energy conversion (Hu et al., 2002; Cogdell et al., 2004; Jones, 2009), the nature of biological electron transfer (Moser et al., 1992), the mechanistic basis of chemiosmotic energy transduction (Crofts et al., 2004a, b), and the structural biology of integral membrane proteins (Deisenhofer and Michel, 1989).

9

PHOTOSYNTHETIC ALTERATIONS OF VIRUS INFECTED PLANTS

9.1 PHOTOSYNTHETIC CHANGES IN INCOMPATIBLE HOST-VIRUS INTERACTION

The most obvious symptom of systemic virus infection is the mosaic pattern of the leaves. Yellowing, chlorosis is also frequent and characteristic sign of the altered photosynthetic activity. Virus infection effects photosynthesis in a complex manner, depending on the particular host-virus combination. The symptoms are basically different in the incompatible or the compatible host-virus interaction.

- Virus infection often causes local lesions in resistant plants. The development of local lesions in hypersensitive reaction (HR) was described in details by electron microscopy, based on the time course of the appearance of symptoms. The ultrastructure of the lesion and the surrounding cell layers and the role of the active zone in the localisation of the infection was highlighted. In tobacco mosaic *tobamovirus* (TMV) infected *N. glutinosa* leaves the size and number of the starch grains increased in the chloroplasts following the infection. One day after the infection plasmolysis occurred, and first the cytoplasm and later the chloroplast membranes completely disappeared. In the remains of the chloroplasts large vesicles were formed, the number of mitochondria increased and their structure changed. Finally, the whole cell structure collapsed and its content disintegrated.
- All these alterations were consequences of the loss of osmotic regulation, and have been related to the so-called 'starch-lesions'. The cells in the lesions contained very few cell organelles; starch grains, disrupted thylakoid membranes, displaced ribosomes. The cells of the active zone showed increased metabolic activity. The number of chloroplasts increased due to their frequent division.

The primary symptom of HR was the change in the permeability of membranes. The oxidative burst, the increase in peroxidase and lipoxigenase activity lead to the brake up of membranes. In the lesions of potato virus *S carlavirus* (PVS) infected *Chenopodium quinoa* Shukla and Hiruki (1975) often found cytoplasmic invaginations containing ribosomes in the chloroplasts and in the stroma, the accumulation of fitoferritin, which was not specific to the PVS infection, because other viruses or certain stress factors cause the same symptoms.

9.2 PHOTOSYNTHETIC CHANGES IN COMPATIBLE HOST-VIRUS INTERACTION

For the synthesis of virions the parasite uses the metabolites and energy of the host plant, which are produced in the course of photosynthesis. It is evident that the physiological processes of the virus and the plant are linked in complex ways, the structure and the function of the chloroplasts are disturbed.

9.3 CHANGES IN THE ULTRASTRUCTURE OF CHLOROPLASTS

- The chlorosis of leaf tissues in the compatible host parasite interactions can be traced back to different causes. The number, size or chlorophyll content of the chloroplasts may decrease. Their shape, inner membrane structure may alter to some degrees, some of their metabolites may accumulate, etc. Often the starch grains enlarge, or accumulate for example in the case of the cauliflower mosaic *caulimovirus* (CaMV) infected cabbage plants. In several other combinations the appearance of large starch grains, the disappearance of thylakoid, stroma and grana membranes and the increase of plastoglobuli is similar to the chloroplast to chromoplast transformation, or to the phenomena of senescence. This idea is underlined by the fact that the application of a citokinin type substance (karbendazim) in the leaves of lettuce infected with beet western yellows *betaluteovirus,* (BWYV) caused the suppression of symptoms. The seasonal rhythm of chloroplast symptoms in abutilon mosaic *bigeminivirus* (AbMV) infected *Abutilon* sp. plants, which can also be correlated with senescence. The microscopic changes reflected the influence of light intensity and season on the lighter/darker mosaic pattern: in summer the inner membrane structure almost completely disintegrated, showed vesicular, tubular or myelin-like structure, while in winter prolamellar body like structures developed in the plastids, which resulted in the reorganisation of grana. According to the authors the virus regulates the chloroplast/ chromoplast transition by switching the organelle's coded 'program' on and off. The effect of virus infection on the development and differentiation of plastids was summarised in the study of cucumber mosaic *cucumovirus* (CMV) infected tobacco leaves. Depending on the time of inoculation, the plastids differentiated irregularly. The ratio of degenerated plastids changed with the time course of symptom development. The myelin-like or tubular structure was not formed by the disruption of the existing lamellae, but by their irregular development. In the case of BSMV infected etiolated barley seedlings we reported not only the accelerated senescence but the inhibition of chlorophyll biosynthesis.

- In tobacco leaves infected with TMV U5 strain it was proved by autoradiography that the chloroplasts do not take part in the biosynthesis of virions, which means that the vesicles do not play any role in virus replication. This was also regarded as a secondary effect in tomato spotted wilt *tospovirus* (TSWV) infected tobacco leaves.
- In contrast, the replication of *tymovirus* genus is related to the chloroplasts. The swelling of the chloroplasts and the formation of small vesicles were reported in Chinese cabbage infected with turnip yellow mosaic *tymovirus* (TYMV). The plastids became globular, the number of vesicles in the plastids increased and they formed clusters covered by endoplasmic reticulum (ER) from the cytoplasmic side. The region between the ER and the plastid envelope became electronlucent, and the ER disappeared. Then the plastids clustered and virions were observable in the zone between the plastids. In the vesicles of the inner surface of the envelope, double stranded nucleic acids were visible, which proved to be the replicative form of viral RNA. Based on these results it was supposed that these vesicles serve as sites of viral replication, then the viral RNA translocates into the cytoplasm where the viral proteins are synthesised, and finally the assembly of the virions is completed. These peripheral vesicles open into the cytoplasm with short necks (channels).
- Similar changes were noted for leaves of wild cucumber (Marah oreganus) infected with wild cucumber mosaic tymovirus (WCMV), where virions were also present in the stroma as well as in the cytoplasmic invaginations. Abnormal vesiculated plastids were observed in susceptible host plants (barley, wheat) infected with barley stripe mosaic hordeivirus (BSMV). The grana of the deformed, swollen plastids had an unusual structure. Between the inner and outer membranes of the envelope single membrane bound vesicles were apparent. In the peripheral vesicles of the chloroplasts and it had observed the replicative form of viral dsRNA, so the plastids could be related to virus synthesis. Symptom development was categorised into time course stages, such as primary, secondary acute (inoculated leaves) and chronic phase (systemic leaves). In yellow tissues of the primary acute phase plastids were often found with translucent stroma, which lacked ribosomes. Their thylakoids were irregularly twisted, forming a tubular reticulum with enlarged grana. In the secondary acute phase, the degradation of plastids continued until the death of the cells. The young plastids of the chronic

phase contained irregular thylakoids, small, few lamellar grana, which sometimes vesiculated. The reason for the development of a mosaic pattern is the clustered layout of the chloroplasts that have been damaged to various degrees. It had thought that in the dividing cells of chinese cabbage infected with TYMV the spontaneously mutating virus strains of different pathogenicity end up in different daughter cells. The symptom development as the result of the different tolerance level of the host cells. As proposed that the virus regulates the phenotype of the plastids through the nuclear or plastidial genome. In contrast, in the case of systemic infection of barley yellow mosaic bymovirus (BYMV) there was a linear correlation between the virus concentration and symptom severity.

- In certain host-parasite interactions the presence of unusually large grana ("supergrana") had been observed. By freeze fracture method replica were made from the granal surface, and based on the number and size of the contained particles the damage of photosystem II (PSII) was determined.

9.4 CHANGES IN THE CHLOROPHYLL PROTEIN COMPLEXES AND CHLOROPLAST PROTEINS

- Virus infected monocotyledons (e.g. barley infected with BSMV or WSMV) contained reduced amount of chlorophyll protein complexes compared to the healthy controls. It is supposed that virus infection inhibits PSII activity selectively by the decomposition of light harvesting antenna complex of PSII (LHCII), based on their findings on the ratio of particles in the surface of the thylakoid membranes in tomato plants infected with TMV. Similarly, *Eupatorium makinoi* leaves infected with tobacco leaf curl *geminivirus* (TLCV) the amount of LHCII decreased in the chloroplasts. The chlorophyll protein complexes and the protein content of the thylakoids in peanut infected with peanut green mosaic *potyvirus* (PGMV) had been altered in parallel to the symptom development. The complexes reduced to the greatest extent were identified by SDS polyacrylamid gel electrophoresis (PAGE). All these components belonged to PSII and LHCII: namely the 47 kD chlorophyll protein, the 23 kD LHCII chlorophyll protein and the a-, b-subunits of the ATP synthase). There were no significant differences in PSI reaction centre, while in LHCI the amount of the chlorophyll protein complexes slightly increased. In our experiments we detected the loss of chlorophyll protein complexes of both photosystems (i.e. PSII and PSI) (unpublished data). In TYMV infected chinese cabbage chloroplasts the antenna complexes (LHCII and LHCI) decreased to the greatest extent, while in TSWV infected *Nicotiana benthamiana* chloroplasts the complexes belonging to the reaction centers (PSII CC and PSI RC) were mostly affected.

- Later in a more sophisticated system (by improved PAGE techniques). Also detected the loss of 14, 18, 19, 23 and 33 kD proteins. The 23 and 33 kD proteins belong to the oxygen evolving complex (OEC) of PSII, and their changes showed a good correlation with the decreased efficiency of O_2 evolution. Scientist determined the 22, 23 and 24 kD proteins of OEC from tobacco leaves infected with CMV separately by using two-dimensional gels. They supposed these three proteins to be the products of the same small multigen family. The amount of the 22 and 23 kD proteins had started to decrease at the time of the appearance of chlorotic spots, while the changes in the amount of the 24 kD and 33 kD proteins which induce more severe symptoms started only after that. To sum up all these results the authors proposed, that the environmental factors do not induce or suppress the distinct elements of certain gene families in the same manner or at the same time. The inhibition of OEC is not a secondary consequence of the symptom development, but it is rather connected with the primary molecular processes of the infection. Among the stroma proteins the loss of the small subunit of ribulose-1,5-bisphosphate- carboxylase-oxygenase (Rubisco) enzyme was reported in the chlorotic tissues of cucumber plants infected with CMV, and in the chloroplasts of tobacco leaves infected with TSWV together with the reduced level of 70 S ribosomes.

- In parallel to the decrease in chlorophyll content a significant increase was measured in the enzyme activity of the chlorophyllase which catalyses the dephytilisation of the chlorophyll molecules. It was interpreted as the consequence of the release of chlorophyllase enzyme bound to the chloroplast inner membrane following the disrupter of the chloroplasts.

9.5 EFFECT OF VIRUS INFECTION ON THE PHOTOSYNTHETIC ELECTRON TRANSPORT

- Chlorotic leaves of tobacco infected with tobacco etch potyvirus (TEV) showed a decreased level of photosynthetic activity compared to control plants. The light reaction (Hill reaction and photophosphorylation) of isolated chloroplasts showed similar efficiency. Hodgson et al. (1989) described the selective inhibition of PSII in spinach infected with TMV. The fluorescence kinetic parameters measured in the thylakoids of mosaic pattern leaves showed considerable deviance. The Fv/Fm ratio (the potential photosynthetic efficiency parameter of PSII) significantly decreased to the values of the uninfected DCMU treated leaves. The O_2-evolution of thylakoids was lower than in the healthy control, and this was the primary result of PSII inhibition. The PSI electron transport was unaffected. The chlorophyll a/b ratio slightly decreased. From the above results the authors concluded that in the symptomatic leaves of virus-infected plants the cyclic phosphorylation and ATP synthesis dominates over the non-cyclic electron transport that results in NADP+ reduction. In TYMV inoculated leaves of chinese cabbage both the cyclic and the non-cyclic phosphorylation is more intensive during the period of active virus synthesis. When the virus concentration reaches its maximum in the plant, in the chloroplasts of systemically infected leaves the previous activity is depressed, the amount of ATP is significantly lower than in the control. Scientist also considered that the ATP formed by photosynthesis as energy source is directly used for the virus protein and nucleic acid synthesis. Kano (1985) drew the same conclusion.

- Net photosynthesis, chlorophyll content and the ratio of chlorophyll a/b decreased in peanut leaves infected with PGMV. The loss of PSII activity was more pronounced than that of PSI. Both cyclic and non-cyclic photophosphorylation was inhibited, although still functioning. Also found higher chlorophyll a/b ratio in the leaves of a *geminivirus* infected *Eupatorium makinoi.* The net photosynthesis measured at saturating light intensity (Pm) and Fv/Fm showed lower values with the increasing Fm. In virus infected plants the proportion of the b reaction centers that have smaller antennae related to the all PSII centers proved to be much higher than in the healthy plants. Energy distribution balance between PSII and PSI broke down.
- In dark-adapted leaves of *N. benthamiana* plants infected with PMMV-S, PMMV-I or with the hybrid viruses constructed of them. detected a considerable increase of the non-photochemical quenching (NPQ) while photochemical quenching (qP) decreased. These changes show the diminished ability for energy capture of PSII reaction centers in virus infected plants, and they function as photoprotective mechanisms to irradiate excess energy.
- In the leaves of *Abutilon striatum* plants infected with AbMV simultaneously with the symptom development the Fv/Fm slightly decreased compared to the controls while the drop of Fo was considerable. In the orange-coloured leaf tissue and in the 'green islands' surrounded by veins NPQ had been continuously reduced or stopped. The disability of the chloroplasts to emit/dissipate excitation energy neither by NPQ, pH-dependent quenching nor by thermal irradiation, leads to the loss of more chlorophylls and to the appearance of more severe symptoms.
- In several virus-host interactions the disturbances of CO_2-fixation and of related metabolic ways, or the alteration in ratio of certain products was reported. At the early stage of systemic TMV infection (virus replication) more CO_2 molecules assimilated by the photosynthetic processes. As the infection had been progressed this tendency inverted: CO_2-uptake was decreasing in the infected tissues. The same changes as well as the increase of respiration were observed in tobacco leaves infected with tobacco etch *potyvirus* (TEV). The enzyme activity of phosphoenol-pyruvate-carboxylase (PEP-carboxylase) increased in the leaves of chinese cabbage infected with TYMV. In contrast, in tobacco leaves infected with TSWV activity loss of PEP-carboxylase was measured, and it was explained as the early senescence induced by the virus infection. In virus infected plants

amino acids translocated from the chloroplasts to the cytoplasm, which means that CO_2 built into amino acids instead of free sugars.

- The primary metabolites of the photosynthetic pathway transformed into glucose, amino acids or organic acids, which are not able to translocate from the site of their production. Sugar and starch accumulation was the result of the disturbed translocation. Therefore, metabolites like soluble carbohydrate cannot get from the healthy leaves into the infected ones. Tomato yellow mosaic *geminivirus* (ToYMV) primarily replicates in the phloem cells of tomato plants causing their pathologic alterations. This way, virus infection inhibits the translocation of assimilates. But in the case of red clover mottle *comovirus* (RCMV), the distribution is not restricted to the phloem of the host plant. Photosynthetic metabolites transformed into starch, amino acids and organic acids, too. Hence starch accumulation was observed in the cells in which virus did not replicate, therefore it could not be a direct or specific effect, but secondary consequences of the upset balance between the sink and source. Disturbances in the translocation of the photosynthetic assimilates may be related to the synthesis of viral movement protein (MP). Scientist analysed the possible relationship between chlorosis, virus replication and starch accumulation in two pathogenically different CMV strains. They concluded that the severity of chlorotic symptoms does not depend on the virus content.
- In the leaves of tobacco infected with the strain inducing strong chlorosis they could not detect more starch than in the control leaves, while in the leaves showing milder symptoms typical pattern of starch accumulation appeared. On the primary leaves of squash cotyledons susceptible to CMV infection chlorotic lesions developed while on the mature leaves yellow mosaic pattern occurred systemically, which contained starch grains. These starch lesions could be divided into four concentric parts based on their structure. In the middle there were cells showing high photosynthetic activity, with a great amount of starch. The next surrounding zone contained mostly starch-free cells with lower photosynthetic activity, then a transitional zone, and the most outer zone contained cells with higher photosynthetic activity, but without a considerable amount of starch. Starch accumulation was the direct effect of the increased photosynthetic activity rather than the result of the metabolite translocation from the non-infected tissues. In the cells of the lesions the physiological changes were connected to the virus replication, which was supported by the fact that positive and negative sense viral RNA accumulated in the first front line of the infection. In the zones of the lesions the activity of the enzymes of the oxidative pentose-phosphate pathway and of the NADP dependent malate dehydrogenase enzyme grew. The spatial distribution of chlorotic symptoms correlated with the increase of glycolysis and respiration, and with the loss of the photosynthetic activity and protein synthesis.

9.6 RELATIONSHIP BETWEEN THE VIRUS AND/OR ITS PRODUCT ACCUMULATION AND THE SYMPTOM DEVELOPMENT

- The build-up of TMV virions was proven by ultrathin microscopic slices of tobacco chloroplasts, or in the chloroplast stroma of Johnsongrass chlorotic stripe *carmovirus* (JoCSV) infected *Sorghum halepense*. Several authors studied the presence of viruses in the etioplasts of tobacco plants. Following infection with TMV U5 strain, in the chloroplasts of the susceptible tobacco plants, the size of the observed virions was 1/3 the normal virion length, so the virus-like particles were probably pseudovirions. It is proved that the *in situ* presence of TMV in the chloroplasts by electron microscopic slides. It still remains as a matter debate, what role the individual viral constituents like the coat protein (CP), or the "movement proteins" (MP), or other possibly toxic product of the viral genome, or the viral nucleic acids play in the development of different symptoms. As described the accumulation of TMV CP and the correlation between the CP concentration and the severity of symptoms. The coat protein was primarily attached to the thylakoid membrane. They proposed that the TMV CP was bound to one of the chloroplast proteins and compared symptoms caused by TMV strains of different pathogenity, and concluded that the impairment of the electron transport was on the reducing site of PSII. Also underlined the hypothesis, that the CP directly inhibited PSII by binding directly onto it.
- The 20, 40 and 60 kD proteins which appeared with the onset of infection in the thylakoid membranes have been shown by immunological detection to be the oligomer forms of the coat protein. Thus, the impairment of the electron transport was directly induced by the CP. Changes in fluorescence kinetics have also been attributed to the attachment of CP to the thylakoids, and later to the aggregation of chlorophylls in the light harvesting antenna complex of PSII. From experiments with coat protein mutant viruses it was evident that even though the virion assembly was inhibited by the modification of the coat protein by gene insertion or deletion, it id not block the replication, or the movement from cell to cell and certain systemisation of the infection. The mutant viruses replicated and moved within the cells in the form of free RNA. The protein products derived from abnormal CP genes led to the development of diverse symptoms, but common among them was the appearance of chlorosis. This observation showed that the CP plays an important role in the initiation of the close physiological interaction between the virus and the host plant. In experiments of Banerjee and Zaitlin (1992) with mutant TMV strains and artificially created chimera viruses had shown that a single nucleotide change in the CP gene caused a drastic change in symptoms.

- The toxicity of CP was proved with the use of genetically modified tobacco plants, into which the potato Y *potyvirus* (PVY) CP gene linked with the transit peptide of the small subunit of Rubisco enzyme was built in, so the virus coat protein was transmitted into the chloroplast. The coat protein like products of the CP mutant TMV strains did not accumulate in the chloroplasts, however still different symptoms were expressed. According to the authors the CP excerted its effect on symptom development outside the chloroplast. Slight and severe symptom producing strains of CMV (where CP did not accumulate in the plastids) were compared based on the rate of CP accumulation. The strain with strong symptoms produced considerably greater concentration of CP within the cells of the host plant, and there was a direct correlation between the concentration of CP and the development of symptoms.
- Viral RNA was detected within the plastids in the case of several virus infections. The presence of these RNAs may indicate the site of viral synthesis, e.g. in TYMV or in BSMV. With regards to the synthesis of *potyviruses* it was possible that the replication takes place within the plastids. During the systemic infection of PVY it was also shown that the CP, the helper component (HC) and the PVY RNA was present in the plastids. The coat protein was present mainly in the thylakoid fraction rather than in the stroma. It is still an open question whether the gene products of the virus RNA are translated in the chloroplasts or are transported from the cytoplasm. TMV L-RNA devoid of coat protein was detected inside the chloroplasts, while S-RNA was not found. The authors therefore doubted the translocation of TMV CP into the plastids, but supposed that the protein is translated by the chloroplast ribosomes from the TMV RNA.
- An answer to this hypothesis was attempted by Banerjee and Zaitlin (1992), who had investigated the *in vitro* mobility of the TMV CP into isolated intact chloroplasts. TMV CP quickly mobilised into the chloroplasts, without the preliminary splitting of any transit peptide as was previously observed *in vitro* when the smaller subunit of the Rubisco enzyme mobilised into the chloroplast. The process also took place in the absence of ATP, so unlike active transport it does not need an energy source. The parallel inoculation with (CMV Y) satellite RNA and CMV strain with weak pathogenity produced the modification of symptoms in tobacco. The authors considered that the chlorosis induced by the satellite RNA was a result of the inhibition of *de novo* chlorophyll synthesis, because the symptoms were solely expressed in the newly developing leaves.
- In the inhibition of photosynthesis and the modification of symptoms not only the coat protein might play a role, but also the different virus MP-s. TMV MP producing transgenic tobacco plants were made in which the photosynthetic activity was lowered and NPQ was higher. The rate of electron transport

decreased. Based on the different chlorophyll fluorescence parameters it seemed likely that the TMV MP did not directly affect any step of the electron transport chain, but caused the specific inhibition of phosphate uptake. The transport of carbon metabolism and CO_2 fixation products into the distal parts of the plant were impaired in these plants). The continuous production of TMV MP affected the plasmodesma size dependent permeability (SEL) of the palisade layer in the mesophyll. It was shown that the product transport through the simplast, the distribution of photosynthetic assimilates and the distal transport processes in the phloem were also affected. In the same experiment the build-up in the concentration of saccharose, glucose and fructose as negative feedback blocked the triose phosphate translocator of the chloroplast. Thus stimulated the starch synthesis and storage, which lead to a net reduction in photosynthesis. The TMV MP therefore affected the photosynthetic activity of plants in a complex way.

- As compared the effect of the potato leafroll *polerovirus* (PLRV) MP with the findings of TMV MP to see how general these changes were. The function of the chloroplasts was altered compared to the control; the maximum rate of photosynthesis was lowered. The Rubisco enzyme was not activated by the substrate CO_2, and thus could not function efficiently. The concentration of Rubisco proteins decreased, the concentration of carbohydrates increased, which in turn inhibited photosynthesis. Unlike the TMV MP producing transgenic plant, the PLRV MP blocks photosynthesis by another process. Based on their findings the TMV MP reduced the rate of sugar transport and allocation through the simplast in the mesophyll cells, while the PLRV MP blocks the charging of the apoplast, and the accumulating sugars take part in the extracellular signalling, which leads to the change in the expression of different genes (pathogenesis related /PR/ proteins, genes of photosynthetic proteins).
- In BSMV infected barley leaves, with the increase in the concentration of RNA dependent RNA polymerase, the rRNA concentration of chloroplasts decreased. This was earlier reported in the chloroplasts of TMV infected tobacco, which was explained by the competition between the virus and the chloroplast for the metabolic products. In the chloroplasts of tobacco leaves infected with TEV the synthesised total protein concentration decreased.

9.7 EFFECT OF VIRUS INFECTION AS A BIOTIC STRESS FACTOR ON PHOTOSYNTHESIS

- Stress has been described as all parameters that suppress the maximum, potential genomial value of growth and reproduction of a plant. Plants adopt to stress conditions in two ways, either by tolerance or avoidance. In the case of tolerance the plant's reaction to a moderate stress is the upregulation, while against a severe stress the downregulation of metabolic processes. At the cellular level the damage in the membranes and the fluorescence of chlorophyll built in the thylakoid membranes indicated the post stress conditions. In the early phase of virus infection both the non-photochemical quenching of fluorescence and the predominantly reduced state of QA the primary electronacceptor of PSII, indicated the development of symptoms and the rapid chloroplast destruction due to photoinhibitory conditions. The build-up of carbohydrates and the brakeup of the equilibrium between synthesis and brakedown probably blocks the gene regulation, which effects the level of chlorophyll-protein complexes and photosynthetic metabolic enzymes.

- The plant's reaction to stress was also determined by other abiotic factors, such as the light intensity or the availability of nitrogen. TMV infection in tobacco plants has multiplied the rate of photoinhibition (which was shown in the high Fo and low Fv/Fm values) especially with low nitrogen concentrations. Similarly, to the abiotic stress processes, stress hormones are also synthesised in virus infected plants. In parallel with the brakedown of chloroplasts in TMV infected tobacco plants, the concentration of abscisic acid (ABA) also rose. In the leaves of healthy control plants, in which drought stress was not induced, ABA was located in the chloroplasts, while virus infection enhanced ABA synthesis, changed its location within the cell and blocked the growth.

10

PHOTOSYNTHESIS IN ALGAE: POTENTIAL AND CHALLENGES

10.1 INTRODUCTION

Algae are a very diverse group of predominantly aquatic photosynthetic organisms that account for almost 50% of the photosynthesis that takes place on Earth. Algae have a wide range of antenna pigments to harvest light energy for photosynthesis giving different types of algae their characteristic colour. Early work done with algae contributed much to what is presently known about the carbon dioxide fixation pathway and the light harvesting reactions. The processes of photosynthesis in algae and higher plants are very similar. From among the three types of carbon dioxide-concentrating mechanisms known in photosynthetic organisms, two types are found in different types of algae. Algae are proposed to play a role in the global carbon cycle by helping remove excess carbon dioxide from the environment. Recently, algae are recognized as a promising biodiesel source due to its efficient absorption and conversion of solar energy into chemical energy.

- Sunlight energy largely exceeds the energy required by anthropic activities, and therefore its exploitation represents a major target in the field of renewable energies. The interest in the mass cultivation of green microalgae has grown in the last decades, as algal biomass could be employed to cover a significant portion of global energy demand. Advantages of microalgal vs. plant biomass production include higher light-use efficiency, efficient carbon capture and the valorization of marginal lands and wastewaters. Realization of this potential requires a decrease of the current production costs, which can be obtained by increasing the productivity of the most common industrial strains, by the identification of factors limiting biomass yield, and by removing bottlenecks, namely through domestication strategies aimed to fill the gap between the theoretical and real productivity of algal cultures. In particular, the light-to-biomass conversion efficiency represents one of the major constraints for achieving a significant improvement of algal cell lines.

- This chapter outlines the molecular events of photosynthesis, which regulate the conversion of light into biomass, and discusses how these can be targeted to enhance productivity through mutagenesis, strain selection or genetic engineering. This review highlights the most recent results in the manipulation of the fundamental mechanisms of algal photosynthesis, which revealed that a significant yield enhancement is feasible. Moreover, metabolic engineering of microalgae, focused upon the development of renewable fuel biorefineries, has also drawn attention and resulted in efforts for enhancing productivity of oil or isoprenoids.

10.2 PHOTOSYNTHESIS IN MICROALGAE

- Oxygenic photosynthesis is the process by which photoautotrophs capture sunlight efficiently and converts it into organic molecules and biomass with an efficiency which is, instead, variable, depending on species and environmental conditions. Oxygenic photosynthetic organisms, namely plants, algae and cyanobacteria, store into biomass the solar energy that reaches the Earth's surface at a rate of 120,000 terawatt/year (TW-y. The current global energy demand of 14.9 TW-y, although it is projected to increase to 23.4 TW-y by 2030, yet falls greater than three orders of magnitude behind the solar energy on Earth. Therefore, exploitation of this potential by culturing photoautotrophs could satisfy at least part of the energy required for anthropic activities.
- Among photosynthetic organisms, cyanobacteria and eukaryotic microalgae are the most promising feedstocks for the sustainable production of bulk bio-based materials such as food, feed, fuel and high-value metabolites; moreover, they can be used for wastewater treatments and in mitigation processes for CO_2-emissions. Algae can grow autotrophically, heterotrophically or mixotrophically in massive cultures for industrial purposes, in either open ponds or closed photobioreactors (PBRs). In contrast with land plants, algae do not require arable land and need far less fresh water for their growth. Moreover, the culture biomass devoid of stems and roots, which consumes metabolic energy, is fully photosynthetically active. Finally, biomass productivity is far less affected by the seasonal cycle. However, while microalgae represent a promising source of valuable bio-based products, an optimization of cultivation technologies is required in order to enhance growth rates and cell densities at saturation, thus making the process profitable.
- Indeed, productivity in photobioreactors is reduced by the inefficient light-to-biomass conversion, that leads to a photosynthetic efficiency significantly lower

than the theoretical maximum of 9–10%, corresponding to ~80 g of biomass/m^2/day or 280 ton/ha/year. In the industrial scale PBRs, algae light conversion yield falls between 3% and 5%. Filling the gap that originates from light-use inefficiency, and that makes the controlled cultivation of microalgae still far from being commercially viable, is therefore essential. Comprehension of the mechanism regulating photosynthesis will allow researchers to identify the targets for genetic improvement and ultimately to enhance biomass yield, thus counterbalancing the costs for cultivation systems and downstream biomass processing.

10.3 REGULATORY MECHANISMS OF PHOTOSYNTHESIS: MICROALGAL SPECIES

- In both green algae and higher plants, the process of oxygenic photosynthesis can be divided into light and dark phases. In the former, photons are absorbed and utilized to drive Linear and Cyclic Electron Transfer (LET and CET, respectively), to form adenosine triphosphate (ATP) and the reduced form of Nicotinamide adenine dinucleotide phosphate (NADPH), which power the Calvin–Benson–Bassham cycle to produce carbohydrates in the dark phase (Figure 1).

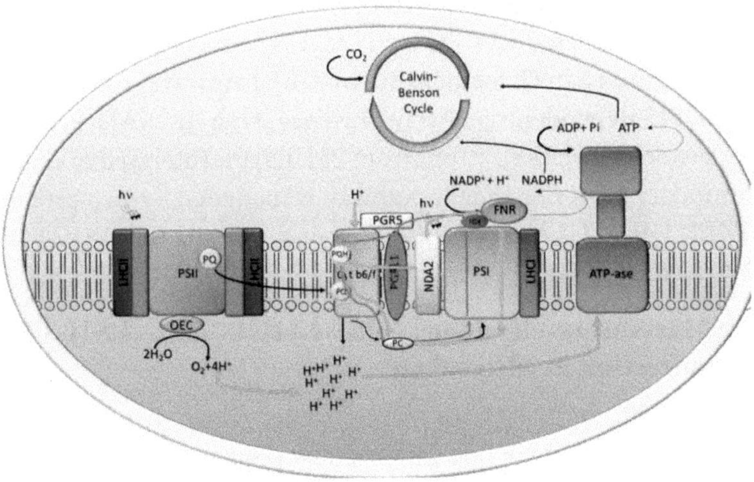

Fig. 1: Schematic representation of photosynthetic electron transport. Arrangement of Photosystem I (PSI), Photosystem II (PSII), cytochrome b6f and adenosine triphosphate (ATP) synthase complexes within the thylakoid membranes is shown. The light-driven water splitting reaction leads to O_2 evolution and originates linear electron transport (LET), indicated with black arrows, from water to nicotinamide adenine dinucleotide phosphate ($NADP^+$), which is coupled to proton translocation from stroma into the luminal side of thylakoids during the light phase. The electrochemical gradient formed is used by the ATP synthase to produce ATP from Adenosine diphosphate (ADP) and Pi in the stroma. The NADPH and ATP formed during the light phase drive the Calvin–Benson–Bassham cycle reactions in the stroma. Two pathways of cyclic electron transport (CET) around PSI are indicated with red (Ferredoxin-dependent pathway) and green (NDA2-dependent pathway) arrows, respectively.

10.4 THE LIGHT PHASE OF PHOTOSYNTHESIS

- The linear electron transport (LET) reaction starts with the water-splitting complex Photosystem II (PSII), that captures sunlight and utilizes excitation energy to oxidize water molecules into protons (H^+) and molecular oxygen (O^2). The electrons removed from water are transferred via the Plastoquinone (PQ) pool to the Cytochrome b6f (Cyt b_6f) complex and then utilized to translocate protons across the thylakoid membrane. The cytochrome f subunit reduces the soluble electron carriers' plastocyanin (PC), the electron donors of PSI. Absorption of photons by Photosystem I (PSI) promotes oxidation of its reaction centre (RC) P700. The electron removed by the oxidation event finally reduces ferredoxin (FDX) and the electron hole in $P700^+$ is filled by electrons from PC, while at the stromal side the ferredoxin $NADP^+$ reductase (FNR) transfers the electrons from FDX to $NADP^+$ to yield $NADPH^+$ H^+. This electron transport is coupled to the build-up of a proton gradient across the thylakoid membrane, with contributions from water splitting and PQH_2 oxidation by the Cyt b_6f. The return of protons to the stromal compartment is coupled to ATP synthesis.
- The ATP:NADPH ratio is regulated by modulating LET and CET, the latter being the reaction which reduces PQ with excess reducing equivalent from FDX or NADPH (Figure 1). In C. reinhardtii, two CET pathways around PSI are suggested: The NADPH-dependent pathway involves type II NAD(P)H-dehydrogenase (NDA2), which recycles electrons from PSI into the intersystem chain via NADPH. The secondary FDX dependent pathway is mediated by two proteins: PROTON GRADIENT REGULATION 5 (PGR5) and PGR5-LIKE PHOTOSYNTHETIC PHENOTYPE 1 (PGRL1) (Figure 1).

10.4.1 Light-Harvesting Systems: PSI-LHCI and PSII-LHCII Super complexes Organization in Microalgae

- Capture of light energy by both photosystems, which drives charge separation in RC and fuels LET and CET, is enhanced by pigment-binding proteins, the light-harvesting complexes (LHC). Various LHCs form the peripheral antenna system in both photosystems. While RC subunits were strongly conserved, the antenna complexes diversified through evolution, yet maintained a common architecture. The most represented member is the major antenna LHCII, a 22 kDa polypeptide which binds 14 chlorophylls (Chl) a and b, and four xanthophylls (Lutein, Neoxanthin, Violaxanthin and, upon high light exposure, Zeaxanthin) (LHC)-like antenna proteins, which were present in a cyanobacterial ancestor, carried out photoprotective functions, while they later evolved into isoforms fulfilling either light-harvesting or energy-dissipative responses.
- The LHC superfamily consists of some 30 proteins, the most conserved being the subunits of PSI and PSII through the Chlorophyta , which have pre-eminently a

light-harvesting role, while the light-harvesting complex stress-related (LHCSR) subunits have an energy-dissipative role, enabling photoprotection in excess light (EL) conditions through the non-photochemical quenching (NPQ) mechanism. In C. reinhardtii, the LHCI subunits, forming the PSI peripheral antenna system, and the monomeric subunits of the PSII supercomplex Lhcb4 (CP29) and Lhcb5 (CP26), are the most conserved antenna proteins. Trimeric LHCIIs, the major antennae of photosynthetic membranes of C. reinhardtii, are encoded by Lhcbm genes (Lhcbm1–9).

10.4.2 PSII-LHCII

- The core complex of PSII is highly conserved in all organisms and consists of 40 different protein subunits. The RC is composed by subunits D1, D2 and cytochrome *b*559 and hosts P680, the PSII RC where the primary charge separation event occurs. Light-dependent transfer of reducing equivalents to PQ leads to $P680^+$ formation. The positive charges accumulated by four events of charge separation drive the water splitting reaction within the oxygen evolving complex (OEC), composed by the extrinsic polypeptides PsbO, PsbQ, PsbP and PsbR. Chl *a*- and β-carotene-binding inner antennae CP43 and CP47 enlarge the light harvesting capacity of the supercomplex. The PSII core is organized into dimers (C_2), which, in turn coordinate a peripheral antenna system.
- In higher plants, this LHC system is made of two layers: The inner, composed by the monomeric LHC proteins CP24, CP26 and CP29, which are bound, respectively, to the CP43 and CP47 core subunits, and the outer layer is made by the trimeric LHCII complexes. In *C. reinhardtii*, the largest PSII-LHCII super complex characterized contains three LHCII trimers (named S, M and N) per monomeric core, and it is characterized by the absence of the monomeric antennae protein CP24. In mosses and higher plants, the N trimer has been substituted for by an additional monomeric LHC, CP24 (Lhcb6), LhcbM1, LhcbM2/7 and LhcbM3, which are the major components of LHCII trimers in the PSII supercomplex of *C. reinhardtii* . Recently, Shen and co-workers reported a cryo-electron microscopy structure of a complete, $C_2S_2M_2N_2$-type PSII–LHCII supercomplex from *C. reinhardtii* at 3.37-A resolution. The high-resolution structure allowed not only locating the LHCII trimers in the complex, but also the plausible energy transfer pathways from the peripheral antennae to the PSII core. Moreover, a number of small core subunits (PsbE, PsbF, PsbH, PsbI, PsbJ, PsbK PsbL, PsbM, PsbTc, PsbW, PsbX and PsbZ) has been elucidated.

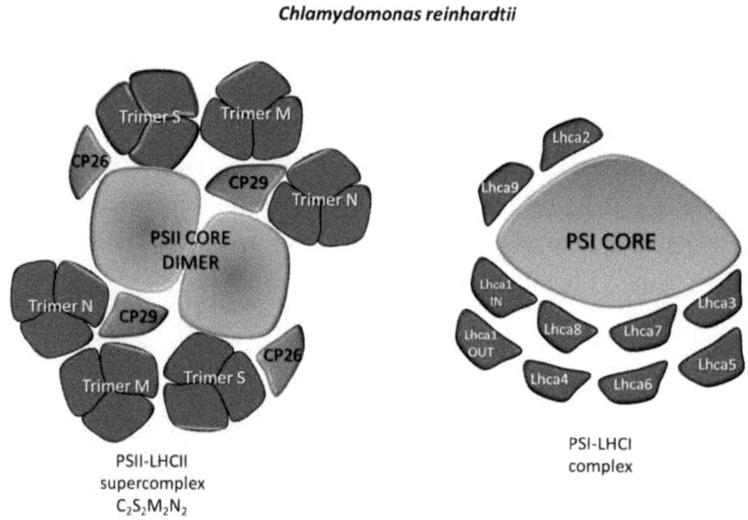

Fig. :2 Supramolecular organization of PSII-LHCII and PSI-LHCI super complexes in the model alga Chlamydomonas reinhardtii. The schematic representations are based upon data from for PSII-LHCII and from for PSI-LHCI. The core complexes of both PSs are shown in light green while the antenna complexes are shown in dark green.

10.4.3 PSI-LHCI

- The organization of PSI-LHCI of C. reinhardtii was investigated by negative stain electron microscopy and single particle analysis, which revealed that the super complex is larger but less stable than that from higher plants. The Lhca1-9 antenna proteins loosely bind to the core, where this can explain the large variation in antenna composition of PSI-LHCI from C. reinhardtii found in the literature.
- The isolation of several PSI-LHCI super complexes with different antenna size allowed to precisely determine the position of Lhca2 and Lhca9 proteins and led to a model of whole PSI-LHCI super complex antenna organization. Moreover, a megacomplex constituted by a Cyt b6f interacting with the PSI-LHCI complex was identified under anaerobic conditions, a treatment that promotes CET in C. reinhardtii. More recently, the structure of C. reinhardtii PSI–LHCI supercomplex has been solved by cryo-electron microscopy, showing that up to ten LHCIs are associated with the PSI core (Figure 2).

10.5 THE DARK PHASE OF PHOTOSYNTHESIS

- In green algae and higher plants, the carbon dioxide reduction occurs in the dark phase of photosynthesis, which also occurs in the light, and is powered by the NADPH and ATP from the light phase. The whole process can be described with the general reaction:

$$CO_2 + 4H^+ + 4e^- \xrightarrow[Enzymes]{2NADPH+H+3ATP} (CH_2O) + H_2O$$

- The entire process of carbon fixation, discovered by Calvin, Benson and Bassham in the early 1950s, requires two molecules of NADPH and three of ATP for each CO_2 fixed into sugars. This energy complement is supplied by the absorption of eight photons in the light phase.

10.5.1 Dark Reactions of Photosynthesis: The Calvin-Benson-Bassham Cycle

- The conversion of CO_2 into sugar (or other compounds) is performed by three distinct phases (carboxylation, reduction and regeneration phases) within the Calvin–Benson–Bassham cycle (CBBc) (Figure 3). During the carboxylation phase, one molecule of CO_2 is added to the 5-carbon sugar ribulose bisphosphate (RuBP) by the enzyme ribulose bisphosphate carboxylase/oxygenase (RuBisCO) to form two molecules of phosphoglycerate (3-PGA).
- The enzyme is controlled by the RuBisCO-activase, which carboxylates a Lys residue in the presence of the substrate CO_2, thus preventing wasteful reaction with O_2 under CO_2-depleted conditions. The subsequent reduction phase catalyses the conversion of 3-PGA into 3-carbon (Triose-P) products glycerhaldeide-3-P (G3P) in two steps, by consuming ATP and NADPH. The regeneration phase restores the initial RuBP reactant from Triose-P, through a complex series of reactions which involves eight distinct enzymes (Figure 3), including transketolase and transaldolase, yielding 5-carbon sugars from 6-carbon plus 3-carbon sugar intermediates, and the sedoheptulose-1,7-bisphosphatase that catalyses the de-phosphorylation of sedoheptulose-1,7-bisphosphate to yield sedoheptulose-7-phosphate.
- Sedoheptulose-1,7-bisphosphatase activity shows a strong correlation with the rate of photosynthetic carbon fixation, thus controlling carbon flux. Under EL conditions, the RuBisCO activity rate becomes limiting, and the rate of synthesis of ATP/NADPH from light reactions exceeds their use by CBBc. Depletion of ADP limits ATPase activity in protons' return to the stroma compartment, leading to lumen iper-acidification and triggering excess energy dissipation reactions. This is further enhanced by NADPH accumulation, since CET activation further acidifies the lumen.

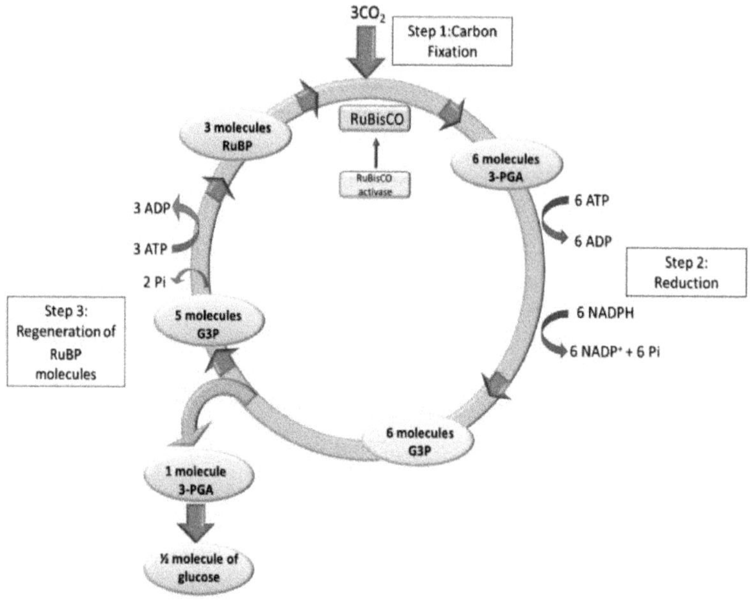

Fig. 3: The Calvin–Benson–Bassham cycle (CBBc) reactions. The CBBc has three stages. In stage 1, the enzyme RuBisCO incorporates 3 CO_2 molecules into the 5-carbon sugar ribulose-1,5-bisphosphate (RuBP) to form 6 molecules of 3-phosphoglycerate (3-PGA). In stage 2, 6 molecules of 3-PGA are converted into 6 molecules of Glyceraldehyde-3-P (G3P) by using 6 molecules of ATP and 6 molecules of NADPH as reducing power. In stage 3, RuBP is regenerated so that the cycle can continue. Stage 3 includes a complex series of reactions combining 3-, 4-, 5-, 6-, and 7-carbon sugar phosphates, which are not explicitly shown in the diagram.

10.5.2 RuBisCO

- In cyanobacteria and plants, along with red, brown and green algae, RuB is CO is found as a large protein complex with a hexadecameric quaternary structures consisting of eight 55-kDa large (L) subunits and eight 15-kDa small (S) subunits (L8S8). The RuBisCO crystal structure was determined on complexes purified from Spinacia oleracea, Nicotiana tabacum and the cyanobacterium Synchronous . In green algae the crystal structure of C. reinahardtii at 1.4 A resolution showed high similarity to the L_8S_8-RuBisCO enzyme assembly from Spinacia oleracea. Since RuBisCO evolved in atmosphere with a higher concentration of CO_2 respect to the current one, it has low affinity for CO_2 and low substrate specificity. Thus, it accepts oxygen as substrate at low CO_2, which leads to the loss of fixed carbon as a consequence of feeding a photorespiratory cycle with phosphoglycolate.
- Although the photorespiratory (C_2) cycle recovers part of phosphoglycolate into phosphoglycerate, this reduces the overall light-to-biomass conversion efficiency. In both green algae and higher plants, the ability to fix CO_2 depends in part upon the properties of RuBisCO: While RuBisCO isolated from a few species of red

algae have three times higher substrate specificity vs. that from C_3 crop species, in most of photoautotrophs RuB is CO is operating at no more than 30% of its capacity under standard atmospheric conditions (21% O_2, 0.04% CO_2).

- Indeed, the chloroplastic abundance of this protein is extremely high. To overcome this drawback, many photosynthetic organisms have developed different systems to increase the level of CO_2 at the catalytic site in order to enhance the carboxylation while disfavouring the oxygenation reaction. Microalgae absorb HCO_3^- ions, which must be converted to CO_2 before the carbon fixation takes place. Moreover, in green algae RuBisCO is compartmentalized into carbon-concentration sub-compartments of the chloroplast, called pyrenoids, which have been purified from C. reinhardtii and shown they consist primarily of RuBisCO. In other algae types including red algae, carboxysomes are present as large molecular architectures including carbonic anhydrase together with RuBisCO, where CO_2 level is increased from carbonic anhydrase activity to limit photorespiration and enhance photosynthetic yield.

10.6 DYNAMICS OF THE PHOTOSYNTHETIC APPARATUS IN RESPONSE TO ENVIRONMENTAL CONDITIONS: PHOTOPROTECTIVE MECHANISMS

- During evolution, photosynthetic organisms are said to have adapted to a wide range of habitats with an extreme variability of light irradiances, water and nutrient abundance and temperature. Abiotic stresses such as drought or nutrient deprivation easily decrease the maximum photosynthetic yield of algae, thus environmental conditions can exacerbate EL stress. In this condition, the energy absorbed exceeds the rate of its utilization by downstream reactions, increases the concentration of Chl-excited singlet states (^{1}Chl *), thus the probability of Chl triplet states (^{3}Chl *) formation together with the release of singlet oxygen ($^1O_2$), a reactive oxygen species (ROS). It comes that a mechanism to dissipate the excitation energy absorbed in excess, is required. Experiencing EL conditions activate the Non-Photochemical Quenching (NPQ) process. This can be experimentally observed as a decrease of fluorescence emitted by PSII upon exposure to over-saturating light. NPQ arises from a number of processes in the thylakoid membranes, and several major components of NPQ can be identified based on the kinetics curves of the relaxation of PSII fluorescence. The fastest component, immediately triggered upon exposure to EL, is the energy-dependent quenching (qE), which relaxes within approx. one minute upon switching actinic light off.
- State transitions (STs) represent changes in the relative antenna sizes of photosystems, however although this fluorescence decline (called qT) has been included in NPQ, it is caused by PSI RC activity, and therefore is of the

photochemical type. An additional quenching component, that rises and relaxes at a longer time scale than qE, is called qZ: This is found in some algae species in which a zeaxanthin-dependent enhancement of NPQ is observed. The slowest component, named qI, develops under long lasting (several hours) high light stress.

- The qE response is dependent on a low lumenal pH and requires LhcsR, Chl-xanthophyll-binding proteins found in eukaryotic algae and mosses and is replaced by the non-pigmented protein PsbS in higher plants. In *C. reinhardtii*, LhcsR proteins are encoded by three highly homologous genes *LhcsR1*, *LhcsR3.1* and *LhcsR3.2*, while PsbS by two closely linked *PsbS1* and *PsbS2* genes. Both PsbS and LhcsR proteins harbour protonatable residues exposed to the luminal side, which detect low pH and activate the heat dissipation of energy absorbed in excess.
- In *C. reinhardtii*, accumulation of gene products involved in qE is induced by signals such as high light, blue light and UV light via increased expression of genes encoding for LhcsR and PsbS. By a forward genetics approach, SPA1 and CUL4 have been identified as components of a putative green algal E3 ubiquitin ligase complex, as critical factors in a signalling pathway that controls light-regulated expression of the dissipative response. The accumulation of two isoforms of LhcsR protein LhcsR1 and LhcsR3 is different. Recently, it has been found that the expression of LhcsR1 protein is constitutive, while the accumulation of LhcsR3 is increased under EL conditions and depends on the activation of the CAS calcium sensor. Upon protonation, *C. reinhardtii* LhcsR subunits switch to a quenching conformation. The dynamics of LhcsR proteins transition between unquenched and quenched conformations has been studied in the moss protein LhcsR1, showing a 50-fold decrease in lifetime from 3.7 ns lifetime to 80 ps.
- The physicochemical mechanisms involved were identified to be dual: (i) the transient formation of carotenoid radical cation, thermally recombining to ground state, and (ii) the energy transfer from a Chl *a* to lutein S1 state, which thermally relaxes to the ground state within approx. 10 ps. Thus, the two types of quenching mechanism reported for plants, as localized, respectively, in two different types of LHC subunits, appear to be both active within the single LhcsR subunit. Under EL conditions, lumen acidification triggers the so-called xanthophyll cycle, which involves the xanthophylls violaxanthin (Vio) and zeaxanthin (Zea), and consists of a light-dependent, rapid and reversible de-epoxidation of Vio to Zea. The reaction is catalysed by VDE (violaxanthin de-epoxidase). This enzyme is luminal in plants where it is activated by acidification, while it is stromatic in *Chlamydomonas*; the xanthophyll cycle of intact *Chlamydomonas* cells is inhibited by the uncoupler nigericin, indicating that the activation of this stromal enzyme also requires the build-up of a pH gradient in EL. The amplitude of qE in

plants correlates with the level of Zea though its binding to specific LHC targets, in *C. reinhardtii* NPQ amplitude is Zea-independent.

- The qT component of NPQ is dependent on ST, i.e., the mechanism of LHCII relocation between PSs, which compensates for PSI/PSII excitation imbalance and optimizes photosynthetic electron transport in response to the light conditions. PSII over-excitation reduces PQ to PQH_2, and activates a thylakoid protein kinase (STT7 in green algae and STN7 in higher plants) which, in turn, phosphorylates LHCII, and leads to its reversible association with the PSI-LHCI complex. In *C. reinhardtii*, most of LhcbM proteins get phosphorylated upon ST, including the monomeric antennae CP29 and CP26, which are recruited as a supplementary antenna for PSI. While this mechanism is widespread in green photosynthetic organisms, in plants the amplitude of ST is lower than in *C. reinhardtii*, possibly indicating differences in the regulation of photosynthetic electron transport. The term "qI" refers to all quenching processes relaxing slowly (>10 min), and comprises multiple processes contributing to the down-regulation, inactivation and damaging of PSII. One of the components of the slowly-relaxing NPQ correlates with the synthesis of Zea, was shown as ΔpH-independent, and is possibly related to the binding of Zea to specific antenna proteins. A second component of NPQ is related to photoinhibition and is enhanced upon prolonged over-excitation. It consists into a light-induced reduction of the quantum yield of PSII, due to the photodamage of the RC protein D1. Thus, quenching relaxation reflects the kinetic of RC repair cycle.
- The NPQ mechanism is highly relevant for the maintenance of the photosynthetic efficiency which contributes to acclimation to the different light environments. The relative contribution of each of the NPQ components changes between organisms and irradiances: qE activates based on sudden increases in light intensity, while ST responds to changes in the light spectrum under low light conditions. Thus, LhcsR protein function is synergically with other photoprotective mechanisms, such as CET and ST in shaping the fast response to environmental conditions. Long-term stresses occur on timescales of days and weeks. Photoacclimation mechanisms to such changes involve a rearrangement at the level of chloroplast protein and lipid composition yielding into an adjustment of the stoichiometry of photosynthetic complexes through the modulation of gene expression and synthesis/degradation of individual chloroplast components. A major component of response to excess light consists into down-modulating the size of the PSII antenna, and enhancing the stoichiometry of the Cyt b_6f complex, ATPase and RuBisCO with respect to PSII RC. In green algae, EL down-regulates LHCII and LHCI genes transcription, while when under limiting irradiance, the opposite response was shown. In *C. reinhardtii* the increase in the Chl *a*/*b* ratio is consistent with decreases in the

amount of both LHCI and LHCII in EL; EL stress also induces an accumulation of proteins involved in the NPQ response, such as LhcsR3 in *C. reinhardtii* .

- The LhcbM isoforms are expressed differentially depending on growth conditions, which suggests a specific role of different LHC complexes in PSII organization and chloroplast photoprotection. LhcbM9 is only expressed in stressing conditions and binds to PSII–LHCII complexes, where it protects PSII by inducing an energy-dissipative state with reduced $^1O_2$ formation. In *C. reinhardtii*, the transcriptional regulation of *LhcbM* genes is mediated by nucleic acid binding 1 protein (NAB1), a cytosolic protein that prevents the translation of LhcbM by sequestering the corresponding mRNAs into translationally silent ribonucleoprotein complexes (Figure 4).
- The relative abundance of NAB1 is regulated by nutrient abundance: Under CO_2 starvation, which hampers the activity of the CBBc, up-regulation of NAB1 promotes an antenna size reduction, thus alleviating the excitation pressure on PSs.

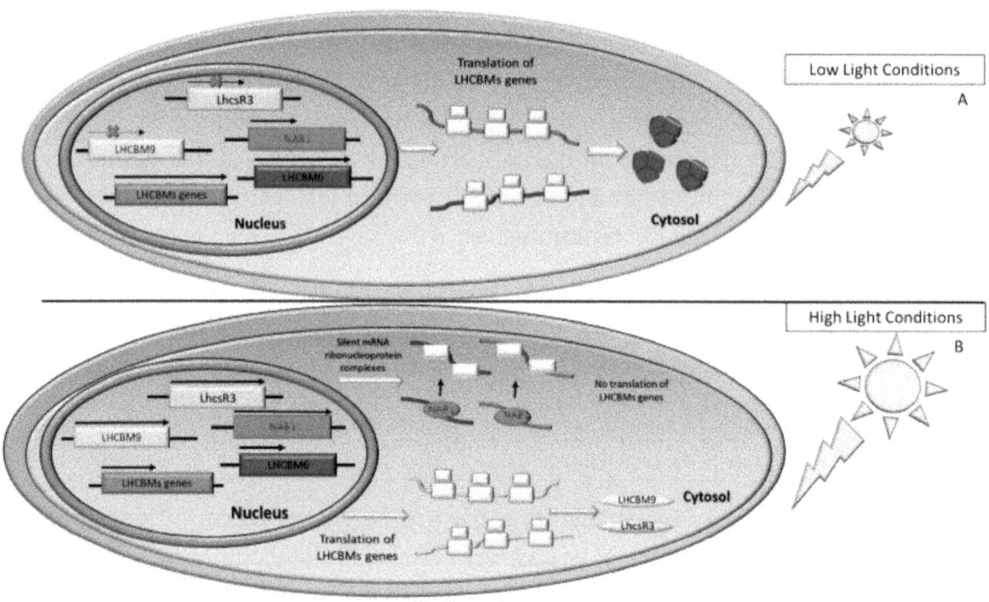

Fig. 4: Scheme of long-term control mechanisms regulating light harvesting antenna size, as described in the model alga C. reinhardtii. (A) In low light conditions, carriers of the photosynthetic electron transport chain are oxidized, and all nuclear genes encoding LhcbMs isoforms associated to the PSII are expressed, except for the isoform 9. LhcbM-encoding mRNAs are translated in the cytosol, then targeted to the chloroplast and inserted in the thylakoid membranes. Under low light conditions, the translational repressor NAB1 is in a less active state. (B) In excess light conditions, ATP and NADPH produced by the light reactions exceed their consumption rate by the CBBc, and the overexcitation of PSII results in the release of reactive oxygen

species (ROS). To alleviate excitation pressure, a remodelling of the antenna system is induced by slowing down the transcription of LhcbM genes. Once the translation of NAB1 is promoted, this subunit interacts with LhcbM-encoding mRNAs to form silent mRNA-ribonucleoprotein complexes. In contrast to all other isoforms, the expression of LhcbM9 and LhcsR3 proteins are induced.

10.7 IMPROVING PHOTOSYNTHETIC YIELD

10.7.1 Light Harvesting Antenna as Target to Reduce Optical Density in Mass Culture

- When grown under mass culture, a condition typical of industrial PBRs, microalgae undergoes a progressive drop in productivity as the cell density gradually increases. This can be mainly ascribed to an inhomogeneous light distribution within the culture, due to its high optical density: In this condition, the surface layers of the culture easily reach the saturation of photosynthesis (and possibly photoinhibition), while the inner layers are light-limited. Such steep gradient in light penetration results in a low productivity of the system. Optimization of the light transmittance within the culture volume was proposed as a strategy to alleviate these constraints.
- A bioengineering approach to decrease the Chl content per cell, thus minimizing the light absorption and enabling a larger fraction of cell suspension to contribute to overall productivity, was first developed in the model alga C. reinhardtii. Truncated light-harvesting antenna (tla) mutants were obtained by random DNA insertional mutagenesis and selection by Chl fluorescence imaging. Mutant tla1 showed a significant reduction of Chl content per cell and a lower functional antenna size of both PSI (−50%) and PSII (−65%) vs. wild type (WT). In batch culture, tla1 cells yielded a higher Pmax at saturating irradiances and higher light-to-biomass conversion efficiency with respect to the WT strain .
- Gene TLA1 was found to participate in the mechanism of Chl antenna size regulation, and indeed its over-expression resulted in a larger antenna size for both photosystems and lower Chl *a/b* ratio with respect to WT, while its down-regulation by RNAi resulted in the opposite phenotype. Strain tla2 was mutated in the gene encoding the chloroplast-localized signal recognition particle (CpSRP) receptor CpFTSY, whose deletion was responsible for a pale-green phenotype and a lower Chl *a/b* ratio than WT. Components of the CpSRP complex, involved in the proper folding of LHCs and targeting of these proteins to the thylakoids, are therefore promising molecular targets to achieve a substantial reduction in Chl antenna size without impairing the photosynthetic electron transport (Figure 5). Moreover, CRISPR-Cas9 technology was recently shown as a reliable approach by which to produce tla mutants. Pale-green mutants were obtained in species other than C. reinhardtii (Figure 5): N. gaditana and C. sorokiniana mutant strains

with truncated antenna were isolated by random mutagenesis and phenotypic selection; once characterized, they showed higher photosynthetic efficiency than WT and improved photoresistance under EL conditions, in both lab-scale and industrial-scale PBRs An additional molecular target expected to affect antenna size was CAO (encoding for Chlorophyllide *a* oxygenase) (Figure 5), encoding for the enzyme responsible for Chl a → Chl b conversion.

- In Chlamydomonas, both insertional knock-out and point mutations on CAO impaired the biogenesis of antenna systems, which were affected in different ways depending on the light conditions. Moreover, CAO expression was modulated by RNAi, which resulted in knock-down mutants showing a lower Chl b content. Therefore, by tuning the Chl b relative abundance, corresponding regulation of antennae size can be obtained, and a reduced optical cross-section improves the growth and photosynthetic rate under high light conditions, without impairing other regulatory mechanisms such as ST and NPQ.

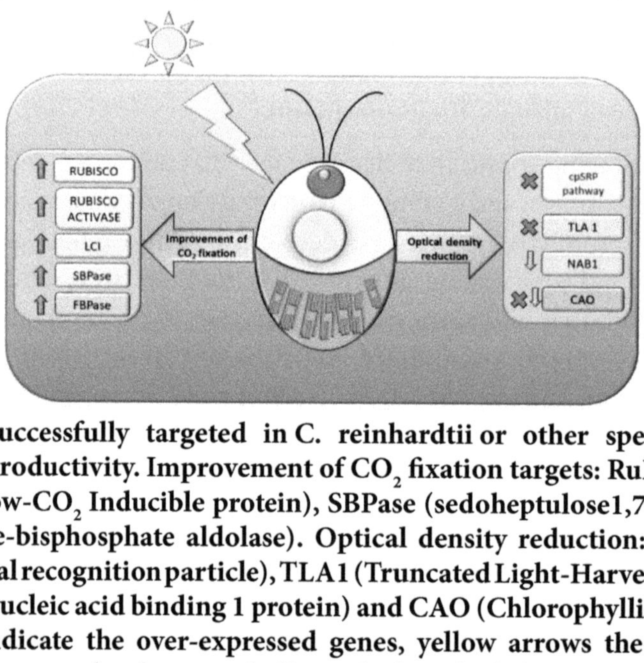

Fig. 5: Genes successfully targeted in C. reinhardtii or other species to improve photosynthetic productivity. Improvement of CO_2 fixation targets: RuBisCO, RuBisCO activase, LCI (Low-CO_2 Inducible protein), SBPase (sedoheptulose1,7-biphosphatase), FBPase (fructose-bisphosphate aldolase). Optical density reduction: cpSRP pathway (chloroplast signal recognition particle), TLA1 (Truncated Light-Harvesting Antenna 1), LhcbM, NAB1 (nucleic acid binding 1 protein) and CAO (Chlorophyllide a Oxygenase). Green arrows indicate the over-expressed genes, yellow arrows the down-regulated genes and red crosses indicate the knocked-down genes.

- Additional perspectives towards enhancing the light use efficiency in algae are likely to be developed in the future based on the emerging functional diversity of individual Lhcm proteins which have been reported to be involved in state1–state2 transitions, NPQ and/or in sustained photoprotection, thus opening the perspective of enhancing such functions selectively in industrial strains. Nevertheless, it is not yet clear how engineering antennas can be combined with the well-established enhanced growth efficiency of truncated antenna strains.

10.7.2 Bioengineering Response to Light Fluctuations and Improving Resistance to Photo-Inhibition

- The capacity to counteract EL stress and avoid photoinhibition clearly provide a carbon-gain advantage and therefore represent an important component of productivity. In particular, responses to fluctuating light conditions are clearly beneficial for photosynthesis since they enhance the 'tracking' of light, thus maintaining high rates of C assimilation, as shown in field crops. In microalgae, the WT strains show impaired growth when excess irradiance induces photoinhibition, since the repair of photodamage requires metabolic energy. Engineering of the response to EL succeeded in mitigating this loss in efficiency: Very high light resistant (VHL-R) Chlamydomonas strains were selected for their ability to grow at irradiances lethal to the control genotype, and found they were affected in the pathways which regulate photoprotective responses, including PSII repair and ROS detoxification.
- The Chlamydomonas WT strain was UV-mutagenized and selected on a lethal concentration of Red Bengal, a photoreactive chemical releasing 1O2; characterization of tolerant strains identified SOR1 as a factor enhancing resistance to photoinhibition. Analogously, UV-mutagenesis and selection under high irradiance (2000 μmol photons m−2 s−1) identified the Light Responsive Signal 1 (LSR1) gene, which conferred improved resistance against exogenous ROS]. Recently, Dall'Osto and colleagues applied two steps of mutagenesis and phenotypic selection to Chlorella vulgaris. First, they selected a strain characterized by a 50% reduction of Chl content per cell and a 30% increased photon-to-biomass conversion efficiency with respect to WT. After a second mutagenesis cycle followed by a selection on Rose Bengal, they selected pale-green genotypes exhibiting higher resistance to singlet oxygen (strains SOR) that showed a further enhancement in biomass productivity with respect to both parental and WT strains.
- Alternatively, to genetic engineering and mutation/phenotypic selection, an alternative approach consists into sampling and evaluating algal biodiversity, particularly in extreme environments which might provide interesting performance when such strains are grown in optimal conditions. An example of this is the case of Chlorella ohadii, a chlorella strain from the Sinai desert, which was reported to exhibit high productivity and the robustness of growth.

10.7.3 RuBisCO as Target to Improve Carbon Assimilation Efficiency

- The rate-limiting step of the CBBc is the fixation of inorganic carbon catalysed by RuBisCO, as the complex has low turnover rate and low substrate specificity. Moreover, it shows affinity for O_2 which leads to futile reactions. The consequences of the wasteful oxygenation reaction are partially alleviated by the photorespiration process which, nevertheless, yields into a partial loss of the CO2, and thus decreases light-to-biomass conversion efficiency. Therefore, the engineering of microalgal strains with enhanced RuBisCO catalytic activity would be crucial for improving the efficiency of solar energy conversion. Some species of red algae express isoforms with high specific activity. Thus, combining positive mutations from different isoforms has been suggested as a way to obtain RuBisCO with the improved Vmax of carboxylation catalysis. A major constraint to this approach is the high intolerance of the catalytic region to mutations, that made sparsely successful direct evolution strategies; nevertheless, some enzymes variants with higher activity have been identified, and their heterologous expression represents a promising approach.

- Other RuBisCO-improved variants were obtained by site directed mutagenesis, targeting either the rbcL gene (RuBisCO large subunit) or the subunit that interacts with Rubisco activase. However, their over-expression in Chlamydomonas failed to enhance the C fixation efficiency. On the contrary, the over-expression of endogenous RuBisCO activase in Nannochloropsis oceanica increased biomass and lipid productivity up to 40% (Figure 5). Consistently, over-expression of RuBisCO in the cyanobacterium Synechocystis enhanced photosynthetic efficiency and fatty acid productivity (Figure 5).
- In Chlamydomonas, a number of strategies were tested to improve carbon assimilation. RuBisCO isoforms with the higher Vmax of carboxylation catalysis were obtained by the PCR-based gene shuffling of the rbcL gene consisting into a restriction of encoding DNA following low fidelity replication and re-ligating into random assembled sequences with enhanced biochemical variability. The site-directed mutagenesis of rbcL resulted in a low-activity RuBisCO variant, which instead triggered a ten-fold higher H_2 production in Chlamydomonas, possibly by increasing the pool of reducing equivalents available to the hydrogenase. An alternative approach would alter the engineering of cyanobacterial CO^{2-} concentrating mechanisms, as a possible route to enhance the RuBisCO operating efficiency. Before the approach delivers potential benefits, characterization of algal HCO^{3-} transporters and carbonic anhydrases, and identification of factors regulating RuBisCO aggregation into the pyrenoids, is required. Recent advances in dissecting the details of pyrenoid biogenesis in Chlamydomonas might guide future redesign of the mechanism, to augment the overall C fixation rate.
- Besides RuBisCO, other CBBc enzymes and accessory proteins have been targeted, e.g., sedoheptulose 1,7-bisphosphatase from C. reinhardtii has been successfully over-expressed in D. bardawil, resulting in a significant enhancement of photosynthetic efficiency (Figure 5). Over-expression of the fructose 1,6-bisphosphatase in Synechocystis enhanced the growth rate with respect

to the control genotype under EL conditions (Figure 5). A strong raise in the photosynthetic productivity of Synechocystis was obtained by over-expressing RuBisCO, sedoheptulose1,7-biphosphatase, fructose-bisphosphate aldolase and trans-ketolase.

- Finally, by the over-expression of Low-CO^2 Inducible (LCI) proteins in C. reinhardtii maintained at high CO^2 concentration, namely under conditions which repress LCI synthesis, an increase of biomass productivity up to 80% with respect to the control genotype was reported (Figure 5).

10.7.4 Engineering of the Lipid Biosynthesis for Renewable Energies Production

- The triose phosphate produced by photosynthesis supports the main metabolic pathways of the algal cell, therefore the enhancement of photosynthetic yield potentially results in the enhancement of lipids, proteins and other high value compounds synthesis. Genetic manipulation approaches can generate strains with desirable commercial traits, by either expressing new biosynthetic pathways or enhancing the yield of a product of interest already present in a given strain.
- The major research targets is the engineering of strains for a significant increase of total lipid accumulation, and/or the optimization of fatty acid chain-length profile, which can be carried out by targeting single or multiple genes involved in the lipid biosynthesis or by down-regulating competing pathways. Saturated and mono-unsaturated C^{14}-C_{20} fatty acids from microalgae are exploited for renewable liquid biofuel production, while the engineering biosynthesis of long chain-polyunsaturated fatty acids (LC-PUFAs), important components of the human diet, might become a viable option in the market of high-value food additives. Metabolic engineering reports for redirecting carbon fluxes toward fatty acid production in microalgae, included the up-regulation of key biosynthetic enzymes:

  (i) acetyl-CoA carboxylase, catalysing the first step for fatty acid biosynthesis, was successfully over-expressed in the chloroplast of P. tricornutum

  (ii) malonyl-CoA ACP transacylase enzyme, which catalyses the formation of malonyl-ACP, was over-expressed in both N. oceanica and Schizochytrium spp., resulting in 31% and 39.6% total lipid increase with respect to the control genotype, respectively;

  (iii) acyl-ACP thioesterases (TE), involved in the last step of fatty acids biosynthesis, was heterologously expressed in different strains of microalgae. In Dunaliella tertiolecta, a combinational expression platform involving plant lauric acid-TE (C12TE) and medium-chain fatty acid-specific ketoacyl-ACP synthase was recently engineered, and resulted in a significant increase in lauric acid (C12:0) and myristic acid (C14:0) accumulation.

- In addition to the strategies for increasing lipid production, fatty acid desaturation could be regulated in order to modify the PUFAs profile, by engineering either the desaturase/elongase pathway or the polyketide synthase pathway. Both endogenous and exogenous desaturases and elongases have been successfully over-expressed in different oleaginous microalgae such as N. oceanica and P. tirocinium . In Chlamydomonas, over-expression of the endogenous CrΔ4FAD, a monogalactosyl-diacylglycerol Δ4-desaturase, resulted in increased production of the specific product hexadeca-4,7,10,13-tetraenoicacid (16:4).
- Microalgal lipid composition differs among species, moreover it is further affected by the cultivation conditions. The main storage lipids in oleaginous microalgae are triacylglycerols (TAG), which are more desirable for commercial-scale biodiesel processing than polar lipids or free fatty acids. The first step of TAG biosynthesis is catalysed by acyl-CoA: glycerol-3-phosphate acyl-transferase, whose over-expression in P. tricornutum was shown to promote the formation of oil bodies, and led to a significant increase in lipid content. A significant increase in lipid productivity was also obtained by up-regulating endogenous lysophosphatidic acid acyltransferase (LPAAT) both in C. reinhardtii and in P. tricornutum, and by over-expressing the Brassica napus LPAAT isoform in Chlamydomonas . An additional rate-limiting steps in the TAG biosynthesis of microalgae is catalysed by chloroplastic (DGAT1) and ER (DGAT2) isoforms of diacylglycerol acyltransferase (DGAT). Over-expression of DGAT1/DGAT2 in P. tricornutum and Nannochloropsis spp. resulted in a significant increase of total lipid accumulation.
- Other enzymes which have been targeted by genetic engineering approaches, and which have succeeded in improving TAG content, included that which was involved in acetyl-CoA synthesis: (i) the E. coli isoform of ACS (acetyl-CoA synthetase) was introduced in Schizochyrium ; (ii) PDC (pyruvate dehydrogenase complex) abundance was increased by down-regulating pyruvate dehydrogenase kinase (PDK) in P. tricornutum ; malic enzyme (ME) and glucose-6-phosphate dehydrogenase (G6PDH) were over-expressed in P. tricornutum .
- In addition to the modification of fatty acid biosynthesis, competitive pathways e.g., starch synthesis have also been manipulated for generating wider metabolic changes. The conversion of phosphoenolpyruvate (PEP) to oxaloacetate by PEPC (PEP carboxylase) preferentially diverts the carbon flow towards protein synthesis, and indeed PEPC down-regulation resulted in the over-expression of pyruvate kinase, which increased the levels of pyruvate and acetyl-CoA. On the other hand, knock-down of a PEPC isoform by either artificial microRNA-mediated technology and CRISPRi in Chlamydomonas, and by RNAi in P. tricornutum, resulted in higher TAG accumulation.

- Another strategy to improve TAG production is the inhibition of specific lipases. In Nannochloropsis oceanica the TAG lipases mutants tgl1 and tgl1;2 increased the TAG content by two-fold in early log phase cells.
- Transcription factors represent a suitable target to be manipulated as an alternative to structural genes, since they proved to be more effective for inducing global metabolic changes. Several transcription factors controlling lipid biosynthetic pathways have been identified, and their over-expression resulted into more productive strains. The up-regulation of a Dof-type transcription factor in Chlamydomonas doubled the cellular content of TAG. Over-expression of bZIP and bHLH transcription factors in both N. salina and N. oceanica improved biomass and lipid productivity. In N. gaditana, 20 putative regulators of lipid production, down-regulated by nitrogen deprivation, were identified by RNA-seq; a strategy of selective knock-out by the Crispr-Cas9 system allowed us to identify a homolog of the fungal Zn (II)2Cys6-encoding gene, which triggered lipid biosynthesis in Nannochloropsis and resulted in a 200% increase of C partition to lipids, without significantly affecting the growth rate.
- More recently, a Myb-like transcription factor Phosphorus Starvation Response (PtPSR) was identified in P. tricornutum; its modulation might represent a good strategy to enhance cell growth and TAG production in limited-phosphorous conditions. A MYB DNA binding protein involved in cell cycle regulation was instead targeted in Chlamydomonas, showing that the mutants devoid of CrCDC5 accumulate more oil and starch with respect to WT]. Integrated, multiomic analysis has been recently performed on different high-lipid productive microalgal strains, in order to pinpoint a range of candidate molecular targets, aimed to enhance oil productivity in different growth conditions.

10.7.5 Endogenous Up-Regulation and Heterologous Expression of Isoprenoid Biosynthetic Pathways in Microalgae

The photoautotrophic nature of microalgae requires the generation of isoprenoids, natural hydrocarbons associated with the photosynthetic apparatus (e.g., carotenoids, plastoquinon, phytol chains) and participating in photon capture and electron transport events. Thus, microalgae own specific isoprenoid biosynthetic pathways that could be manipulated by genetic engineering to increase the natural capacities for the generation of isoprenoids, or to promote heterologous isoprenoid production. These class of molecules, due to their intensive color, fragrance and antioxidant properties, are in high demand for various applications in animal feed, medicines and nutraceuticals, pest control agents, cosmetics and pigments.

Regarding the endogenous isoprenoids, carotenoids represent commercially successful products from microalgae: β-carotene from D. salina and astaxanthin from H. pluvialis are high-price products mainly used in aquaculture, and as food colouring agents and nutraceuticals. As for zeaxanthin, hyper-accumulating D. salina strains were identified, while a ΔZEP Chlamydomonas strain with a far higher zeaxanthin content than WT (56-fold) was shown to have a commercial potential in the production of eggs fortified with carotenoids. In H. pluvialis, the endogenous phytoene desaturase was modified by site-directed mutagenesis, enhancing both the resistance to the herbicide norflurazon and the astaxanthin productivity. Significant enhancements in carotenoid productivity were obtained by either over-expressing the β-carotene ketolase or expressing the endogenous phytoene desaturase in the chloroplast under the control of the psbA promoter.

In Chlamydomonas, the over-expression of C. zofingiensis ZEP resulted in an astaxanthin production, reaching values of 0.5 mg g−1 DW. To increase the production of fucoxanthin, employed as an antioxidant agent, the 1-deoxy-d-xylulose 5-phosphate synthase and the phytoene synthase were over-expressed in Phaeodacthylum.

Two pathways lead to synthesis isoprenoid products: The 2-c-methyl-D-erythritol 4-phosphate/1 deoxy-D-xylulose 5-phosphate (MEP/DOXP) pathway and the mevalonate pathway (MVA). These are localized in different cell compartments but generate the same 5-C precursors, isopentenyl diphosphate (IPP) and dimethylallyl diphosphate (DMAPP). The MVA pathway was lost during evolution in the Chlorophyta and Eustigmatophyceae, while it was conserved in other microalgae, e.g., P. Tricornutum maintained the cytosolic MVA pathway, thus representing a choice species for isoprenoid engineering and heterologous terpenoid production.

In Phaeodacthylum, lupeol synthase isoforms from A. thaliana and L. japonicus were expressed to produce lupeol, an anti-inflammatory triterpene. Moreover, co-expression in the diatom of CYP716A12 from M. truncatula and its corresponding reductase MtCPR, was aimed to further functionalize lupeol in betulin for the production of the antiretroviral, antimalarial agent betulinic acid. In Chlamydomonas, the fragrance patchoulol was produced by expressing the patchoulol synthase from Pogostemon cablin , while the synthesis of bisabolene, a sesquiterpene compound identified as a promising jet fuel candidate, resulted by both over-expressing the bisabolene synthase from Abies grandis and down-regulating both the geranylgeranyl diphosphate (GGPP) synthase and the squalene synthase.

Finally, several non-native diterpenoids were successfully produced in C. reinhardtii, by the heterologous expression of diterpene synthases targeted to the algal chloroplast, to convert native GGPP into the diterpenoid products casbene, taxadiene and 13R(+) manoyl oxide.

Notably, besides diatoms and green microalgae, Cyanobacteria represent the most targeted photosynthetic organisms for isoprenoid production, especially terpenoids. Cyanobacteria are, indeed, promising platforms for biofuel production, and the knowledge on isoprenoid biosynthetic pathways regulation is instrumental in over-expressing key enzymes toward a higher productivity. To enhance isoprene synthesis, different plant isoprene synthase (IS) genes were expressed in Synechococcus elongatus, in particular the IS from Eucalyptus globulus was responsible for the most significant productivity enhancement among the isoforms tested. Moreover, limonene is a terpenoid evaluated as a promising alternative fuel. The limonene synthase from Mentha spicata was successfully expressed in different species such as Synechococcus elongatus PCC 7942, Synechocystis sp. PCC 6803 and Synechococcus sp. PCC 7002.

10.8 AT A GLANCE

- Nowadays the technical advance in genetic engineering made it possible to engineer algal strains for enhancing both biomass productivity and the yield of high value products from
- microalgae. The major efforts are focused on obtaining strains with higher photosynthetic efficiency in order to decrease the unitary cost of algal biomass. The best results have been obtained by enhancing the homogeneity of light distribution within photobioreactors and by decreasing their susceptibility to photodamage. Besides the obvious strategy of targeting antenna systems, photoprotection was also enhanced by increasing photochemical quenching by boosting CO2 fixation through over-expressing RuBisCO activity, and other rate-liming enzymes in the CBB cycle.
- Besides productivity, microalgae exploitation targets high value compounds. First, oleaginous strains, especially diatoms, were engineered to optimize the fatty acid chain-length profile for biodiesel production. Secondly, isoprenoid biosynthesis pathway has been engineered to redirect carbon flux towards specific and high values products for chemical industry and/or as biofuels. While the overall strategies have been well defined, limitations derive mainly from the low number of species that can be engineered to some extent besides the model species Chlamydomonas reinhardtii and the efficiency of the expression of the transgenes in different species.

11

METABOLITE TRANSPORT ACROSS THE CHLOROPLAST ENVELOPE OF C_3-PLANTS

11. 1. INTRODUCTION: AN OVERVIEW

During photosynthesis, energy from solar radiation is used to convert atmospheric CO_2 into intermediates that are used within, as well as outside, the chloroplast for a series of metabolic pathways. These intermediates have to be exchanged between the chloroplast and the cytosol. The envelope membrane of chloroplasts contains various specific translocators that are involved in these transport processes. The basic properties of several of these translocators have been extensively studied in the past. The elucidation of the molecular structure of some of these translocators during the last years opened the way to study the function of particular translocators in more detail. This r focuses on the progress achieved in this field with emphasis on the characterization of different classes of phosphate translocators and on translocators that are specific for dicarboxylates, adenylates and hexoses.

Communication between chloroplasts and the surrounding cytosol occurs via the plastid envelope membrane. The outer envelope possesses pore forming proteins (porins) that allow the nonspecific diffusion of hydrophilic solutes. The inner envelope membrane contains various metabolite transporters that mediate the exchange of metabolites between both compartments. The carbon fixed during the day can be exported from the chloroplasts into the cytosol for the synthesis of sucrose which is subsequently allocated to hetero-trophic organs of the plant such as roots, seeds, fruits or tubers. When the rate of sucrose biosynthesis and export falls below that of CO_2 assimilation, the fixed carbon is retained in the chloroplasts and directed into the biosynthesis of assimilatory starch. Remobilization of starch during the following dark period and export of the starch breakdown products ensures a continuous supply of photosynthates to heterotrophic tissues in the dark. Two chloroplast translocators are involved in these processes:

- The triose phosphate/phosphate translocator (I'PT), a member of the phosphate translocator family, that exports the fixed carbon in form of triose phosphates and 3-phosphoglycerate from the chloroplasts in exchange for inorganic phosphate. This translocator is present in photosynthetic tissues only.
- A glucose translocator exporting the products of amylolytic starch breakdown. This chapter focuses on the characterization of the different classes of phosphate translocators, the glucose translocator, and on chloroplast translocators involved in nitrogen assimilation and amino acid biosynthesis.

11.2 PLASTIDIC PHOSPHATE TRANSLOCATORS

11.2.1 The Triose Phosphate-3-Phosphoglycerate-Phosphate Translocator

11.2.1.1 Structure and Function

The triose phosphate-3-phosphoglycerate-phosphate

Abbreviations: DiT1—2-oxoglutarate/malate translocator; DiT2—glutamate/malate translocator; GS — glutamine synthetase; OEP —outer envelope protein; PEP —phosphoenolpyruvate; 2-PGA—2- phosphoglycerate; 3-PGA — 3-phosphoglycerate; PPT —phosphoenolpyruvate/phosphate translocator; TPT — triose phosphate/3-phosphoglycerate/phosp hate translocator; TrioseP —triose phosphates; VDAC — voltage-dependent anion channel translocator (TPT) mediates the export of the fixed carbon (in the form of trioseP and 3-PGA) from the chloroplast into the cytosol, where it is converted into other substances, e.g. sucrose and amino acids (see Fig. 1). The inorganic phosphate released during these biosynthetic processes is reimported into the chloroplast via the TPT for the formation of ATP catalyzed by the thylakoid ATP synthase. Sucrose and amino acids are the main products for the supply of heterotrophic tissues with photo-assimilates. They are actively loaded into the sieve element/companion cell complex by specific H7 symporters. The isolation and elucidation of the molecular structures oftransporters for sucrose and various amino acids have been recently achieved using a yeast complementation system. Transport mediated by the TPT was first inves-tigated in intact chloroplasts and later, after purification and cloning of the corre-sponding cDNA, in artificial membranes containing the recombinant expressed protein (see below). In its functional form, the TPT is a dimer composed of two identical subunits. As substrates, the TPT accepts either inorganic phosphate or a phosphate molecule attached to the end of a three-carbon chain, such as trioseP or 3-PGA. C^{3}-compounds with the phosphate molecule at C-atom 2 (phosphoenolpyruvate, 2-PGA) are only poorly transported (see below). Under physiological conditions, the substrates are transported via a strict 1:1 exchange. Transport proceeds via a ping-pong type of reaction mechanism, i.e. the first substrate has to be transported across the membrane and then leave the transport site before the second substrate can be bound and transported to the opposite direction. In intact chloroplasts, unidirectional transport of phosphate can be observed but with a V that is two to three orders of magnitude lower as max compared to the antiport mode. Using the reconstituted system in which the concentrations of phosphate in both the internal and the external compartments are accessible to experimental variations, it could be demonstrated that the transport activity of the reconstituted TPT can reach values that exceed those measured for an antiport mode by at least one order of magnitude. It is suggested that transport under these conditions proceeds by a mechanism different from the antiport mode, probably by a (channel-like) uniport mechanism. Evidence for ion channel properties of the TPT is provided, (i) by a decrease of the activation energy for phosphate transport from 46 kJ/mol (antiport mode) to 18 kJ/mol (uniport mode), a value that is in the range observed for ion channels and, (ii) by measuring the TPT-mediated unidirectional transport by the patch-clamp technique. It can be concluded from these electrophysiological experiments that the PT can behave as a voltage-dependent ion channel, preferentially permeable to anions, as well as an antiporter. Different classes of transporters might share common structural motifs and may have arisen from a common ancestor. A small structural change within the translocation pathway might then allow transport via an ion channel mode or might result in strong coupling of substrate binding with conformational changes as observed for transporters operating in the antiport mode.

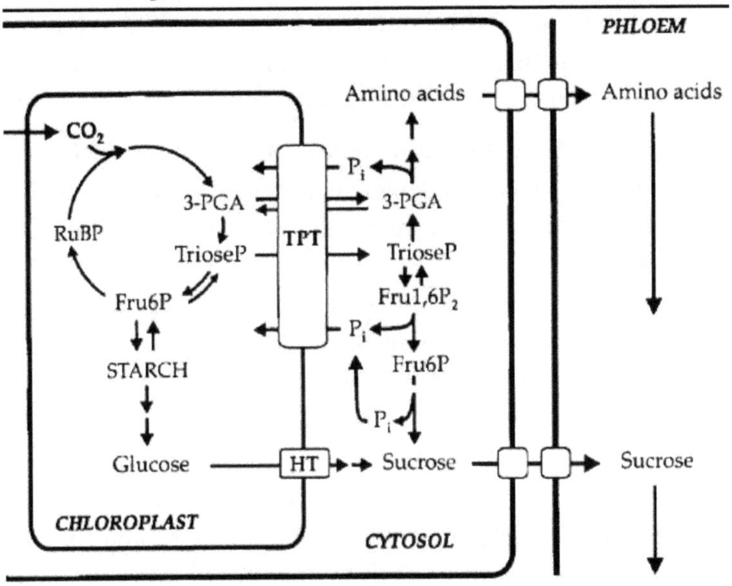

Fig. 1: Transport processes involved in the export of fixed carbon from chloroplasts. 3-PGA, 3-phosphoglycerate; Fru6P, fructose 6-phosphate; Fru1,6P2, fructose 1,6-bisphosphate; HT, hexose translocator; Pi, inorganic phosphate; RuBP, ribulose 1,5-bisphosphate; TPT, triose phosphate/3-phosphoglycerate/phosphate translocator; TrioseP, triose phosphates.

The spinach TPT was the first plant membrane transport system for which the primary sequence could be determined. Meanwhile, TPT-sequences from various plants are available, e.g., those from Arabidopsis, pea, potato, maize, Flaveria and tobacco that all have a high similarity to each other. All TPTs are nuclear-encoded and possess N-terminal transit peptides (about 80 amino acid residues) that direct the adjacent protein to the chloroplasts. Import of the translocator into chloroplasts is driven by ATP and depends on the translocation machinery of the envelope membrane. The mature part of these transporters consists of about 330 amino acid residues per monomer, is highly hydrophobic, and contains information (envelope insertion signals) for the integration of the protein within the inner envelope membrane. It is composed of five to seven hydrophobic segments in an a-helical conformation that traverse the membrane in zig-zag fashion connected by hydrophilic loops. The TPT thus belongs to the group of translocators with a 6 + 6 helix folding pattern, as is the case for mitochondrial carrierproteins. Based on a tentative model for the arrangement of the TPT in the membrane, it is probable that all twelve a-helices take part in forming a hydrophilic translocation channel through which the substrates are transported across the membrane. According to the proposed three-dimensional structural model of the TPT, an asymmetry in the structure of the substrate binding sites on either membrane side is suggested. This structural asymmetry was verified experimentally by demonstrating that only the cytosolic side of the translocator is accessible to specific inhibitors, namely pyridoxal 5 '-phosphate and 4,4 '-diisothiocyanostilbene-2,2 '-disulfonate. Bisubstrate initial velocity studies of the reconstituted translocator further showed that the TPT possesses a higher affinity (about five-fold) toward its substrates on the cytosolic side compared to the stromal side.

The availability of the TPT-cDNA also opened the way to produce the corresponding protein in heterologous expression systems. Since prokaryotic systems failed to express the hydrophobic translocator protein, yeast cells (Schizosaccharomyces pombe) were used. It could be shown that the TPT was produced in yeast cells in a functional state representing about 1% of the total protein. To facilitate the isolation of the TPT from the cells, a recombinant protein was engineered containing a C-terminal tag of six consecutive histidine residues (His6-tag). This was initially achieved by overlapping PCR; however, in the meanwhile, various commercially available vectors can be used for introducing different kinds of tags. Despite the low level of expression, the TPT-His6 protein could be purified to apparent homogeneity by a single chromatography step on metal-affinity columns. It could be demonstrated that the recombinant protein possessed transport characteristics almost identical to those of the authentic chloroplast protein.

11.2.1.2 Physiological Role of the TPT

The TPT is an important link between metabolism in the chloroplast and the cytosol. It can be calculated that only about 10% of the total transport activity of the TPT can be used for (productive) net triose export in order to provide the carbon skeleton for further biosynthetic processes. Since both the cytosol and the stroma contain triose P, 3-PGA and inorganic phosphate competing for transport in either direction, it appears feasible that most of the TPT activity is used for catalyzing (non-productive) homologous exchanges.

From subcellular metabolite concentrations in intact spinach leaves, it has been proposed that the TPT can exert a kinetic limitation during sucrose biosynthesis in vivo (Gerhardt et al., 1987). We have assessed the role of the TPT on photosynthetic metabolism by means of antisense repression of its corresponding mRNA. Transgenic potato (Solarium tuberosum L.) plants were created that showed a dramatically reduced expression of the TPT at the RNA level. In all transgenic plants analyzed, the maximum reduction of both the amount and the activity of the TPT was about 2530%. Under ambient CO, and light conditions, there was no significant effect on photosynthetic rates (Table 1), growth, and tuber development. However, even under ambient conditions, the reduction of the TPT activity resulted in a marked perturbation of leaf metabolism. Most remarkably, the content of 3-PGA was greatly increased compared to the corresponding value in wild-type plants and this increase was found to be almost entirely due to an increase of the stromal 3-PGA content with the cytosolic values remaining almost unchanged. Also, the sum of organic phosphates (3-PGA plus sugar phosphates) was about twice as large in the transformants as in the wild-type plants. This should result in a large decrease of the stromal content of inorganic phosphate since the TPT mediates a strict counter-exchange of substrates. Thus, reduction of the TPT activity in vivo resembles the situation of chloroplasts performing photosynthesis under phosphate limitation: the increase in 3-PGA (and the 3-PGA/phosphate ratio) results in a large increase of starch synthesis due to the allosteric activation of the ADP-glucose pyrophosphorylase (AGPase). Indeed, both at the end of the day and end ofthe night the starch content in transformants was much greater than in wild-type plants (Table 1). It appears that due to the restriction of the TPT, the daily assimilated carbon is mainly maintained within the plastids in which it is directed into the accumulation of starch (Table 2). Since there was no obvious effect on plant growth, the transformants were obviously able to efficiently compensate for their deficiency in TPT activity. A detailed analysis of the transgenic potato plants indeed showed that these plants mobilized and exported the major part of the daily accumulated carbon during the following night period. By contrast, wild-type plants generally export the major part of the fixed carbon during ongoing photosynthesis (Table 2).

Table 1. Antisense repression of the triose phosphate/phosphate translocator (TPT) in transgenic potato plants. Rates of phosphate transport and photosynthesis, and starch levels in wild-type and transformant leaves. Data from Riesmeier et al. (1993)

			Plants	
	Wild-type	**TPT 1**	**TPT 7**	**TPT 39**
Uptake of [3213] phosphate into chloroplasts (% of V)	= 100	72	69	83
Photosynthetic rate, ambient conditions (mmol 02 per m2 per h)	17.5	21.9	14.3	19.8
Starch (end of the day) (patom C per mg Chl.)	761	1310	2101	1336
Starch (end of the night) (patom C per mg Chl.)	330	654	1291	505

The question arises how the mobilization and export of photoassimilates is achieved in the antisense TPT plants during the night. Mobilization of starch and export of the resulting breakdown products can, in principle, either be performed via the TPT (after phosphorolytic starch breakdown resulting in trioseP) or via a hexose translocator (after amylolytic starch breakdown resulting in hexoses, see Fig. 1). Due to the reduced activity of the TPT, the mobilized carbon cannot be exported (at least not at high rates) from the chloroplasts as C_3-compounds that are end-products from phosphorylytic starch breakdown. Obviously, the transformants follow the hydrolytic starch breakdown pathway and preferentially export the mobilized carbon as hexoses via the hexose translocator. This is in line with the observation that transgenic antisense TPT plants from tobacco showed increased rates of amylolytic starch mobilization which process is accompanied by a higher transport capacity for glucose across the envelope membrane. It can be concluded that plants can efficiently compensate for their deficiency in TPT activity provided that a carbon sink (i.e., starch) can be generated during photosynthesis which can be mobilized during the following night period. Interestingly, transgenic potato plants with a reduced ability to synthesize assimilatory starch (e.g., by anti-sense repression of the AGPase) also show no effect on growth and productivity. Export of the daily fixed carbon via the TPT is obviously so efficient in these plants that heterotrophic tissues can be supplied sufficiently with reduced carbon. However, if both starch formation and the activity of the TPT are reduced, the corresponding transformants show a dramatic phenotype. On the one hand, the transformants are not able to export sufficient amounts of fixed carbon during the day (due to antisense repression of the TPT) and on the other hand, they do not have a carbon store which they could use during the dark period (due to antisense repression of the AGPase).

11.2.2 The Phosphoenolpyruvate/Phosphate Translocator

11.2.2.1 Structure and Function

Another type of phosphate translocator has been described recently. This transporter, named PPT (phosphoenolpyruvate/ phosphate translocator) was initially identified in non-green tissues, but it is also present in photosynthetic tissues. As in the case of the TPT, the PPTs are nuclear-encoded and consist of a transit peptide (about 80 amino acid residues) and a mature part (about 320 amino acid residues). Sequence comparisons between the TPT and the PPTs from different plants and plant tissues showedthat the PPTs represent a second class of phosphate translocators (Table 3). Within each group of transporters, the mature proteins are highly homologous to each other. In contrast, the homologies between members of the TPT family and the PPT family, respectively, are only about 35%.

Table 2. Antisense repression of the triose phosphate/phosphate translocator (TPT) in transgenic potato plants. Effect of antisense repression of the TPT on the accumulation and export of photoassimilates. TPT39, antisense TPT potato plant.

	Plants		
	Wild-type	**TPT 39**	
I. Amount of assimilate			
Assimilated during the day	1085	975	Matom c-mg chl.-I
Accumulated during the day	466 (= 43%)	866 (= 89%)	Matom c-mg chl.-I
2. Rate of assimilate export			
During the day	52 (= 57%)	9(= 11%)	Matom cmg ch1,-1h
During the night	39 (= 43%)	72 (= 89%)	Matom cmg chl.-I

Table 3. Homologies (in %) between the chloroplast triose phosphate/phosphate translocators (TPT) and the phosphoenolpyruvate/ phosphate translocators (PPT) from different plants.

		TPTs from				PPTs from	
	Spinach	**Cauliflower**	**Maize**	**Cauliflower PPT3**	**Arabidopsis PPT12**	**Tobacco PPT10**	**Maize PPT1**
Spinach TPT	(100)	88	85	35 35 34	34	35	35
Cauliflower TPT		(100)	81	(100)	34	36	37
Maize TPT			(100)		35	36	35
Cauliflower PPT3					93	77	75
Arabidopsis PPT12					(100)	80	73
Tobacco PPTIO						(100)	76 (100)
Maize PPT 1							

To study the transport characteristics of the cloned PPT proteins, the cDNA from the cauliflower PPT3 clone was fused to a DNA fragment encoding a tag of six consecutive histidine residues (Hiss tag) and was then subcloned into a yeast expression vector. As outlined above, this procedure enables the one-step purification of the produced transporter to apparent homogeneity by metal-affinity chromatography for the subsequent analysis of the transport characteristics by reconstitution experiments.

Figure 2 shows the substrate specificities of the purified PPT and the TPT, both purified from yeast cells, as determined by measuring phosphate transport into proteoliposomes that had been preloaded with the indicated counter-substrates. Both classes of phosphate translocators greatly differ in their transport characteristics. The PPT protein transports inorganic phosphate preferentially in exchange with PEP. This is in sharp contrast to the transport characteristics of the TPT. This transporter accepts either inorganic phosphate or C_3-compounds that are phosphorylated at C-atom 3 i.e., triose phosphates and 3-phosphoglycerate (see above). These substrates, however, are only poorly transported by the PPT protein. Both types of transporters strictly rely on the presence of an exchangeable substrate within the vesicles (i.e., they function as antiparty systems) and do not transport hexose phosphates like glucose 6-phosphate at all.

11.2.2.2 Physiological Role of the PPT

Do plastids rely on externally produced PEP or can PEP be generated inside the organelles? Work from several laboratories has shown that the conversion of hexose phosphates and/or triose phosphates via glycolysis cannot proceed further than to 3 phosphoglycerate due to the absence (or low activities) of phosphoglycerate mutase and/or enolase in most plastids. These plastids depend therefore on the provision of externally produced PEP. This fact has been overlooked to date, or it has been assumed that the transport of PEP into chloroplasts is facilitated by the TPT. This, however, appears unlikely because PEP is only poorly accepted as a substrate by the TPT (Fig. 2) and has, under physiological conditions, to compete with inorganic phosphate, triose P, and 3 phosphoglycerate for binding to the TPT. Hence, the presence of a PPT that bypasses the TPT is the most likely alternative for an efficient provision of the chloroplasts with PEP.

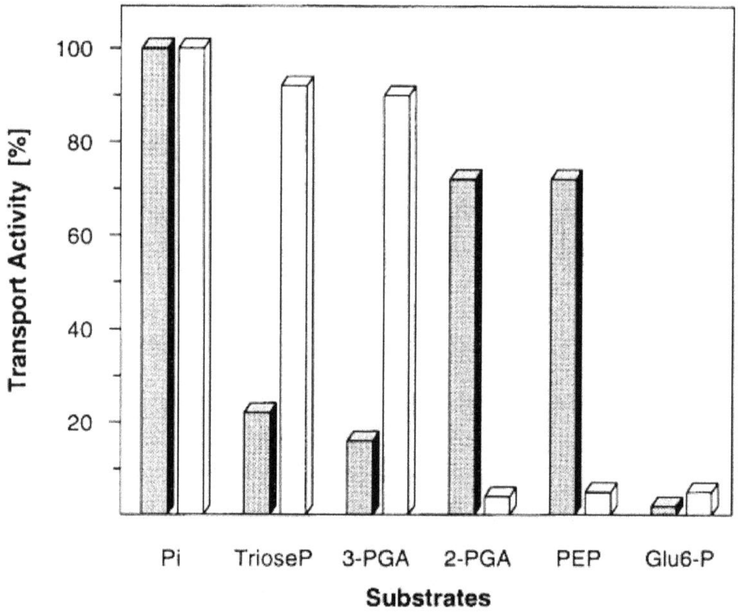

Fig. 2: Substrate specificities of the TPT and the PPT. The cauliflower PPT3-His6 protein and the spinach TPT-His6 protein were expressed in Schizosaccharomyces pombe cells and purified from these cells by metal affinity chromatography. The recombinant proteins were reconstituted into liposomes that had been preloaded with 25 mM substrates as indicated. Phosphate transport activity was measured as described by Fischer et al. (1997) and is given as a percentage of the activity measured for proteoliposomes preloaded with inorganic phosphate. The 100% exchange activities (Etmol mg-1protein x min-1) were 1.5 (PPT; dark gray, left), and 0.85 (TPT; light grey, right). 3-PGA, 3-phosphoglycerate; PEP, phosphoenolpyruvate; Pi, inorganic phosphate; TrioseP, dihydroxyacetone phosphate; Glu6-P, glucose 6-phosphate.

Several processes inside the plastids rely on PEP (see Fig. 3). This compound is an immediate substrate for the shikimate pathway which leads to the biosynthesis of aromatic amino acids and to a large number of secondary compounds that are important in plant defense mechanisms and stress responses. It has also been shown that the plastid-located fatty acid biosynthesis can be driven by exogenous pyruvate. Plastidic acetyl-CoA can subsequently be formed by the action of either acetyl-Co A synthetase or via the plastidic pyruvate dehydrogenase complex. In those plastids missing a complete pathway from 3-phosphoglycerate to pyruvate, PEP could be imported into the organelles via the PPT and pyruvate kinase could subsequently convert PEP into pyruvate.

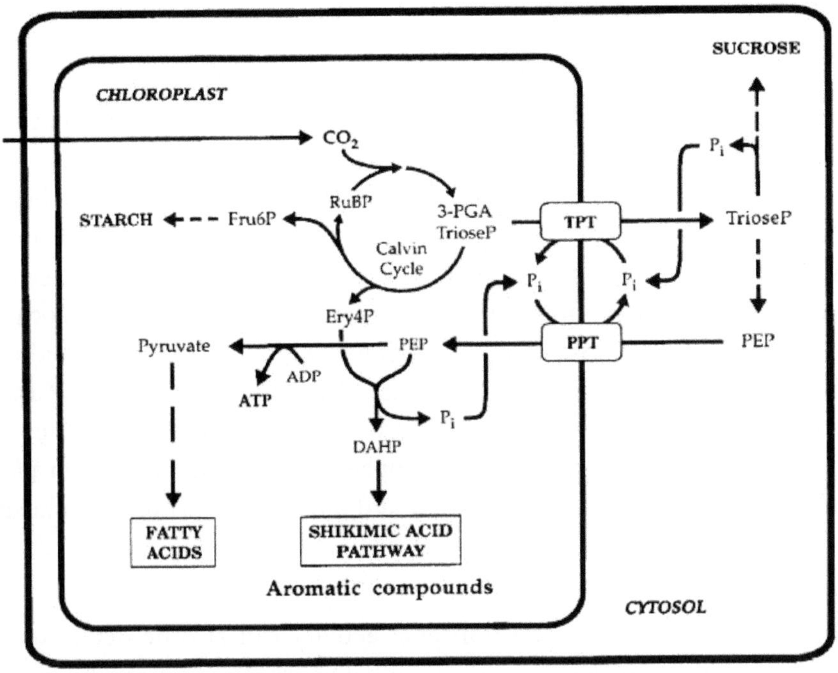

Fig. 3: Proposed function of the PPT protein in photosynthetically active tissues. The combined action of the TPT and the PPT results in the supply of the organelle with PEP generated from photosynthetically fixed carbon. 3-PGA, 3-phosphoglycerate; DAHP, 3-deoxy-D-arabino-heptulosonate-7-phosphate; Ery4P, erythrose 4-phosphate; Fru6P, fructose 6-phosphate; RuBP, ribulose 1,5-bisphosphate; TrioseP, those phosphates.

Figure 3 outlines the proposed physiological function of the PPT in chloroplasts. The fixed carbon is exported from the chloroplast by the TPT and partially converted into PEP in the cytosol. The PPT can the provide the chloroplast stroma with PEP for the shikimate pathway. In addition, PEP can be converted into pyruvate as a precursor for fatty acid biosynthesis, concomitantly providing the plastids with ATP The latter issue is probably important in chloroplasts during night or in non-photosynthetic tissues. Because inorganic phosphate is used as a counter-substrate by both the TPT and the PPT, the combined action of both translocators would result in an exchange of triose phosphate with PEP without net phosphate transport.

With the identification of a phosphate translocator specific for PEP, it is proposed that plastids contain a set of phosphate translocators with different structures but overlapping substrate specificities. Chloroplasts, or at least a subtype of chloroplasts, contain members of both the TPT and the PPT family. The TPT is present only in photosynthetic tissues, whereas PPT-specific transcripts can be detected in both leaves and non-green tissues although they are more abundant in non-green tissues. The presence of different phosphate translocators with different substrate specificities would allow the uptake of any particular phosphorylated substrates even in the presence of high concentrations of other phosphorylated metabolites, which would otherwise compete for the binding site of a single phosphate translocator.

11.2.3 The Hexose Phosphate/Phosphate Translocator

It has been shown that chloroplasts from guard-cells contain a phosphate translocator that also transports glucose 6-phosphate. Stomatal opening is associated with the production of malate as a counterion for potassium. Malate can derive from starch mobilization, the products of which are exported from the chloroplasts and used for malate formation. For the biosynthesis of starch, guard-cells have to import reduced carbon. It has been proposed that trioseP and/or 3-PGA, imported via the chloroplast phosphate translocator, can be used for the formation of hexose phosphates. However, guard-cells are devoid of fructose 1,6-bisphosphatase activity,the key enzyme for the conversion of trioseP into hexose phosphates, and therefore, rely on the provision of hexose phosphates for starch biosynthesis. In this respect, guard-cell chloroplasts behave like non-green plastids that also lack fructose 1, 6- bisphosphatase activity and have to import hexose phosphates for starch biosynthesis. Both organelles obviously contain a phosphate translocator that can import hexose phosphates in exchange with inorganic phosphate.

A hexose phosphate/phosphate transport activity can also be detected in spinach chloroplasts after feeding of detached leaves with glucose. The chloroplasts of these leaves are photosynthetically activebut contain large quantities of starch. As is the case for heterotrophic plastids in sink tissues, the precursor for starch biosynthesis is imported into the chloroplasts in the form of glucose 6-phosphate by a novel hexose phosphate/ phosphate translocator. Glucose-feeding has obviously induced a switch in the function of the chloroplasts from carbon-exporting source organelles to carbon-importing sink-organelles. These experiments show that the physiological state of a tissue can greatly alter its biochemical properties and demonstrate the high flexibility of plant metabolism.

We have recently identified a translocator that mediates the transport of glucose 6-phosphate in exchange with either inorganic phosphate or triose phosphates (glucose 6-phosphate/phosphate trans-locator, GPT). Glucose 6-phosphate imported into the plastids via the GPT can thus be used either for starch biosynthesis, during which process inorganic phosphate is released, or as a substrate for the oxidative pentose phosphate pathway, yielding triose phosphates. The GPT shares about 38% identical amino acids with members of the TPT and PPT families and thus represents a third group of plastidic phosphate antiporters. The GPT gene is predominantly expressed in heterotrophic tissues but is most probably also responsible for the hexose phosphate transport activities in chloroplasts.

11.3 TRANSPORT OF GLUCOSE

The amylolytic degradation of starch at night results in the formation of glucose that can be exported from the chloroplasts by a glucose translocator and converted to sucrose in the cytosol (see Fig. 1). As mentioned above, transgenic anti-sense TPT potato plants export carbon during the night preferentially via this translocator to compensate for their deficiency in TPT activity. Evidence that the preferred route for exporting carbon from chloroplasts proceeds via the glucose translocator comes from studies on a high starch Arabidopsis mutant. This mutant is able to degrade starch but obviously unable to export the products of starch degradation. These are therefore redirected into the synthesis of starch leading to the observed starch-excess phenotype. The chloroplasts possess a functional TPT but obviously lack a functional chloroplast glucose translocator for carbon export. Little is known about this translocator that had first been described in the late seventies. The availability of starch-excess mutants will presumably open the way to isolate the corresponding gene, e.g., by map-based cloning.

11.4 DICARBOXYLATETRANSLOCATORS

The photosynthetically fixed carbon that is exported from the chloroplasts via the TPT as triose P and 3-PGA can also be used for the formation of carbon skeletons (2-oxoglutarate) via processes occurring in the cytosol and in the mitochondria. 2-oxoglutarate can then be imported into the chloroplasts for the fixation of ammonia that derives from nitrate reduction or photorespiration. Fixation of ammonia within the chloroplast is achieved via the stromal glutamine synthetase (GS2)/glutamate synthase cycle. Glutamate synthesized during this cycle is the key compound of nitrogen metabolism in plants. It can be subsequently exported into the cytosol (Fig. 4). Two different dicarboxylate antiport systems are involved in this process. Both translocators mediate an exchange with malate:

- A2-oxoglutarate/malate translocator (DiT1) that transports 2-oxoglutarate into the chloroplasts in exchange with malate and that does not transport glutamate or aspartate and
- A glutamate/malate translocator (DiT2) exporting glutamate in exchange with cytosolic malate. DiT2 can be considered as a general dicarboxylate. Imported OAA can also be used for transamination to aspartate which is, in turn, the precursor for the biosynthesis of various amino acids. The molecular structure of the OAA translocator is yet unknown.

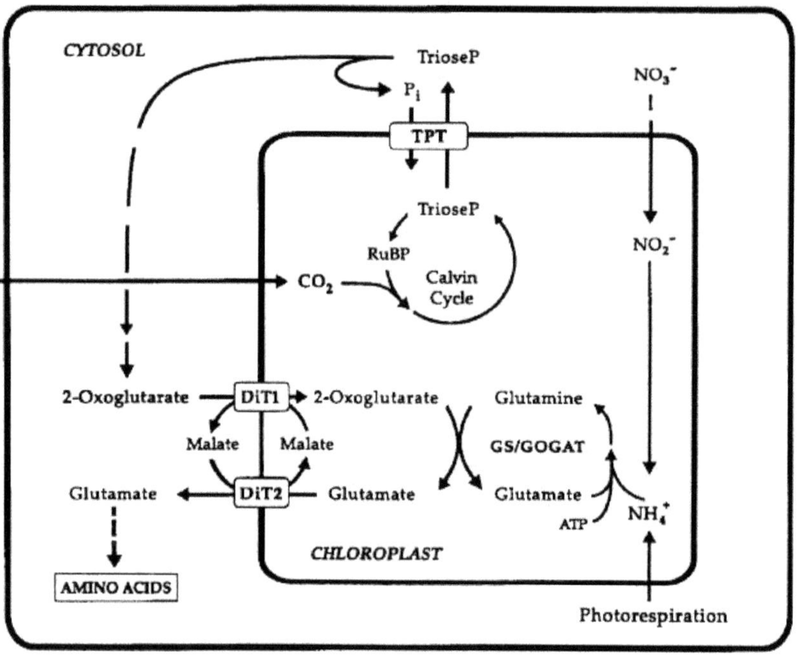

Fig. 4: Two plastidic dicarboxy late translocators are involved in ammonia assimilation. DiT1, 2-oxoglutarate/malate translocator; DiT2, glutamate/malate translocator; GS/ GOGAT, glutamine synthetase/glutamate synthase; 11,, inorganic phosphate; RuBP, ribulose 1,5- bisphosphate; TrioseP, triose phosphates.

Mutants of Arabidopsis deficient in dicarboxylate transport activities are not viable under photo-respiratory conditions. The defect is probably caused by the general dicarboxylate translocator (DiT2). However, attempts to clone the corresponding gene have failed so far.

The identification of the 2-oxoglutarate/malate translocator (DiT1) as a 45 kDa component of the inner envelope membrane has recently been described. Subsequently, a cDNA clone was obtained that codes for the precursor protein with a relative molecular mass of 60 kDa. Interestingly, there is no detectable homology to any other known protein, not even to the mitochondrial 2-oxoglutarate/malate carrier with transport characteristics similar to DiT1.

The mature part of DiT1 contains 13-14 hydrophobic segments; 12 of them are long enough to span the membrane as transmembrane a-helices. These helices are obviously arranged in such a way that two clusters of six helices each are separated by an intervening hydrophilic loop. This transmembrane topology with an 12-helix motif resembles that of other plasmamembrane transporters from prokaryotes and eukaryotes that presumably all function as monomers. The recently identified plastidic ADP/ ATP translocator also belongs to this type of transporter (see below). All other organellar transporters, including those from mitochondria, belong to the second group of transporters having 5-7 transmembrane helices like the chloroplast phosphate translocators. These translocators probably function as homodimers with a '6 + 6' transmembrane helix pattern. Thus, the inner envelope membrane possesses two different types of transporters which may all have, in their functional state, a comparable number of membrane-spanning segments.

To verify the identity of the cloned gene, the full-length cDNA sequence was expressed in the fission yeast Schizosaccharomyces pombe. Yeast transformants harboring the DNA coding for DiT1 indeed displayed reconstituted malate transport activities at rates exceeding those of the endogenous transporters by at least two orders of magnitude. It could be further shown that substrate specificities of the recombinant protein were almost identical to the translocator purified from envelope membranes). The activity of the recombinant protein is thus in accordance with its proposed function as a supplier of the chloroplasts with 2- oxoglutarate for ammonia assimilation.

It has been shown that chloroplasts contain, in addition to DiT1 and DiT2, a specific translocator for glutamine with glutamate as the preferential counter substrate. Glutamine is not transported by DiT1, and by DiT2 only at low rates. It has been proposed that the glutamine translocator is involved in the refixation of photorespiratory ammonia by mediating uptake of glutamine formed by the cytosolic isoform of glutamine synthetase (GS 1). However, it has been shown that the presence of GS 1 is exclusively restricted to the cytoplasm of phloem companion cells whereas the chloroplast-located GS2 is localized in mesophyll cells in which it is involved in nitrogen assimilation. Since the phloem companion cells do not contain photosynthetically active chloroplasts, uptake of GS 1 - produced glutamine into the plastids is not possible. Although the location of the glutamine/glutamate translocator has not yet been determined, it is suggested that this translocator resides in the mesophyll. In these cells, glutamine, produced by the stromal GS 2, could be exported from the chloroplasts in exchange for cytosolic glutamate. This would allow a supply of assimilated nitrogen to the cell.

Chloroplasts are also able to transport oxaloacetate (OAA). Import of OAA into chloroplasts can be closely linked to an export of malate. This malate/ OAA shuttle plays an important role in the transfer of reducing equivalents from the chloroplasts into the cytosol. In principle, OAA can be transported by DiT1 and DiT2. However, this appears unlikely because the concentrations of OAA in both the cytosol and the stroma are at least one order of magnitude lower than that of malate and other dicarboxylates with which substrates OAA has to compete for binding to the translocator. Measurements of the uptake of OAA into chloroplasts of C--as well of C_4-plants revealed the presence of a specific translocator with a high affinity for OAA. This translocator is able to transport OAA even in the presence of a large excess of other dicarboxylates Imported OAA can also be used for transamination to aspartate which is, in turn, the precursor for the biosynthesis of various amino acids. The molecular structure of the OAA translocator is yet unknown. C_3-chloroplasts also contain an antiport system for monocarboxylates, i.e.

glycerate and glycolate, which are both intermediates of the photorespiratory cycle. This transporter will be dealt with in the chapter on photorespiration.

C_3-chloroplasts also contain an antiport system for monocarboxylates, i.e. glycerate and glycolate, which are both intermediates of the photorespiratory cycle. This transporter will be dealt with in the chapter on photorespiration.

11.5 TRANSPORT OF NUCLEOTIDES

Plant cells contain a highly active ADP/ATP translocator (AAT) in the inner membranes of mitochondria that exports the ATP generated by oxidative phosphorylation. Heldt (1969) was the first to show that chloroplasts are also capable of transporting ADP and ATP in a counterexchange mode. The proposed physiological function of this translocator in mature leaves is to supply chloroplasts with mitochondrial ATP during the night whereas nongreen plastids depend on the supply with ATP as the driving force for biosynthetic processes, e.g., starch and fatty acids. A detailed analysis ofthe kinetic constants of the AAT showed that ATP and ADP are preferentially transported. AMP and other nucleotides showed only a poor interaction with the translocator. It has been claimed that ADP-glucose can also be transported via the AAT into both chloroplasts and amyloplasts. However, the affinity of AAT toward ADP-glucose is 2 orders of magnitude lower than toward ATP. Because ADP-glucose has to compete with ATP and ADP for binding to the translocator, the amount ofADP-glucose transported by the AAT cannot be of any physiological significance. Nevertheless, a mechanism for the transport of ADP-glucose is required in the endosperm of some cereals. It has been shown recently that in these tissues the AGPase is mainly present in the cytosol but not in the plastids. The cytosolically formed ADP-glucose is presumably transported into the plastids for starch biosynthesis via an ADP-glucose/ADP antiporter which is distinct from the AAT. Most probably, the Btl protein, the cDNA of which has already been cloned, serves this function. The corresponding Btl mutation results in an accumulation of ADP-glucose and a reduction of starch biosynthesis in maize, probably reflecting the lack of the ADP-glucose/ADP translocator.

The molecular structure of the plastidic AAT remained elusive until Kampfenkel et al. (1995) happened to isolate a partial cDNA clone from A. *thaliana* that exhibited similarities to the ADP/ATP translocase from the bacterium *Rickettsia prowazekii*. Isolation and characterization ofthe full-length clone revealed no homology to the known mitochondrial adenylate translocators. The corresponding hydrophobic membrane protein (AATP1) possesses an N-terminal region with typical features of plastidic transit peptides. Interestingly, the AATP1 protein contains 12 potential transmembrane helices and thus belongs to the group of transporters with a 12 helix motif as is the case for DiT1 (see above). The transit peptide of AATP1 was able to target the adjacent protein to isolated chloroplasts. In purified chloroplast envelope membranes, theAATP1 protein (62 kDa) couldbe detected immunologically. Expression of the AATP1-cDNA in heterologous systems and subsequent transport measurements demonstrates the identity of AATP1 as an ATP translocator. AATP1-specific transcripts could be detected both in photosynthetic and nongreen tissues. Taken together, the results suggest thatAATP1 represents the plastidic adenylate translocator that is present in all plant tissues.

11.6 OTHER TRANSLOCATORS

It is well established that chloroplasts are the major site of the synthesis of many amino acids including those of the aspartate and the pyruvate family (Bryan, 1990). In addition, the plastid-located shikimate pathway leads to the formation of the aromatic amino acids phenylalanine, tyrosine and tryptophan. The obvious reason for the chloroplast-located biosynthesis of amino acids is that reduced nitrogen and reducing equivalents, required for some of the biochemical reactions, are produced within the chloroplasts via nitrite reductase and photosynthetic electron transport, respectively, and carbon skeletons (pyruvate (phosphoenolpyruvate), OAA, 2-oxoglutarate) can be imported from the cytosol via specific translocators. The synthesized amino acids are either used for the plastid-located protein biosynthesis or are, in the main part, exported into the cytosol for intercellular nitrogen transport. Unfortunately, almost nothing is known about plastidic translocators that are involved in these transport processes.

11.7 CHANNELS IN CHLOROPLAST ENVELOPE MEMBRANES

So far, translocator proteins of the inner chloroplast envelope as the major osmotic barrier between the chloroplast and the cytosol have been dealt with. Both the inner and the outer envelope membrane also contain channels and pores to mediate the flux of ions across these membranes. The outer envelope membrane appears to be permeable to low molecular weight solutes, mediated by the chloroplast porin. Porins are voltage-dependent channels (VDACs) that are also present in the outer membrane of bacteria and mitochondria. In planar bilayers containing reconstituted chloroplast outer envelope membrane proteins, a large conductance channel (A @720 pS, in 100 mM KC1) was found. Heiber et al. (1995) reported the presence of a slightly cation-selective high conductance channel (A @ 1.14 nS) in the outer envelope membrane. Weak anion and cation-selective large conductance pathways (A @ 512 pS and A @ 1016 pS in 100 mM KC1) have also been detected in the chloroplast envelope membrane by direct patch-clamping of chloroplasts from the green alga Nitelopsis obtusa. It has been suggested that the weakly anion-selective channel with a conductivity of A 512 pS may be related to the mitochondrial VDAC.

Although the identities of these electrophysiologically characterized channels are not yet known, primary sequences of one class of porins from non-green plastids were recently obtained. These porins showed only a 20-25% identity to the known mitochondrial porins from eukaryotes. However, the plant porins contain 16 antiparallel transmembrane Li-strands as is the case for porins from other organisms. The plant porin could be functionally expressed in bacterial cells and the recombinant protein revealed, after reconstitution into planar lipid bilayers, single channel conductances and voltage dependencies almost identical to the authentic protein, 3 and 1.5 nS in 1 M KC1. These values are similar to those obtained for plant mitochondrial porins indicating that both non-green plastids and mitochondria possess one type of closely related porin proteins. This conclusion is corroborated by direct alignments of porin sequences from plant mitochondria.

More recently, the primary sequences of two chloroplast outer envelope membrane proteins were obtained (OEP16, OEP24) that both possess channel-like activities but do not show similarities to bacterial or mitochondrial porins. Both proteins act as selectivity filters of the outer envelope membrane allowing the passage of amino acids (OEP16) andphosphate, dicarboxylates or adenylates (0EP24), respectively. It is suggested that the outer envelope membrane may contain various channel-like proteins with different selectivities and that the combined action of these channel proteins defines the permeability properties of the chloroplast outer envelope membrane.

Ion channels are also present in the inner envelope membrane. It has already been mentioned that the TPT can function as an anion-selective channel. Ion channel activity of the TPT, however, can only be observed if the concentrations of the monovalent ions are much higher than those of the physiological substrates. Recently, a weakly anion-selective high conductance channel with a single channel conductance of A 525 pS has been discovered in proteoliposomes containing purified inner envelope membranes from spinach chloroplasts. A similar weak anion-selective high conductance channel (A w 540 pS) was observed by Heiber et al. (1995) in proteoliposomes containing inner and outer envelope membranes from spinach chloroplasts. To prevent uncontrolled dissipation of ion gradients across the inner envelope membrane, these porin-like channels have to be highly regulated.

The inner envelope membrane also contains a chloride channel that might be responsible for the rather high chloride permeability of this membrane. Similar anion-selective channels have been observed in the chloroplast membranes from green algae.

The presence of a potassium channel in the inner chloroplast envelope has been demonstrated by Berkowitz and coworkers. This K'-channel revealed an inward rectifying IN-relationship, was blockedby millimolar concentrations of ATP and activated by Mg^{t-}. The proposed physiological function of the K+-channel in photosynthetic ion fluxes is outlined in Fig. 5. Light-driven FL-transport into the thylakoids can be counterbalanced by a release of Mg^{2}+-ions from the thylakoid lumen. Elevated levels of Mg" and alkalization of the chloroplast stroma are both required for full activity of the photosynthetic carbon cycle. Stromal Mg' activates the envelope K+-channel leading to an increased K+-influx that is electrically balanced by H+-efflux via the envelope Fr-ATPase. On the other hand, the K^{-}-channel is blocked by increasing concentrations of photosynthetically produced ATP. This effect results in a decreased H+-efflux from the chloroplasts and an acidification of the chloroplast stroma. It is proposed that the opposing modulation of the IC-channel by Me-ions and ATP play a major role in the pH-stasis across the chloroplast envelope and the regulation of photosynthesis.

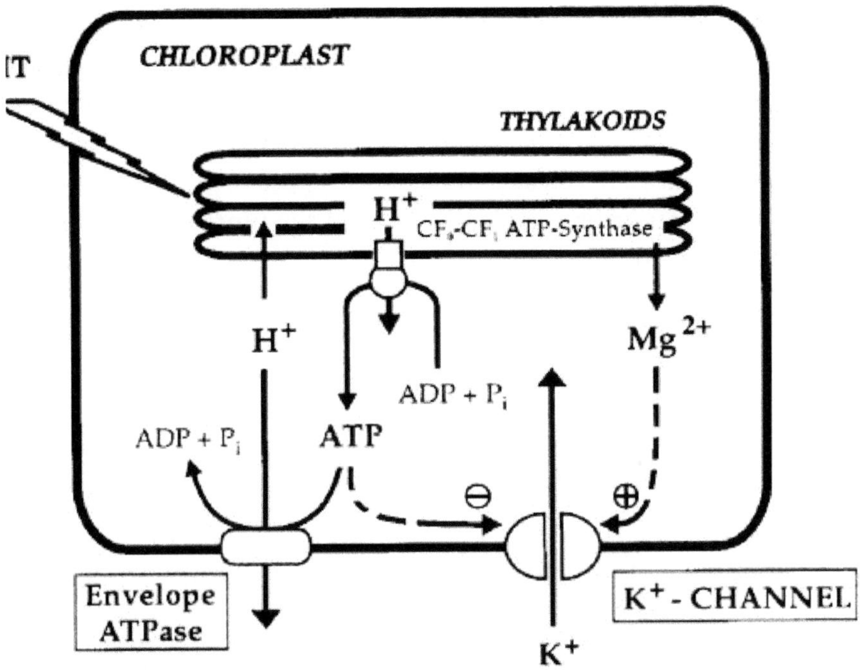

Fig. 5: Proposed function of the 1C+-channel of the chloroplast inner envelope membrane in energy-dependent fluxes of Fr and K+-ions. Modulation of the K+-channel activity by photosynthetically generated ATP and by Mg2tions takes part in the pH-stasis across the chloroplast envelope during photosynthesis. For details, see text. P,, inorganic phosphate.

11.8 AT A GLANCE

- Much progress has been achieved during the last years in studying translocators at biochemical and, more recently, molecular levels. When the corresponding gene is available, its cell- and tissue-specific expression can be studied (e.g., by RNA gel blot analysis, by in situ hybridization, or by expression studies of the corresponding promoter fused to a reporter gene).
- Transgenic plants with a controlled expression of the transporter protein can be used to elucidate the precise role of these translocators in plant metabolism. On the other hand, our knowledge of the identity of other plastidic translocators is still rudimentary. There are multiple reasons for this. First, the biochemical identification and isolation of non-abundant translocators is difficult. Second, the identification of unknown plastidic transporters appears not to be possible using the heterologous complementation system of yeast mutants. As mentioned above, this system has been used extensively to isolate transporters of the plasma membrane. Third, the primary sequences of plastidic translocators known to date

revealed only limited homologies to transporters known in other systems. Thus, genes coding for plastidic translocators are rarely obtained by screenings of DNA libraries with heterologous probes.

- In the foreseeable future, the Arabidopsis and rice genomic sequencing programs will provide the sequences of complete higher plant genomes. In addition, 30,000 Arabidopsis ESTs (Expressed Sequence Tags) tagging about half of the expected Arabidopsis genes are already available. It appears evident that future work will then concentrate on the identification of the functions of genes coding for putative envelope translocators.

12

CONTROL OF PHOTOSYNTHESIS, ALLOCATION AND PARTITIONING BY SUGAR REGULATED GENE EXPRESSION

12.1 INTRODUCTION

- Plant tissues can be divided into those that export (source) and those that import (sink) carbon and nitrogen. Mature green leaves, seed storage tissue during early post germinative growth and sprouting tubers are examples of source tissues. Sink tissue can be sub-divided into storage sinks, such as developing seeds and tubers which accumulate various proportions of carbohydrate, proteins and lipids; and metabolic sinks such as meristems and roots which accumulate little or no storage material, but rely on net import of metabolites for growth and development. However, this classification into sink and source organs is not static and depends on genetically fixed developmental processes such as the transition of developing seeds and young leaves from sink to source organs during maturation. It also depends on changes in environmental conditions. The ability of a plant to respond to environmental changes is necessary in order to maximize growth, biomass and reproductive capacity when essential resources such as light, CO^2, water and nitrogen are abundant and to compensate when there is a deficit in one or more of these resources. Coordinated development and interaction of source and sink organs is critical in order to achieve this. One of the key processes that needs to be coordinated between source and sink tissue is the photosynthetic production of carbohydrates with their mobilization, allocation and utilization. This acclimation (maintenance of the supply and demand for photosynthetic assimilates in appropriate balance) in response to fluctuating exogenous factors is achieved by short-term

- Abbreviations: ADP — adenosine diphosphate; AGP-S — ADP-glucose pyrophosphorylase; AMP—adenosine monophosphate; AMPK—AMP-activated protein kinase; ATP — adenosine triphosphate; CAB — chlorophyl alb binding protein; cAMP —cyclic adenosine monophosphate; CaMV35S—cauliflower mosaic virus 35S promoter; Cra — catabolite repressor/activator; CRP —cAMP receptor protein; FBP — fructose 1,6-bisphosphate; F 1P —fructose 1-phosphate; HXK — hexokinase; PC — plastocyanin; PEP — phosphoenolpyruvate; PTS — phosphotransferase system; RBCS — ribulose bisphosphate carboxylase small subunit; Rubisco — ribulose-1,5-bisphosphate carboxylase/oxygenase; SNF 1 — sucrose non-fermenting protein kinase; VSP — vegetative storage protein modulation of enzyme activities and other proteins along with more long-term changes in expression of genes whose products are involved in the production, mobilization, allocation and utilization of photo-assimilates. These genes are controlled by a variety of factors including developmental, hormonal and environmental signals as well as by carbohydrates, either directly or indirectly, since the products of these genes ultimately modify carbohydrate levels.
- A variety of genes, whose products are involved in diverse metabolic pathways and cellular functions, are either induced or repressed depending on the availability of soluble sugars in higher plant cells. This metabolic regulation of gene expression can be viewed at two levels. Firstly, at the cellular level, changes in gene expression result in a modification of anabolic or catabolic processes in response to nutrient availability. This mechanism is important in the control of primary carbon metabolism and nutrient homeostasis within the cell. For example, in mesophyll cells when soluble sugar levels are high there is feedback inhibition on photosynthetic gene expression which can actually override the well documented induction effect of light.
- At the other extreme, catabolic processes are induced at the level of gene expression when soluble sugar levels are depleted in whole leaves or mesophyll protoplasts. Secondly, when viewed at the whole plant level it becomes apparent that sugar regulation of gene expression plays an important role in source-sink interactions. Productivity in source tissues needs to match demand from sink tissues. Sugars play an integral role in achieving this as they act as signals both for repression of genes coding for photosynthetic enzymes and for induction of genes coding for sink-specific enzymes, such as sucrose synthase, extracellular invertase and granule bound starch synthase. Since changes in the intracellular levels of carbohydrates result in changes in expression of a variety of genes, then any factor, either local or remote, that affects carbohydrate levels will affect gene expression. Such factors include environmental conditions such as light, temperature, availability of macronutrients.. In this chapter we will discuss the variety of factors that bring about changes in carbohydrate status in photosynthetic cells

and the effect these changes have on gene expression. We will then draw upon the extensive literature in metabolic regulation of gene expression in prokaryotes and lower eukaryotes in order to highlight possible mechanisms that may be involved in the sugar regulated gene expression in plant cells.

12.2 CARBOHYDRATE REGULATION OF GENE EXPRESSION IN SOURCE AND SINK TISSUES

Photosynthetic gene expression is subject to feedback control by metabolites that signal carbohydrate status of photosynthetic cells. To appreciate this control at the whole plant level it is essential first to consider the factors that regulate carbohydrate status in photosynthetic tissues. Carbohydrate status will be a function of the net production of carbohydrate and the partitioning of assimilated carbon to sink tissues. Therefore, any factor which results in a change in either of these parameters will change the carbohydrate status and so effect a change in photosynthetic gene expression. For example, light intensity, temperature and CO_2 availability will directly affect the rate of photosynthesis and so the production of photoassimilates. Availability of essential nutrients such as nitrate or phosphate, will affect the demand for carbon-skeletons for biosynthesis in both source and sink tissues.

12.2.1 Carbohydrate Control of Gene Expression in Source Tissues

- A number of model systems have been used to demonstrate that elevated levels of soluble sugars result in repression of photosynthetic gene expression via a feedback control mechanism. In several cases this control has been demonstrated to occur at the level of transcription. The first example of this type of control came from work using a maize mesophyll protoplast transient expression system in which it was demonstrated that transcription from seven maize.
- Photosynthetic gene promoters were repressed by sugars or acetate whereas other promoters (e.g. CaMV35S) were not significantly affected. In Arabidopsis, the light induction of the ribulose bisphosphate carboxylase small subunit gene (RBCS) in dark-adapted seedlings is inhibited by growth on 2% sucrose or glucose. In rapeseed cell culture 2% sucrose inhibits the light dependent accumulation of the chlorophyll alb binding protein (CAB).
- Sucrose is also able to repress the early, light independent activation of a plastocyanin gene during early Arabidopsis seedling development. Sugar repression of photo-synthetic genes has also been demonstrated in a variety of other experimental systems including transgenic tobacco expressing invertase in the apoplast resulting in elevated levels of intracellular sugars, a Chenopodium cell culture, and in spinach plants that had been cold-girdled to decrease export of photoassimilates from leaves.

- The repression of photosynthetic genes by high-carbohydrate-asso-ciated metabolic factors represents a basic genetic mechanism. It should be noted that genes encoding other enzymes including the plastid isoforms of aldolase, trio se-phosphate isomerase and phosphogluco-mutase, that have a dual role in photosynthesis and respiratory metabolism, are not subject to carbo-hydrate repression and activities of glycolytic enzymes remain unaltered or increase.

12.2.2 Influence of Carbohydrate Control in Sink Tissues on Gene Expression in Source Tissues

- Carbon partitioning to sink tissues depends on the plants developmental program which largely dictates the temporal and spatial distribution of storage and metabolic sinks. Superimposed on this are fluctuations in partitioning in response to changing environmental conditions since many forms of environmental stress can restrict availability of essential resources. In general, genes involved in allocation and accumulation of storage reserves show increased expression when carbohydrate levels are high. Expression of these genes decreases when carbohydrate levels decrease.
- One of the key mechanisms regulating carbon partitioning to sink tissue appears to be the metabolic regulation of two families of genes encoding enzymes responsible for the initial metabolism of sucrose. Sucrose is the major form of carbohydrate that is transported in many plants. The genes encoding the two enzymes known to be responsible for sucrose cleavage in the phloem and importing tissues of plant cells, namely sucrose synthase and invertase, are themselves regulated by carbohydrate status. Both of these enzymes are encoded by a gene family in maize and in both families one class of genes is up-regulated by elevated sugar levels (SUCROSE SYNTHASE1 and INVERTASE2), whereas a second class of genes in the same family is repressed by sugars and upregulated when sugar levels fall (SHRUNKEN1 and INVERTASE1). Interestingly, the two classes of genes from each of these families are also differentially expressed during development. Those genes that are induced by sugars are expressed in sugar-importing tissues and those that are repressed by sugars are induced during reproductive development. Carbohydrate regulation of sucrose synthase and invertase genes from various other species including potato, rice, bean, Chenopodium rubrum and Arabidopsis have also been reported.
- The metabolic regulation of sucrose synthase and invertase genes therefore provides a possible mechanism for the regulation of import of carbon into sink tissues. This regulation in itself will effect changes in the carbohydrate status in photosynthetic tissues and in so doing ultimately give rise to changes in photosynthetic gene expression. Expression of extracellular invertase following wounding or bacterial infection is correlated with repression of photosynthetic

gene expression. It has been postulated that the resulting increased hydrolysis of apoplastic sucrose to glucose and fructose may play an important role in coordinating the plants defense response by inducing certain genes encoding for proteins such as proteinase inhibitor II and chalcone synthase, and repressing others such as those involved in photosynthesis.

- The majority of sink genes that respond to sugar related signals are also regulated by a number of other mechanisms including developmental, stress or environmental signals. This leads to a complex regulatory network in which specific genes are responding to a variety of signals. Further details of metabolic regulation of gene expression in higher plants can be found in recent reviews and more specifically carbohydrate control of photo-synthetic gene expression has also been reviewed.

12.3 INFLUENCE OF OTHER METABOLITES ON SUGAR REGULATED GENES

- Photosynthetic acclimation is suggested to be caused by accumulation of carbohydrates in the source leaves resulting in feedback inhibition of photosynthesis. Therefore, any developmental or environmental factor which results in increased levels of carbohydrate in source tissues could cause a repression of photosynthesis. Photosynthetic acclimation has been observed in some instances when plants are grown at elevated CO_2.
- Acclimation of tomato plants to elevated levels of CO_2 in which an initial increase in photosynthesis is followed by a long term decrease has recently been linked to sugar repression of photosynthesis. Exposure of tomato plants to either elevated CO_2 or sucrose resulted in decreased levels of nuclear encoded mRNA for photosynthetic enzymes, whereas repression of chloroplast genes was less pronounced. The similar effects of elevated CO^2 and sugars has led to the proposal that the initial increase in photosynthesis after exposure to elevated CO^2 results in accumulation of sugars (if sink strength is insufficient) and this, in turn, results in repression of nuclear encoded photosynthetic genes. Deficiency of nitrogen or phosphorous results in accumulation of carbohydrates in mature leaves and roots within hours after withdrawal of these nutrients. This implies that nitrogen or phosphorous deficiency limits utilization of carbohydrate more than it limits photosynthesis. In nitrogen-deficient shoots and roots of tobacco seedlings, levels of hexose phosphates and 3-phosphoglyceric acid are elevated and upon feeding of exogenous sucrose, protein, Rubisco and chlorophyll content is decreased in shoots as nitrogen is recycled.
- Transcripts encoding enzymes involved in nitrate assimilation and starch synthesis increase in response to elevated levels of soluble sugars in photosynthetic tissues. Sucrose can substitute for light in eliciting an increase in steady state levels of

nitrate reductase mRNA in dark-adapted green Arabidopsis plants. In the same experiment, sucrose was shown to override the light induction of the RBCS gene. Both light and exogenous sucrose repress accumulation of asparagine synthetase and increase nitrate reductase mRNA levels in Arabidopsis. The partial elimination of sucrose repression of asparagine synthetase expression by amino acids suggests that nitrogen:carbon ratio rather than carbon alone is responsible for the metabolic regulation of at least this carbohydrate regulated gene. Control of gene expression by metabolites may therefore regulate major fluxes of carbon and nitrogen in plants. Allosteric regulation of the starch biosynthetic enzyme ADP-glucose pyrophosphorylase by phos-phate and 3-phosphoglyceric acid is a well established mechanism for regulation of starch biosynthesis. An additional level of control appears to act at the level of gene expression.

- Genes encoding ADP-glucose pyrophosphorylase (AGP-S) and granule bound starch synthase have been shown to be induced by high levels of sugars in sink tissues. A similar mechanism appears to operate in source tissues. Feeding glucose to autotrophic cell cultures of Chenopodium or cold-girdling of spinach leaves results in a rapid increase in the AGP-S transcript. However, other factors must also be involved in the large stimulation of starch accumulation after adding glucose or inhibiting export since ADP-glucose pyrophosphorylase activity does not increase significantly. The decrease in activity of the plastidic (but not the cytosolic) isoform of starch phosphorylase when glucose is added to Chenopodium cells suggests that carbohydrates may also regulate the breakdown of starch. Transgenic tobacco plants with very low nitrate reductase activity have been used to provide evidence that nitrate acts as a signal to induce organic acid metabolism and repress starch metabolism in tobacco. Addition of nitrate to these plants results in an increase in transcripts for phosphoenolpyruvate carboxylase, cytosolic pyruvate kinase, citrate synthase, and NADP-isocitrate dehydrogenase even though the nitrate is not being reduced and entering metabolism proper. The same treatment results in a decrease of the ADP-glucose pyrophosphorylase transcript and enzyme activity, and starch decreases in the leaves and roots. This work has led to the proposal that nitrate acts as a signal to initiate coordinated changes in carbon and nitrogen metabolism.

12.4 HORMONES AND SUGAR REGULATION

- Animal systems provide the best examples of how sugars and hormones can interact to regulate a given process. One such example is the glucose-induced insulin secretion from pancreatic cells. To trigger the secretion of insulin, glucose has to be phosphorylated by glucokinase and processed by glycolysis and the Krebs cycle. The consequent increase in ATP relative to ADP is thought to mediate a glucose induced depolarization of the P-cell membrane possibly by inhibiting the activity of metabolically regulated potassium channels resulting in an influx of calcium into the cell. The increase in intracellular calcium activates the exocytosis of vesicular insulin.

- This regulation of insulin secretion by glucose availability is modified by other processes including the inhibitory and stimulatory action of a large number of hormones and neurotransmitters thus allowing a high plasticity of the response reaction. Do similar interactions between sugars and hormones exist in plants as have been demonstrated in animal systems? Plant hormones have been shown to have an effect on carbohydrate metabolism in both source and sink tissue. For example the activities of cytosolic sucrose phosphate synthase and fructose-1,6-bisphosphatase, which determine to a great extent the amount of sucrose in a leaf, are affected by exogenous application of hormones. Gibberellin application to soybean leaves increases sucrose phosphate synthase activity while the application of abscisic acid inhibits fructose-1,6-bisphosphatase activity.
- A recent report of work carried out on a Chenopodium rubrum cell suspension culture has shown that cytokinin treatment results in induction of a cell wall invertase gene (CHU) and a hexose transporter gene at the mRNA level. The increase of cell wall invertase transcripts corresponds to higher steady state protein level and enzymatic activity of an invertase.
- Cytokinins did not affect the expression of two putative intracellular invertases (CIN2 and CIN3), sucrose synthase or two other hexose transporters present in the cell suspension culture showing the specificity of this response. The increased transcription of CINI and the hexose transporter results in higher rates of sugar uptake in induced compared to non-induced cells. CINI mRNA is also induced by cytokinin in tissues of C. rubrum plants thus indicating a physiological role for this hormone in the control of carbohydrate partitioning in plants. Since cell wall invertase is also induced by glucose and 6-deoxyglucose in the same C. rubrum cell suspension culture it would appear that sugars and hormones are involved in the induction of this enzyme at the level of gene expression.
- In contrast to the induction effect of cytokinin, physiological concentrations of ethylene were found to reduce the enzyme activity of cell wall invertase by 25-47% and this was paralleled by repression of the corresponding gene. Dual control by sugars and hormones has also been reported for rice a-amylase and the soybean vegetative storage protein (VSP) gene. The rice a-amylase gene is repressed by sugars and abscisic acid and can be induced by gibberelin, whereas the activation of VSP gene expression by sugars and methyljasmonate is inhibited by auxins. The possibility that sugar sensing and ethylene signal transduction pathways are integrated was recently highlighted by a report from Sheen's laboratory on the characterization of Arabidopsis gin (glucose insensitive) mutants. Arabidopsis

gin mutants were isolated due to their ability to germinate and grow on 333 mM glucose.

- It was revealed that Arabidopsis mutants that produce significantly increased levels of ethylene (eto: ethylene overproduction) or mutants which show an ethylene over-production phenotype due to a defect in the ethylene signal transduction pathway (ctr: constitutive triple response) behave like gin mutants on 333 mM glucose. Furthermore AVG (L-a-(2- 238 Ian A. Graham and Thomas Martin aminoethoxyvinyl)-glycine), an inhibitor of ethylene biosynthesis, reverts the gin phenotype to wild type in the eto mutants and in at least one of the gin mutants (gin1.1). However, CAB] gene expression, which is insensitive to glucose in the gin 1.1 mutant, is sensitive to glucose in the ethylene mutants as it is in wild type. This work therefore suggests that ethylene may be playing an important role in the response of germinating seedlings to high levels of glucose whereas it does not appear to be involved in the carbohydrate repression of CAB gene expression.

12.5 SUGAR SENSING

All organisms regulate the flux of carbon through key metabolic pathways in order to balance carbohydrate metabolism with nutrient uptake, cellular osmotic potential, growth and physiology. In microorganisms, sugar sensing enables cellular metabolism to be adjusted in response to changes in carbon source availability in the external environment. Glucose is the preferred carbon source for the majority of microorganisms. The availability of glucose acts as a specific signal to repress genes encoding proteins involved in the utilization of alternative carbon sources. Our understanding of sugar sensing and sugar signal transduction pathways is much more advanced in bacteria and yeast than it is in plants or animals.

12.5.1 Distinct Sugar Sensing Mechanisms Exist in Evolutionary Divergent Prokaryotes

Carbon catabolite repression is a universal phenom-enon in prokaryotes. Multiple mechanisms of control exist within a single cell and distinct mechanisms exist between evolutionary divergent species. The classic picture of catabolite repression in prokaryotes is thatinvolvingacAMP-dependentmechanism.InE.colithisinvolvesaphosphoenolpyruvate (PEP) dependent phosphotransferase system (PTS) that serves multiple functions one of which is sugar sensing. This system transports, phosphorylates and detects a variety of sugars. In so doing, it initiates sugar metabolism and regulates a number of physiologically important processes including the phenomena of inducer exclusion, inducer expulsion and catabolite repression. The PTS consists of several phosphoryl transfer proteins (Enzyme I, Hpr, IIA, JIB, and IIC) that must be sequentially phosphorylated before sugar substrates can be transported and concomitantly phosphorylated. Enzyme I and Hpr are general (non-sugar-specific), soluble, cytoplasmic components of the system while the

IIA, IIB and IIC components are sugar-specific membrane bound bifunctional permeases/ kinases. Adenylyl cyclase is believed to be allosterically activated by the phosphorylated form of the glucose loaded IIA protein (I . When active, it synthesizes cAMP, which binds to the cAMP receptor protein (CRP). The cAMP-CRP complex binds directly to DNA to activate transcription. Catabolite repression results when glucose in the medium causes a dephosphorylation of HA.8k-P, there by deactivating adenylyl cyclase. This inturn, causes the cytoplasmic cAMP concentration to diminish which causes the cAMP-CRP complex to dissociate from the DNA, thus resulting in catabolite repression. In recent years, cAMP-independent catabolite repression mechanisms have been delineated in bacteria. Recent studies have shown that E. coli and Salmonella typhimurium also utilize other mechanism which are completely independent of the cAMP-CRP complex to control transcriptional initiation of genes encoding the key enzymes of the major pathways for carbon metabo-lism. These pathways include the Embden-Meyerhof and Entner Duodoroff glycolytic pathways (subject to catabolite activation) and the Krebs cycle, glyoxylate shunt and gluconeogenic pathways (subject to catabolite repression). The pleiotropic transcriptional regulatory protein that controls expression of these central metabolic genes has been termed the' Catabolite represser/activator' (Cra). In vitro Cra-binding experiments have shown that micromolar concentrations of fructose 1-phos-phate (F 1 P) or millimolar concentrations of fructose 1,6 bisphosphate (FBP), but not other catabolites or gluconeogenic intermediates, displaced the Cra protein from the promoters ofall Cra regulated genes examined. Cra exerts a negative effect on glycolytic genes and a positive effect on genes encoding Krebs cycle, glyoxylate shunt and gluconeogenic enzymes. Under glucose-rich conditions, the concentrations of FBP and PIP increase and these interact with the Cra protein and act to dissociate it from target genes thus mediating catabolite activation ofglycolytic genes and catabolite repression of Krebs cycle, glyoxylate shunt and gluconeogenic genes. In regulating such a variety of genes, the Cra protein effectively controls the direction of carbon flux through metabolic pathways in enteric bacteria. In evolutionary distinct species, still other mechanisms have evolved to control carbon catabolite repression. For example, in Bacillus subtilis and other Gram-positive bacteria cAMP is absent and instead, intracellular carbon metabolites such as fructose-1,6-bisphosphate (FBP) and gluconate-6- phosphate, the common intermediates of the Embden-Meyerhof and Entner Duodoroff pathways, respec-tively, promote catabolite repression. The mechanism of repression involves a metabolite activated, ATP-dependent protein kinase that phosphorylates a serine residue in HPr, the small phosphocarrier protein of the PI'S. HPr-(ser-P) then binds to various protein targets (transcription factors, carbohydratepermeases and hydrolytic enzymes) to regulate their activities. It is evident, then, that several different catabolite repression and inducer control mechanisms have evolved in prokaryotes, many which are independent of cAMP. Some of these mechanisms may be operative in eukaryotic microorganisms, higher plants and animals. The variety of cAMP independent mechanisms for catabolite repression, particularly those which involve universal metabolite intermediates as effector molecules, provide models for similar mechanisms in higher plants. If this is the case then the rapid accumulation of sequence data for model plant species such as Arabidopsis and rice along with already complete genome sequences for some bacteria (eg Bacillus subtilis, Haemophilus in, luenzae and Mycoplasma genitalium) should allow identification of putative regulatory components from plants in the near future. The task will then be to prove functional significance of these genes and proteins in plants.

12.5.2 Sugar Sensing in the Yeast Saccharomyces Cerevisiae

As in E. coli, carbon catabolite repression in S. cerevisiae has been found to involve a number of distinct regulatory mechanisms. Addition of fermentable sugars to S. cerevisiae grown on non-fermentable carbon sources causes a variety of short-term and long-term regulatory effects. The regulatory switch is characterized by a rapid transient activation of cAMP synthesis, mobilization of the storage carbohydrate trehalose and induction of genes encoding glycolytic enzymes and ribosomal proteins.

Furthermore, high-affinity sugar transport and key gluconeogenic enzymes are rapidly inactivated and transcription of a variety of genes encoding respiratory, gluconeogenic and other enzymes is repressed. Several signaling pathways are involved in mediating the above phenomena of which the classic glucose repression pathway and the Ras-cAMP pathways are the best studied. Sensing of the available sugar by both these signaling pathways requires transport and phosphorylation but no further metabolism. S. cerevisiae contains two homologous hexokinases, PI (Hxkl) and PII (Hxk2), that can phosphorylate both glucose and fructose, and a single glucokinase (Glk1), only capable of phosphorylating glucose. Activation of the Ras-cAMP pathway by glucose or fructose is dependent on sugar phosphorylation by any of these three hexose kinases. In contrast, glucose-dependent transcrip-tional repression exerted through the main glucose repression pathway is primarily dependent on the hexokinase PII. Deletion of the hexokinase PII gene abolishes glucose repression, whereas deletions of the hexokinase PI or glucokinase genes have no effect. Overexpression of hexokinase PI, but not glucokinase, relieves the requirement for PII in establishing glucose repression suggesting a function of the hexokinases in triggering catabolite repression that is separate from enzyme activity. A recent report has demonstrated that the main glucose repression pathway involves two distinct mechanisms: an initial rapid response that can be mediated through any one of the three S. cerevisiae hexose kinases; and a second, long-term repression response that specifically requires hexokinase PII for glucose repression and either hexokinase PI or hexokinase PII for fructose repression. Both the hexokinase PI and the glucokinase genes are themselves repressed by glucose and fructose. However, the repression of the hexokinase PI gene by fructose is transient which is in agreement with the preference of PI for fructose as substrate and its role in long-term fructose repression. These results demonstrate that sugar sensing and establishment of catabolite repression in S. cerevisiae is controlled by an interregulatory network, involving all three yeast hexose kinases and the Ras-cAMP pathway. As discussed below, it would now appear that at least part of the hexokinase sensing mechanism may have been conserved and adapted for sugar sensing in higher plants. Characterization of the phenotypes of additional S. cerevisiae mutants have resulted in the identification of two signaling pathways which operate downstream of sugar sensing. Although one of these pathways is responsible for repression and the other for de-repression of genes that are subject to carbon catabolite control there is substantial crosstalk between the two. Two protein complexes have been identified in the repression pathway. One of these (REG2/REG1/GLC7 type 1 protein phosphatase) mediates a signal by regulating a protein phosphatase activity but no substrate for this activity has as yet been identified. The second complex (SSN6/TUP1) functions as a general repressor of transcription through modulation of chromosome structure. Interaction of the SSN6/TUP1-complex with DNA is mediated by the MIG1 DNA binding protein, which somehow directs binding of the complex to specific sites in the promoter region of genes that are repressed by glucose. Derepression of gene expression in the absence of glucose is necessary to allow alternative carbon sources to be utilized. This derepression process involves another regulatory pathway containing at least two protein complexes. The first complex (SNE1/ SNF4) mediates a signal through activity of the SNP1 protein kinase of which there are plant and mammalian homologues (see below). The second complex (SWI/SNE1) consists of at least ten proteins which modulate chromatin conformation and affect expression of a variety of genes. Targeting of this complex to specific gene promoters is mediated by transcription factors. It is not yet known how the SWI/SNF complex receives information from the SNP]. complex in the derepression pathway. The SNP1 complex also directly affects the repression pathway by inactivating the MIG1 repressor protein. Similarly the GLC7 protein phosphatase complex from the repression pathway inhibits the activity of the SNP]. kinase complex. Our understanding of the mechanisms outlined above, although detailed relative to our knowledge in plants, is still far from complete. One important gap still remaining to be filled is how the metabolite signals involved are sensed and transduced into the regulatory pathway. As discussed above, hexokinase has been implicated as the sugar sensor triggering the repression pathway but a mechanism of action still has to be defined.

An alternative mechanism involving AMP:ATP ratios for signaling intracellular nutritional status to the de-repression pathway has recently emerged through biochemical studies of the S. cerevisiae SNF1 kinase (Wilson et al., 1996). The mammalian AMP-activated protein kinase (AMPK) is a homologue of SNP]. kinase. AMPK is made up of three subunits and all ofthese have homologues that are part of the SNP]. complex. AMPK is activated by the elevation of the cellular AMP: ATP ratio, which occurs during cellular stress in mammalian cells. The mechanism of activation involves allosteric regulation by AMP, promotion of phosphorylation by an upstream protein kinase (AMPK kinase) and inhibition of dephosphorylation of AMPK. The proposed mode of action in mammalian cells is that AMPK functions as a 'fuel gauge'. When AMPK detects a low fuel situation it protects the cell by switching off ATP-consuming pathways (eg. fatty acid synthesis and sterol synthesis) and switching on alternative pathways for ATP generation (eg. fatty acid oxidation). SNP1 is rapidly activated by phosphorylation on removal of glucose fromS. cerevisiae, and can be inactivated by protein phosphatases and reactivated by mammalian AMPK kinase, thus demonstrating that the mammalian AMPK and S. cerevisiae SNP1 protein kinase cascades are highly conserved (Wilson et al., 1996). The only apparent difference in these two complexes is that the SNF1 complex is not regulated allosterically by AMP in vitro. However, multiple levels of control by AMP:ATP ratio exist for the mammalian complex and a similar situation may exist for the yeast complex. Based on this work, Wilson et al. (1996) put forward a new model for the role of SNP1 protein kinase in carbon catabolite repression. This model proposes that adenine nucleotides, rather than an interaction of glucose with hexokinase, are the metabolic signals which indicate the lack of glucose in the medium. These signals trigger activation of SNP1 kinase and de-repression of glucose repressed genes. AMP K/SNE1 homologues also exist in plants but their function has not yet been defined. SNF1 homologues from rye and tobacco can complement the snfl mutation in S. cerevisiae. Unlike the plant homologue, the mammalian AMPK does not complement snfl regulatory mutants. Additional SNF1 homologues have been cloned from barley, wheat and Arabidopsis . A tobacco SNF1 homologue, NPK5, constitutively activates expression of the glucose repressible SUC2 gene in S. cerevisiae snfll mutants. However, expression of NPK5 does not rescue mutants disrupted in the SNF4 gene, the product of which associates with SNF1 and is necessary for maximal activity of the kinase in yeast. These results indicate that NPK5 can substitute for SNF1 in the SNF1/SNF4 complex in S. cerevisiae snfl mutants (Muranaka et al., 1994). Carbohydrate starvation of wheat seedlings results in expression ofthe SNF1 homologue WPK4 (Sano and Youssefian, 1994). Light and cytokinins also induce WPK4 gene expression which suggests that this particular kinase may be involved in more than one signal transduction pathway. An AMPK specific peptide based kinase assay has been used to identify AMPK-like activities in extracts from avocado, potato, carrot, pea, oilseed rape, wheat and cauliflower. Immunological studies indicate that the cauliflower kinase is the homologue of the rye and tobacco SNF1-like protein kinases (Ball et al., 1995). Like the S. cerevisiae SNF1, the cauliflower kinase is not sensitive to AMP. In all other respects, including being able to inactivate mammalian acetyl-CoA carboxylase and HMG-CoA reductase, it is similar to the mammalian AMPK. The physiological substrate for the plant AMPK/SNF1 homologue is not yet known but it has been shown that the cauliflower kinase inactivates bacterially synthesized HMG-CoA reductase from Arabidopsis by phosphorylating the protein at the same site that is phosphorylated in the mammalian HMG-CoA reductase. Much work still has to be done to establish the in vivo role(s) of the plant SNF1 homologues. It is possible that this family of protein kinases plays a role in signal transduction of metabolic status in plant cells. Alternatively, or as well, these kinases could play an important role in the post-translational control of key metabolic enzymes as is the case in mammalian cells. Two such enzymes in plant cells are sucrose phosphate synthase which catalyzes a key step in sucrose biosynthesis in source tissues and nitrate reductase which catalyzes the first step in nitrogen assimilation. In spinach leaves both these enzymes are regulated by phosphorylation by multiple protein kinases. One of three nitrate reductase kinases detected in extracts from spinach leaves has identical properties to other SNF1-like enzymes. It is, therefore, likely that plant SNF1 homologues do play an important role in regulating metabolic processes in source tissue through post-translational control of key metabolic enzymes. Experiments involving antisense and over-expression of plant SNF1 homologues are currently underway in several laboratories and these along with specific kinase assays and monitoring expression of appropriate genes under different metabolic conditions should provide useful information on the role ofthe plant kinase. In addition, identification of plant homologues to other components in the SNF1 signal transduction pathway, such as SNF4 or a possible upstream kinase, along with their functional analysis would provide additional evidence for a conserved mechanism of carbon catabolite repression between plants and yeast.

12.5.3 Sugar Sensing in Plants—the Case for Hexokinase

Studies on the sugar mediated repression of the glyoxylate cycle and photosynthetic genes has led to the suggestion that sugar sensing in plants involves a hexokinase. In cucumber cell cultures, the malate synthase and isocitrate lyase genes are repressed in the presence of glucose, fructose or sucrose and de-repressed when intracellular levels of these sugars fall below a critical threshold. A cucumber mesophyll protoplast transient expression system was used to demonstrate that these same genes are de-repressed at the transcriptional level upon carbohydrate depletion. The glucose analogues 2-deoxyglucose and mannose, which are phosphorylated by hexokinase but not further metabolized by the glycolytic pathway, cause repression even though intracellular amounts of sucrose, glucose and fructose fall below the critical levels normally associated with de-repression. In contrast to glucose, fructose or sucrose, 3-0-methylglucose, an analogue of glucose that is taken up by the cells but is not phosphorylated by hexokinase, does not result in repression of malate synthase and isocitrate lyase. These results implicate hexokinase or events associated with the hexokinase reaction in the initial sugar sensing step. However it should be noted that experiments with hexose analogues alone do not rule out the possibility of a stereospecific hexose receptor 242 Ian A. Graham and Thomas Martin operating independently ofhexokinase. Using a C. rubrum cell culture, Krapp et al. (1993) demonstrated that 6-deoxyglucose or 3-0-methyl-glucose (both imported into the cell but not phosphorylated by hexokinase) had no effect on RBCS gene expression, whereas glucose at the same concentration caused severe repression. It was concluded from this work that transport of carbo-hydrates does not provide the signal for regulation of RBCS and that phosphorylation of glucose is necessary. Using a maize mesophyll protoplast transient expression system, Jang and Sheen (1994) demonstrated that 2-deoxyglucose causes repression of photosynthetic gene promoters whereas 3-0- methylglucose has no effect on expression. The importance of hexokinase in sugar sensing was further demonstrated by feeding experiments using manno heptulose, an apparently specific inhibitor of hexokinase, which was shown to block the repression effect of glucose and 2-deoxyglucose (Jang and Sheen, 1994). However, more recent work using a C. rubrum cell culture has reported that both D-glucose and 6-deoxyglucose treatments result in a similar decrease in RBCS mRNA levels and an increase in the mRNAs for a sink specific cell wall invertase and for phenylalanine ammonialyase, a key enzyme of defense response. An explanation for these conflicting reports is not apparent. Cloning of two Arabidopsis genes, At HXK 1 and AtHXK2 has allowed overexpression and antisense experiments to be carried out in order to establish the in vivo involvement of hexokinase in the response of seedlings to elevated levels of glucose. Wild type seedling growth is inhibited in the presence of 333 mM glucose whereas antisense seedlings with reduced levels of hexokinase show less inhibition of growth. Photosynthetic genes such as RBCS and CAB are also more strongly repressed by glucose in wild type than in antisense plants. Overexpression of the hexokinase in transgenic seedlings results in increased sensitivity to 333 mM glucose relative to wild type. The authors' interpretation of these results together with other results using the yeast hexokinase, is that hexokinase is a sugar sensor. Decreasing hexokinase protein decreases the capacity of seedlings to sense glucose hence they grow better on 333 mM glucose and do not exhibit the same degree of glucose repression of photosynthetic gene expression as is observed in wild type. Increased levels of hexokinase lead to increased sugar sensing capacity and increased sensitivity to 333 mM glucose. Another interpretation is that 333 mM glucose is toxic to seedlings, possibly because it results in a massive flux of hexose sugar into the hexose phosphate pool. This might lead to a sequestration of phosphate which could disturb the balance of cellular metabolism. A third possibility is that high concentrations of hexose phosphates in the cell have deleterious effects. Decreasing hexokinase activity will decrease this flux whereas increasing hexokinase activity will increase the flux. The crucial experiment which supports the view that hexokinase is acting as a sugar sensor involved overexpressing the S. cerevisiae HXK2 gene (which is implicated in sugar sensing in S. cerevisiae) in Arabidopsis. If the yeast hexokinase were to exhibit sugar sensing in Arabidopsis then transgenic seedlings should be more sensitive than wild type to 333 mM glucose. In fact, the opposite phenotype was observed in which the transgenic seedlings were less sensitive to 333 mM glucose than wild type even though the total hexose phosphorylation activity was significantly greater than that in wild type seedlings. The interpretation given for this is that the yeast hexokinase competes effectively for the same substrate as the endogenous glucose sensing hexokinases resulting in reduced sensitivity to glucose. The hexokinase under- and over-expressing plants were also used to investigate the role of hexokinase in the regulation of sugar inducible nitrate reductase. Arabidopsis lines over-expressing

At HXK show higher steady state nitrate reductase mRNA levels in response to glucose than wild type seedlings, thus suggesting that hexokinase is also important in signaling expression of sugar-inducible genes. However, other effects such as increased flux of hexose sugars into glycolysis cannot be ruled out.

12.5.4 Sugar Transport and Signaling

The generation of a signal by metabolism of a sugar is one way in which carbohydrate status is perceived. An alternative or additional mechanism could involve intra- or extracellular receptors. In the simplest case, a DNA binding protein might act as a sensor and transmit the signal directly to the respective gene as in the case of some hormone induction in mammalian cells. Alternatively, the receptor could be part of a sugar transport system (possibly similar to the PTS system in E. coli) in which increased or decreased levels of sugars lead to a chain of signaling events which ultimately cause a sugar specific response. The yeast SNF3 and RGT2 genes encode putative glucose transporters with high similarities to mammalian, yeast and plant glucose transporters, but differ from them in having a long C-terminal extension. This extension may be located in the cytoplasm and could serve as a signaling domain Mutational analysis has shown that both SNF3 and RGT2 may act as sugar sensors and are involved in the induction of glucose induced HXT gene regulation in yeast. Furthermore this analysis has shown that SNF3 and RGT2 might act as a pair of glucose receptors recognizing either high (RGT2) or low (SNF3) levels of extracellular glucose. The authors suggest that as in the case of the bacterial PTS system the generation of the sugar signal is coupled to the transport of this sugar and that the components of the sugar signaling pathway are present in the cell even in the absence of the sugar. In this model, the activity of the glucose sensor becomes the limiting factor of the signaling system. Despite the cloning of a number of sugar transporter genes from plants a sugar transporter/sensor protein equivalent to the yeast SNF3 or RGT2 has not yet been identified in plants. Experiments using non-metabolizable sugar analogues provide evidence for at least two different signal perception mechanisms in plants. As discussed above, source-related genes, e.g. RBCS and CAB, can be repressed by 2-deoxyglucose, or mannose but not by 3-O-methylglucose (Jang and Sheen, 1994). In contrast, the more sink metabolism-related sucrose synthase, invertase and class I patatin genes are induced by 3-0-methylglucose, 6-deoxyglucose or 2-deoxyglucose. In addition, the induction of a sugar regulated chimeric patatin class I/uidA (B33GUS) gene in transgenic Arabidopsis plants by the glucose analogs 6- deoxyglucose and 3-0-methylglucose shows that at least some sugar signaling pathways in an intact plant system do not require sugar phosphorylation. The lack ofinduction ofpatatin gene expression in transgenic potato tubers with increased carbohydrate levels due to AGPase or sucrose synthase antisense repression is taken as evidence to indicate that the flux rather than the concentration of sugars is sensed in the case of induction of patatin gene expression. However, the demonstration that the sugar-inducible promoter of the potato proteinase inhibitor II gene is not induced by 6-deoxyglucose in transgenic tobacco suggests that at least two mechanisms may operate in parallel to regulate genes that are induced by sugars. The suggestion that hexose sensing may be associated with the endomembrane system has come from recent work on transgenic tobacco plants expressing yeast-derived invertase targeted to the apoplast and vacuole. Targeting of this enzyme to the apoplast or vacuole results in increased levels of monosaccharides in the cells and an associated induction of defense genes and repression of CAB genes. Expression of the yeast invertase in the cytosol of tobacco plants results in equally elevated levels of sugars but does not have any effect on defense or photosynthetic genes. Bacterial fructosyltransferase converts sucrose into fructans with a concomitant release of equimolar amounts of glucose. Expression of a bacterial fructosyltransferase gene in the endomembrane system of plants results in a bleached phenotype which suggests accumulation of hexose sugars may be resulting in repression of genes encoding enzymes and proteins in the photosynthetic apparatus.

12.5.5 Sucrose as a Direct Signal

Sucrose is the main transport sugar in most plant species and it is also an important storage compound. Due to the fact that extracellular invertases hydrolyze sucrose to fructose and glucose, it is difficult to prove a direct effect of this molecule in feeding experiments. However, sucrose specific induction of the patatin promoter has been reported.

12.6 SIGNAL TRANSDUCTION

Specific components of the signal transduction pathways involved in regulation of plant genes by sugars have not yet been identified. As discussed above (Section V.B.) one family of protein kinases that may play a role in the transduction of sugar signals are the plant homologues of the AMPK/ SNF1 protein kinase which have been shown to be essential for post-translational control of key biosynthetic enzymes in mammalian cells and de-repression of genes that are subject to carbon catabolite repression in yeast. As in other systems, phosphatases, kinases, calcium and calmodulin appear to be involved in the transduction of sugar signals in plants.

12.7 SUGAR RESPONSE ELEMENTS IN GENE PROMOTERS

In contrast to the detailed characterization of various photosynthetic gene promoters in relation to their response to light, there have been few reports characterizing their response to carbohydrates. Sheen (1990) used the maize protoplast transient expression system to demonstrate that sucrose repression of a number of maize photosynthetic genes was mediated by positively acting promoter elements upstream of the TATA box, but no specific consensus sequences were defined. More recently, transient expression experiments with protoplasts from primary leaves of Phaseolus vulgaris have been used to identify promoter sequences involved in sucrose repression of the P. vulgaris RBCS2 gene. A region of this promoter from position —203 to —187 contains sequences resembling elements involved in the sugar stimulation of transcription of several plant and mammalian genes. These include the sucrose response elements (SURE elements) of the patatin gene promoter (Grierson et al., 1994). The SURE elements are similar to elements defined in the sporamin and 13-amylase genes of sweet potato both of which are induced by sucrose. The —203 to —187 region of the RBCS2 gene implicated in sugar repression also has remarkable similarity to carbohydrate response elements (ChoRE) found in the promoters of mammalian genes induced by sugars, one of which is the L-type pyruvate kinase gene. Whether the RBCS2 —203 to —187 promoter region is sufficient to mediate a sucrose response, or whether any functional similarity exists with the elements described above, requires further research.

12.8 SUGAR SENSING MUTANTS

A number of laboratories have designed various genetic screens in the model plant Arabidopsis thaliana in an effort to isolate mutants in the signal transduction mechanisms involved in sugar mediated repression and induction of gene expression. In the majority of cases young seedlings have been screened as these respond to exogenous sugars in various ways and it is also possible to look at large mutagenized populations. One such screen has exploited the fact that intracellular sugar concentrations increase dramatically in Arabidopsis seedlings grown on high concentrations of sucrose and growth limiting amounts of nitrate compared to when seedlings are grown on high concentrations of sucrose and sufficient amounts of nitrate. The increase in intracellular sugars results in an increase in anthocyanins and a decrease in chlorophyll. Using high sucrose, low nitrate conditions a number of carbohydrate insensitive (cai) mutants have been isolated which do not show the same decrease in chlorophyll and increase in anthocyanins as in the wild type but do show the same increase in intracellular sugar levels. As mentioned previously, growth of Arabidopsis seedlings on 333 mM glucose results in developmental arrest and this is overcome in plants that have been antisensed for hexokinase I or hexokinase II (Jang et al., 1997). These same conditions have been used to isolate glucose insensitive (gin) mutants which are less sensitive to 333 mM glucose in the growth medium (Zhou et al., 1998). The Arabidopsis plastocyanin (PC) gene is transiently activated in both the light and dark during early seedling growth and this expression can be repressed by sucrose. A transgene screening approach using a PC gene promoter-luciferase reporter gene construct has been used to isolate Arabidopsis mutants that are disrupted in the sugar repression of the plastocyanin PC gene during early seedling growth. Sucrose uncoupled (sun) mutants show no or reduced repression of the transgene and, importantly, the endogenous PC, CAB and RBCS mRNA levels are similarly insensitive to sucrose repression. Studies of the affect of 2 mM 2-deoxyglucose on the mature rosette stage Arabidopsis have shown that whole plant photosynthesis, plastocyanin gene expression and total extractable Rubisco activity is unaffected in one mutant sun. This treatment results in reduced photosynthesis, PC gene expression and Rubisco activity in wildtype Arabidopsis. Many photosynthetic genes that are subject to sucrose repression are also subject to light regulation via phytochrome. The presence of sucrose in the growth medium results in altered PHYA responses in the sun6 mutant compared to wildtype. This suggests that the mechanisms of sugar signaling and light signaling are interrelated. Arabidopsis mutants disrupted in the induction of specific genes by sugars have also been isolated. For example, the /3-amylase gene is induced by sugars in Arabidopsis and low level beta amylase (lba) mutants have been isolated which do not show this induction response. In another example the patatin class I promoter, which is induced by sugars in potato, was transferred to Arabidopsis where it was found to be similarly regulated. Transgenic Arabidopsis harboring a patatin promoter-fl-glucuronidasetransgene were used to isolate mutants showing a reduced sucrose response (rsr). The demonstration that the promoter of the potato tuber storage protein is regulated by sugars in the same manner in Arabidopsis demonstrates a high degree of conservation between the mechanisms underlying sugar signaling in plants. Characterization of Arabidopsis mutants will allow us to identify the key regulatory components involved in sugar signaling. One application of this knowledge will be manipulation of sugar signaling and resource allocation in crop species in order to increase yields of current plant products and to over-ride regulatory mechanisms that may inhibit high yields of genetically engineered novel plant products.

12.9 AT A GLANCE

- Plants contain several sugar signaling cascades, one of which is important in source tissues, possibly involving hexokinase, and another which is independent of hexokinase possibly located at the plasma membrane in sink tissues. The components of a hexokinase independent sugar signaling pathway could be as in yeast a sugar sensing transporter protein or simply a sugar sensing protein as the first component of a signaling cascade.
- The analysis of Arabidopsis mutants disrupted in sugar induced responses should help in understanding how plants sense their sugar status and coordinate sink and source metabolism. We have given many examples of how metabolites regulate gene expression in other organisms in an attempt to emphasize the variety of mechanism that may operate in plants. It is reasonable to predict that at least some of these mechanisms will be conserved across phyla and indeed a precedent for this has already been set in the case of the regulation of heat shock genes. In higher plants various mechanisms appear to have either evolved or been conserved and adapted throughout evolution to meet the physiological requirements of source and sink tissues. Future efforts to manipulate production in source tissues, and ultimately plant yield, will require a thorough understanding of the regulatory mechanisms involved.

13

STARCH METABOLISM IN LEAVES

13.1 INTRODUCTION: AN OVERVIEW

- Starch is the most abundant carbohydrate reserve in plants: it has been estimated that in excess of 109 tonnes are produced annually (Galliard, 1987). Two types of starch can be distinguished according to function: reserve starch and transitory starch. Reserve starch accumulates in organs of perennation and dispersal, and is localised in non-green plastids, called amyloplasts.
- Transitory starch is formed during photosynthesis in the chloroplasts of many higher plants, and serves as an important store of carbohydrate for periods of darkness when photosynthesis is not possible. In this chapter we will discuss the occurrence, importance, structure, synthesis, and degradation of transitory starch. We will pay particular attention to the physiological significance of transitory starch and discuss current ideas on how starch metabolism is regulated at the biochemical level.

13.2 THE OCCURRENCE AND FUNCTION OF TRANSITORY STARCH

- Abbreviations: 3-PGA — 3-phosphoglycerate; AGPase —ADPglucose pyrophosphoiylase; GBSSI — granule-bound starch synthase I; PGM—phosphoglucomutase; Pi —inorganic phosphate; SSI — starch synthase I; SSII — starch synthase II; SSIII — starch synthase III; TPT — triose-phosphate translocator portrayed as a strong diurnal rhythm, in which starch is synthesized in relatively large amounts during the day and almost entirely degraded during the night. This is certainly the case in some well-studied experimental systems. Leaves of pea grown under an 18-h photoperiod, Arabidopsis grown under a 12-h photoperiod and spinach grown under a 9-h photoperiod all degraded 85%

or more of the accumulated starch in the subsequent dark period. However, the extent of starch accumulation and the magnitude of the diurnal change vary considerably. First, different species allocate very different proportions of their newly-assimilated carbon to starch.

- Many species do not accumulate starch in their chloroplasts under normal growth conditions. Sucrose or—in the Gramineae—fructans are stored in the vacuoles of mesophyll cells in these species. In other species, starch synthesis can account for anything from a few percent to a very substantial proportion of the newly-assimilated carbon. For example, leaves of annual meadow grass (*Poa annua* L.) partitioned most of their assimilated carbon into sucrose and fructans, and less than 2% into starch whereas in leaves of sugar beet 40% of assimilated carbon was partitioned into starch. Second, the extent of the diurnal change in starch content changes as the leaf ages in some species but not in others. The starch content of tobacco, soybean and cotton leaves increases as leaves age, and the magnitude of the diurnal change decreases, but there is little change in the diurnal pattern of starch accumulation and degradation through development in Arabidopsis leaves. Third, the extent of starch accumulation may vary from one cell type to another within a single leaf. The parenchymatous bundle sheath of barley leaves contains four distinct types of photosynthetic cell, each of which displays different patterns of starch accumulation and degradation. The relative amounts of starch in the two palisade layers and the spongy mesophyll layer of soybean leaves changed considerably during the photoperiod. Fourth, starch content and the extent of the diurnal change are highly dependent upon the conditions in which the plant is grown—for example, the length of the photoperiod, the photon flux density and the night temperature. Deficiencies of nitrogen and potassium and elevated levels of carbon dioxide usually result in increased accumulation of starch in leaves. Attack by fungal and viral pathogens may also lead to accumulation of starch. This accumulation may be specifically in or around infected cells, or generally throughout the leaves of an infected plant. The importance of transitory starch in those plants in which it accumulates is demonstrated by the phenotypes of mutant and transgenic plants in which accumulation is prevented or its magnitude is altered. Mutations in Arabidopsis and tobacco that eliminate activity of plastidial phosphoglucomutase or ADPglu-cose pyrophosphorylase prevent accumulation of starch in the leaves. Growth of these mutants is indistinguishable from that of wild-type plants under long photoperiods. In the tobacco mutants, sucrose is synthesized and exported during the light period at a higher rate than in wild-type plants, compensating for the lack of sucrose synthesis from starch at night. However, when grown with long dark periods this compensatory effect is not adequate to support growth rates comparable with those of wild-type plants: all of the mutants exhibit reduced growth rates.

- Indeed, a mutant of Arabidopsis lacking plastidial phosphoglucomutase also exhibited a late-flowering phenotype in days shorter than 16 hours. Transgenic plants in which the extent of accumulation of transitory starch has been altered also show compensatory changes in the diurnal pattern of sucrose export from the leaf. Potato plants in which the activity of the triose-phosphate translocator (TPT) in the chloroplast envelope has been reduced through the expression of antisense RNA have increased rates of transitory starch synthesis and reduced rates of sucrose synthesis and export during the day. This is compensated for by high rates of starch degradation leading to increased sucrose synthesis and export at night. Potato plants in which the synthesis of transitory starch is decreased through expression of an antisense RNA for AGPase display increased rates of sucrose synthesis and export during the day, compensating for the reduction in sucrose synthesis from starch at night. Consideration of conditions which promote the accumulation of starch in leaves has given rise to two opinions about its function. First, transitory starch synthesis may occur primarily as an overflow for newly-assimilated carbon when assimilation exceeds the demand for sucrose. The sequestration of carbon as insoluble starch allows high rates of assimilation to be maintained when the capacity to store and to export sucrose has been reached. Recent studies on transgenic plants with reduced rates of transitory starch synthesis have demonstrated a positive correlation between the capacity for starch synthesis and the rate of photosynthesis at elevated carbon dioxide concentrations.
- In intact leaves of some species, sugar beet and maize the rate of starch synthesis is low in relation to that of sucrose during the early part of the day and rises later, in parallel with a rise in the sucrose content of the leaf. Further experimental support for the idea of transitory starch as an overflow is provided by the general observation that restriction of the export of carbon from the chloroplast or from the leaf leads to increased rates of starch synthesis. For example, starch synthesis is increased in transgenic potato plants in which activity of the TPT or a sucrose-proton translocator involved in phloem loading has been reduced, and in plants in which 208 Richard N. Trethewey and Alison M. Smith phloem loading has been impaired by expression of high activities of invertase in the apoplast. Starch synthesis is also increased in leaves of Clarkia mutants with decreased activity of cytosolic phosphoglucoisomerase and hence a reduced rate of sucrose synthesis. In spinach leaf discs the rate of starch relative to sucrose synthesis increased as the sucrose content of the discs increased. Second, transitory starch may provide a source of carbon for growth during the following night. This view is supported by the compensatory changes in the pattern of sucrose synthesis and export that occur in mutants unable to synthesize transitory starch, and by the fact that growth of these plants is restricted under short-day regimes. The importance of

transitory starch to the growth of the plant is clearly dependent upon the general growth conditions: an inability to synthesize transitory starch is likely to restrict growth to a greater extent in conditions when carbohydrate supply is limiting growth (for example in high nitrogen) than in conditions where some other factor is more important in limiting growth. The pattern of starch synthesis in several species is also not consistent with a role simply as an overflow mechanism.

- For example, starch synthesis can occur concurrently with sucrose synthesis at low rates of photosynthesis. In Clarkia xantiana Gray (Onagraceae) rates of starch and sucrose synthesis were approximately equal under both saturating and limiting light, even though the rate of photosynthesis was three times greater in saturating than in limiting light. The ratio of sucrose to starch synthesis in Phaseolus was constant over a wide range of light intensities and carbon dioxide concentrations except where the rate of photosynthesis was very low, when sucrose synthesis was preferentially maintained. In a number of species [soybean, spinach, maize, pangola, sugar beet, and potato the proportion ofnewly-assimilated carbon that is partitioned to starch varies inversely with the length of the photoperiod. These observations, and extensive studies of the allocation of newly-assimilated carbon to starch in leaves of sugar beet, have led Geiger and colleagues to suggest that partitioning to starch may be regulated by an endogenous circadian rhythm which ensures that the amount of starch synthesized during the day is commensurate with the demand for carbon during the following night. Overall, it seems likely that transitory starch is synthesized both as an overflow fornewly-assimilated carbon and as a source of carbon during the night. The relative importance of these two roles depends on the species and the growth conditions. The ways in which the partitioning of newly-assimilated carbon into starch may be regulated.

13.3 THE STRUCTURE OF TRANSITORY STARCH

- Very little is known about the composition and structure of the starch granules in chloroplasts: almost all research on starch structure has been on the starches of non-photosynthetic storage organs. We will summarise the present understanding of storage starches and discuss whether this picture is likely to be applicable to transitory starch. Storage starch granules vary enormously between species in size and shape but are similar in important features of composition and structure. They consist of two distinct types of glucose polymer. About 20-30% of their mass is made up of amylose, a predominantly linear molecule consisting of between 500 and 10,000 a(1-4) linked glucose residues. The remaining 70-80% of the granule is amylopectin, a polymer consisting ofrelatively short chains of a(1 -4) linked glucose residues, joined by a(1-6) linkages.

- Microscopical studies and X-ray diffraction and scatter analyses indicate that granules of storage starches are semi-crystalline, containing concentric rings of alternating crystalline and amorphous lamellae with a repeat distance of 9 nm. This level of organization is a function of amylopectin rather than amylose molecules since it occurs in the storage starches of both wild-type plants and mutant plants unable to synthesis amylose. The existence of a 9-nm repeat has been explained by the fact that the distribution of chain lengths within the amylopectin molecule is polymodal, with maxima at chain lengths of approximately 12-16, about 40 and about 70 glucose residues. It is believed that the amylopectin chains of 12-16 glucose residues are organised into clusters. Chains of 40 glucose residues span two clusters, and chains of 70 glucose residues span three clusters. Within clusters, adjacent chains form double helices which pack together in ordered arrays, giving rise to the crystalline lamellae (Fig. 1).

◉ The amorphous lamellae are the zones in which a large number of all-6 linkages occur, hence few double helices are formed. The crystalline and amorphous lamellae form concentric zones, usually several hundreds of nm wide, within the starch granule. These semi-crystalline zones alternate with amorphous zones in which the arrangement of the amylopectin molecules is not understood (Fig. 1). It is thought that one semi-crystalline and one amorphous zone are laid down per day—this repeat structure is often termed a 'growth ring' but definitive evidence that this is the case is lacking. The transitory starch granules of chloroplasts are generally discoid in shape and less than five microns across. Where it has been studied, considerable variation has been found in the size of transitory granules within a single leaf. For example, most of the granules in a sunflower leaf were 0.2 to 2.5 microns across, but about 3% of the granules were 4 to 5.5 microns across. In Pellionia leaves granules in the spongy mesophyll were much larger than those in the palisade layer.

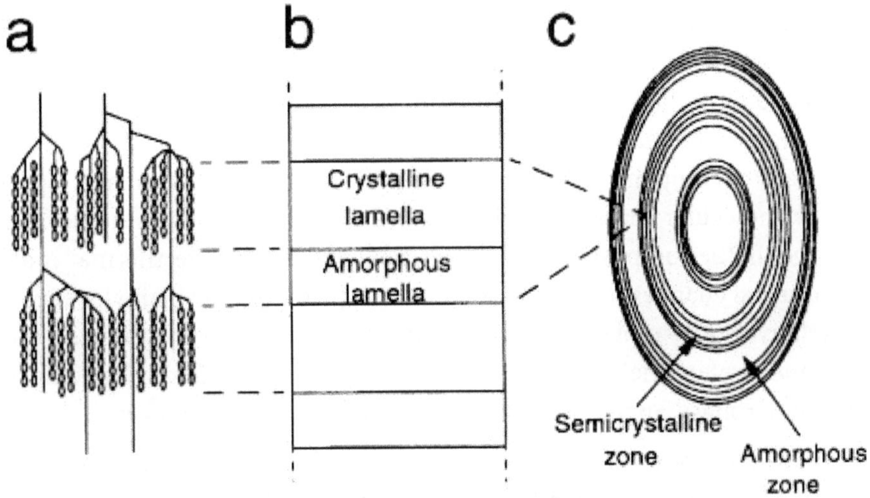

Fig. 1: Organization of starch polymers within a granule of storage starch. (a) 'Cluster' arrangement of chains in an amylopectin molecule. Adjacent chains within a cluster

form double helices. The free (non-reducing) ends of the chains point towards the periphery of the granule. (b) Crystalline lamellae, formed by regular packing of the double helices in the clusters, alternating with amorphous lamellae representing inter-cluster regions with high concentrations of a(1,6) linkages. The periodicity of the crystalline-amorphous repeat is 9 nm. (c) Cross-section of a starch granule, showing alternating concentric zones (`growth rings') of semi-crystalline material—consisting of the alternating crystalline and amorphous lamellae shown in B—and amorphous material. The zones of semi-crystalline material are several hundred nm in width.

- There is general agreement that transitory starch granules contain both branched polymers of high molecular mass and relatively unbranched polymers of lower molecular mass, corresponding approximately to the amylopectin and amylose components of storage starch. The ratio of amylose to amylopectin is lower in transitory than in storage starch, and the size and chemical structure of both of these components probably differs between transitory and storage starch. Around 16% or less of the starch of leaves of tobacco, sunflower, rice, pea and potato is amylose, whereas this value is almost invariably 20% or more for storage starches. The amylose content of the transitory starches of tobacco and cotton leaves increases as the starch content of the leaves increases through development . The amylose of transitory starch is reported to be of lower molecular mass and more branched than that of storage starches. The amylopectin of transitory starch has a polymodal distribution of chain lengths which approximates to that of storage starches. However, a detailed study of the distribution of lengths of the shorter chains of amylopectins from the transitory and storage starches of pea revealed considerable differences between the two. The amylopectin of the transitory starch had a very strong polymodal distribution of chain lengths in the range of 5-40 glucose residues, with maxima at lengths of 12, 15 and about 21 glucose residues. This pattern was observed regardless of the age and growth conditions of the plants. It may be a common feature of transitory starches: a similar distribution is observed for the amylopectin of the transitory starch of Arabidopsis. Increases in the starch content of leaves during the light period are generally accompanied by decreases in the ratio of amylose to amylopectin, and in the molecular mass of amylopectin. Taken together with the developmental changes in the composition of leaf starches described above, this may indicate that the diurnal changes in starch content occur primarily in an amylopectin fraction of relatively low molecular mass which is preferentially synthesized during the day and preferentially degraded at night. Amylose, and amylopectin of higher molecular mass, appear to accumulate gradually as the leaf ages.
- Very little is known about the organization of amylose and amylopectin within granules of transitory starch. In general it appears that these granules lack the relatively high level of crystalline order seen in storage starches. However, like the

starch of storage organs, the starch of pea leaves is at least in part semi-crystalline with a lamellar repeat of 9 nm. The starch of tobacco leaves contained amorphous and semi-crystalline zones analogous to the growth rings of storage starches, and the starch of spinach leaves was reported to consist of an amorphous outer 'mantle' and an inner, more crystalline core. The mantle was more readily degraded during the night than the core. The mantle may represent the amylopectin of relatively low molecular mass that is preferentially turned over on a diurnal basis. The core, and the material organised into growth rings in tobacco starch, may represent material that accumulates as the basal level of starch in these leaves increases with leaf age. The material in transitory starches that is not turned over on a diurnal basis may thus resemble storage starch in that it contains amylopectin molecules of high molecular mass and considerable amounts of amylose and it is probably semi-crystalline.

13.4 THE SYNTHESIS OF TRANSITORY STARCH

13.4.1 Pathway

- Starch in chloroplasts is synthesized from fructose 6-phosphate derived from the reductive pentose phosphate pathway. Fructose 6-phosphate is converted to glucose 6-phosphate and then to glucose 1-phosphate via the enzymes phosphoglucoisomerase and PGM. The substrate for the synthesis of the starch polymers, ADPglucose, is synthesized from glucose 1-phosphate and ATP via the enzyme AGPase (EC 2.7.7.27) in a reaction which also produces pyrophosphate. This reaction is rendered effectively irreversible in vivo by the presence in the chloroplast of an alkaline inorganic pyrophosphatase, which hydrolyzes the pyrophos-phate produced to phosphate. Evidence that this pathway is exclusively responsible for starch synthesis in leaves comes from mutant plants with reduced or undetectable activities in leaves of either plastidial PGM or AGPase.
- Both classes of mutant are either unable to synthesize transitory starch or have a severely restricted capacity to do so. For example, peas carrying a mutation at the rug3 locus, tobacco and the Arabidopsis mutant TC7 all lack activity of the plastidial isoform of PGM, and are unable to synthesize transitory starch. Arabidopsis plants carrying mutations at the adg2 and adgl loci have 5% of wild-type and undetectable activities of AGPase, respectively. adgl mutants are unable to make transitory starch, and adg2 mutants accumulate only 40% of the amount of starch in leaves of wild-type plants. ADPglucose is the substrate for starch synthase, which adds glucose residues to the non-reducing ends of glucan chains via all-4) linkages. The a (1-6) linkages that create the branch points in the starch polymers are introduced by starch-branching enzyme (SBE), which transfers linear glucans from the end of a linear chain to the side of the same or an adjacent chain. The present understanding of the nature and properties of

AGPase, starch synthase and SBE in leaves, and the roles of starch synthase and SBE in determining starch structure, are considered below.

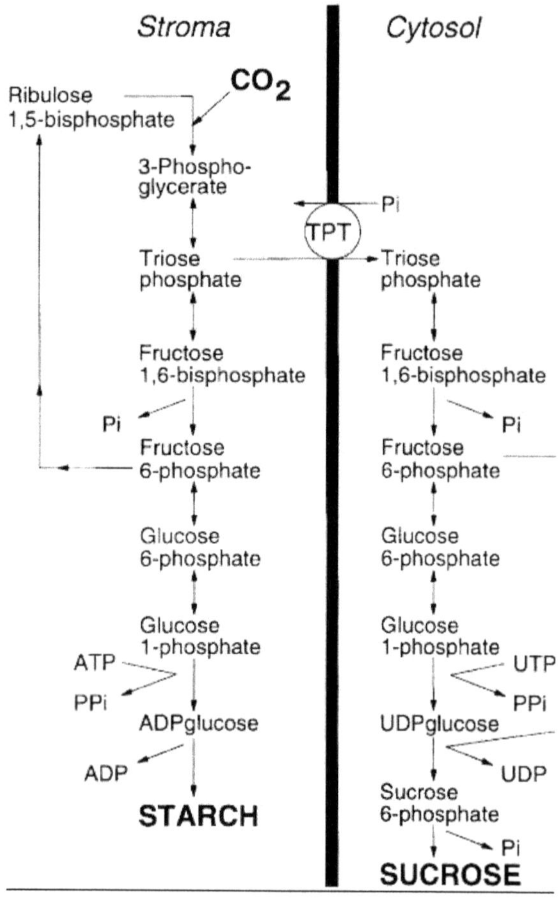

Fig. 2: Pathways of starch and sucrose metabolism in leaves. TPT, triose-phosphate translocator.

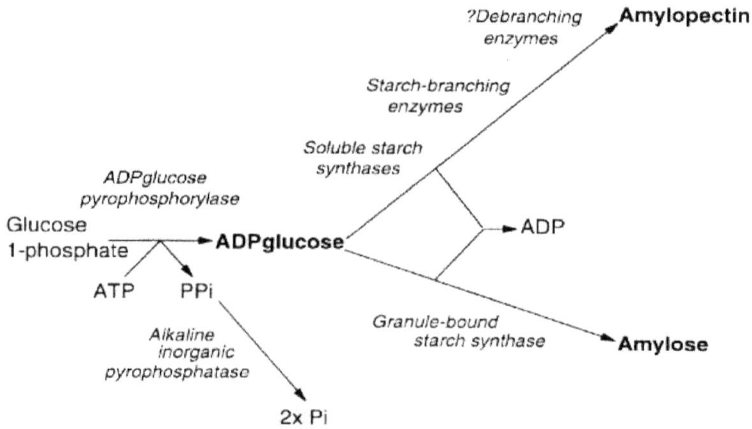

Fig. 3: Pathway of starch synthesis in leaves. PPi, inorganic pyrophosphate; Pi, inorganic phosphate.

13.4.1.1 ADPglucose Pyrophosphorylase

- The activity of AGPase in leaves is potentially subject to regulation both allosterically by metabolites, and at the levels of transcription and translation. Our description will concentrate on aspects of its regulation that are of relevance to its role in the regulation of starch synthesis (Section IV.B). AGPase from leaves is allosterically regulated by a number of metabolites central to chloroplast metabolism. It is activated most strongly by 3- phosphoglycerate (3-PGA): for example the enzymes from spinach and barley leaves are activated about 25-fold and 13-fold with A05 values of 30 pM and 5 pM, respectively but fructose 6-phosphate, fructose 1,6-bisphosphate, phosphoenolpyruvate and 2-phosphoglycerate also activate the enzyme from some sources. Inorganic phosphate (Pi) is a potent inhibitor, with 10.5 values of 45-64 pM and 25/2M for the enzymes from spinach and barley leaves respectively. The presence of 3-PGA weakens and alters the nature of the inhibition by Pi, causing the Pi inhibition curve to change from hyperbolic to sigmoidal. The activity of the enzyme is thus extremely sensitive to changes in the ratio of 3-PGA to Pp and this has important consequences for its regulation in the chloroplast (see Section IV.B). AGPase is a heterotetramer of two types of subunit—known as large and small subunits—that are related in sequence. The precise roles of the two subunits in determining the properties ofthe enzyme are not yet fully understood, and both appear to be capable of binding substrates and effectors. Studies of the catalytic properties of large and small subunits expressed separately and together in the bacterium Escherichia coli are consistent with the view that the small subunits form a basic catalytic unit, the regulatory properties of which are modified by association with a large subunit.
- Many species of plant contain small multigene families encoding one or both of the subunits. The small subunit is evolutionarily highly conserved while the large subunit shows much more divergence of sequence between species, and between members of multigene families within a species. Analysis of transcripts indicates that members of large and small subunit families are differentially expressed between organs, and during the development of individual organs. The expression of subunit genes has been studied in most detail in potato. One gene encoding a small subunit and two different genes encoding large subunits are expressed in potato leaves. Transcript 212 Richard N. Trethewey and Alison M. Smith from the large-subunit gene agpS2 was more abundant in young sink leaves than in source leaves, and was weakly induced by incubation of detached leaves in sucrose. Only small amounts of transcript from the second large-subunit gene, agpS, were detectable in attached leaves, but levels increased dramatically upon incubation of detached leaves with sucrose. The agpS and agpS2 genes appear to be expressed in different tissues within the leaf. A promoter-a-glucuronidase

fusion system indicated that the agpS gene is expressed in stomata and the starch sheath surrounding the main veins, but apparently not in the mesophyll cells. It seems likely that the agpS2 gene is expressed in mesophyll cells. Amounts of protein ofAGPase subunits in potato leaves do not reflect directly the amounts of transcript, implying that there is strong post-transcriptional regulation of expression. Whereas levels of small-subunit transcript increased during the photoperiod, and levels of both small and large subunit transcripts increased in response to incubation with sucrose, there were no corresponding changes in amounts of the proteins or in AGPase activity. Activity and amounts stayed constant or actually fell both during the photoperiod and in response to incubation with sucrose. The significance of this regulatory complexity is not yet clear.

- It is not known whether the transcriptional and post-transcriptional regulation of expression actually leads to changes in the subunit composition of AGPase in the mesophyll cells of the leaf during development, on a diurnal basis, or in response to carbohydrate status or environmental conditions. If the subunit composition does change, this may lead to changes in the regulatory properties of the enzyme. It is likely tha tAGPases with different subunit compositions will be differently sensitive to allosteric activation and inhibition. AGPase is generally accepted to be an exclusively plastidial enzyme, and there is compelling evidence for some plant organs that this is the case. However, recent work has demonstrated that endosperms of barley and maize have distinct plastidial and extra-plastidial forms of the enzyme. In both tissues most of the AGPase activity is contributed by the extra-plastidial form, and in maize endosperm analysis of a mutant (brittle2) lacking the extraplastidial activity indicates that it provides much of the AD glucose for starch synthesis in wild-type endosperm. There is at present no evidence that leaves contain an extra-plastidial AGPase. In leaves of barley, immunological studies suggested that the AGPase was attributable to chloroplastic proteins likely to be the same as those responsible for the minor, plastidial activity in the endosperm.

13.4.1.2 Starch Synthase

- The starch synthase activity of plant organs is composed of several, distinct isoforms, each encoded by a different gene. Some of these isoforms are soluble, ie in the stroma of the plastid, some are tightly bound to the starch granule, and some are present in both of these locations. For a few starch-synthesising storage organs, most or all of the isoforms have been identified. However, their kinetic and regulatory properties are poorly understood, and in vitro studies have shed little light on their roles in the synthesis of the starch polymers. Recent studies of mutant and transgenic plants with altered isoform compositions are starting to elucidate these roles. Although some of the first descriptions of starch synthases were made on the enzyme from leaves studies of starch synthase in leaves have been less extensive and less detailed than those in storage organs.

- We will summarise the present understanding of the enzyme in storage organs, then examine the extent to which this picture is applicable to leaves. Starch-synthesising storage organs of wild-type plants all contain a highly-conserved class of isoform of starch synthase that is tightly bound to the starch granule. This class is known as granule-bound starch synthase I. Analysis of mutant and transgenic plants in which GBSSI activity has been reduced or eliminated (for example the waxy mutants of cereals, the lam mutant of pea, the amf mutant of potato, and potatoes expressing antisense RNA for GBSSI) provides evidence consistent with the idea that GBSSI is specifically responsible for the synthesis of the amylose component of the starch . These mutants have a normal starch content, but the starch consists almost entirely of amylopectin. Exactly how GBSSI is able to synthesize an almost linear glucan. Transitory Starch Metabolism in the same compartment and at the same time as the synthesis of a highly branched glucan is unclear. One possibility is that it is buried within the starch granule in such a way that it is accessible to its substrates—ADPglucose and glucans of low molecular mass—but its product is not accessible to SBEs. It is also not clear whether GBSSI contributes to the elongation of amylopectin chains in vivo. It is certainly capable of incorporating glucose residues from ADPglucose into long chains of amylopectin in isolated starch granules.

- Other isoforms of starch synthase—those responsible for amylopectin synthesis —are less well understood. Three classes, distinguished by features of their predicted amino-acid sequences and/or their molecular mass and antigenic properties, have been identified so far. Starch synthase I (SSI) has been identified in potato and the endosperms of rice and maize, SSII in potato, pea embryo and cereal endosperms, and SSIII in potato, pea embryo and maize endosperm. It is likely that these three classes are of widespread occurrence, and between them they may account for starch synthase activity other than that accounted for by GBSSI. However, their relative importance differs from one type of storage organ to another. For example, SSII accounts for 60-70% of the soluble activity in pea embryos, and SSIII was a minor component of the activity. A mutation that eliminates SSII (at the rug5 locus) dramatically reduces the average chain length of amylopectin in the embryo, indicating that this isoform plays a specific role in determining amylopectin structure.

- In contrast, SSII accounts for about 10% and SSIII for 85% of the soluble starch synthase activity in a potato tuber. Large reductions in SSII activity in the tuber, achieved through expression of antisense RNA, had very little effect on amylopectin structure. It is thus not possible at present to make generalizations about the nature and specific roles of isoforms of starch synthase other than GBSSI. There is good evidence that leaves, like storage organs, contain multiple isoforms of starch synthase, and that these are both soluble and granule bound. Multiple, soluble forms of the enzyme have been resolved by ion-exchange chromatography and gel electrophoresis from spinach, rice and maize leaves, and granule-bound activity has been reported for starch from, for example, geranium and rice leaves. These reports have generally not identified the proteins actually responsible for the activities, hence it is not possible to determine unequivocally the nature and the number of gene products that contribute to the activity. Nonetheless, there is evidence that the isoforms of starch synthase present in leaves of cereals differ from those in the storage organs of the same plant.

- For example, Taira et al (1991) reported that the granule-bound activity and the amylose content of the starch of rice leaves were unaffected by the waxy mutation that eliminates GBSSI and amylose from the endosperm. Leaf starch contained no proteins antigenically related to the GBSSI of the endosperm. In maize, the chroma-tographic behaviour and kinetic properties of the soluble starch synthase activity of leaves and developing endosperm differed considerably, leading to the suggestion that leaves contained only one of the two isoforms distinguished biochemically in endosperm.
- Detailed study of pea revealed that both the granule-bound and soluble starch synthase activities of leaves are attributable to isoforms different from those in the starch-storing embryo. Leaf starch contains an isoform of starch synthase (GBSSIb) closely related in kinetic and antigenic properties and in N-terminal sequence to the GBSSI class, but encoded by a different gene from the GB SSI of the embryo. Most of the soluble activity in the leaf is contributed by an isoform antigenically related to the SSIII class. Although a minor isoform of the SSIII class is also present in the embryo it is not clear whether the leaf and the embryo SSIII-like isoforms are the same gene product. The SSII isoform that contributes 60-70% of the soluble activity of the embryo contributes only about 10% of the soluble activity of the leaf.
- The precise contributions of different isoforms of starch synthase to the synthesis of amylopectin in transitory starches has not been assessed. It is nonetheless possible that at least some of the differences in the structure and organization of amylopectin molecules between transitory and storage starches are caused by differences in the isoform composition of the starch synthases responsible for the synthesis of these two sorts of starch. However, other factors are also likely to contribute to the differences in the structure and organization of amylopectin molecules. The structure of storage starches is determined over many days, during which time the relative levels of expression of different isoforms of both

starch synthase and starch-branching enzyme probably change. Transitory starch, in contrast, is mostly the immediate product of the isoforms of starch synthase and starch branching enzyme present in the chloroplast during the preceding photoperiod. The structure of any starch present at the start of the photoperiod will be strongly influenced by the actions of starch-degrading enzymes during the previous night. The isoforms of starch synthase responsible for the synthesis of amylose in leaves are not well understood. In potato leaves, amylose synthesis is almost certainly a function of the same GBSSI isoform of starch synthase responsible for amylose synthesis in the tuber. A mutation in the gene encoding GBSSI eliminates amylose from both tuber and leaf starches). However, as discussed above, leaf amylose is not synthesized by the GBSSI isoform of the storage organ in pea and rice. The lam mutation that eliminates amylose from the starch of pea embryo has no effect on the amylose content of leaf starch. A second isoform of the GBSSI class, GBSSIb, may be responsible for amylose synthesis in pea leaves, but definitive proof that this is the case is lacking. It is possible that the amylose of the leaf is at least in part a product of isoforms of starch synthase other than those of the GBSSI class.

13.4.1.3 Starch-Branching Enzyme

- Like starch synthase, SBE exists as multiple isoforms that are much better described for storage organs than for leaves. The isoforms of storage organs fall into two distinct classes (known as A and B), distinguished by features of their amino-acid sequences. All of the storage organs so far examined—including those of rice, pea, maize and potato—possess both A and B isoforms, and these appear to account for all of the activity of SBE in the organ. In the endosperm of maize, two different A isoforms as well as a B isoform are present and rice endosperm also appears to contain multiple forms of the two types of isoform. The A and B isoforms have distinct properties. Detailed studies of these isoforms from maize, both in vitro and when expressed in Escherichia coli, show that the A isoform (maize isoform II) prefers more highly-branched substrates and preferentially transfers shorter branches than the B isoform.

- More limited studies on the A and B isoforms of other species indicate that the properties of the maize isoforms define general differences between the A and B classes. There is evidence from pea, rice and maize that leaves also contain A and B isoforms. The isoforms of pea leaves are antigenically related to those of the embryo, and the effects of a mutation that eliminates the A isoform combined with the results of immunoprecipitation experiments provide good evidence that the same A and B isoforms account for all of the SBE activity of both the leaf and the embryo. The two main forms of SBE purified from maize and rice leaves have kinetic, antigenic and chromatographic properties similar to the A and B isoforms of the endosperm although the relative contributions of the A and B isoforms to the total SBE activity probably differs between leaf and endosperm. Biochemical evidence, and the effects on leaf and endosperm SBE of mutations (at the amylose-extender loci) that lie in genes encoding an A isoform (maize isoform IIb) indicate that leaves of maize and rice may contain a different A isoform from the endosperms of these species. The fact that starch-synthesising organs have two isoforms of SBE with different properties has led to speculation that the two isoforms create. Transitory Starch Metabolism 215 different lengths or at different intervals along the amylopectin chains during amylopectin synthesis in vivo.

- For example, it has been suggested that the A isoform creates the chains of 12-16 glucose units that lie within the amylopectin clusters and the B isoform creates the longer chains that span clusters. Studies of the effects of the mutation that eliminates the A isoform in peas show that this idea requires refinement. Although the mutation increases the average chain length of amylopectin in both embryos and leaves, it has little effect on the polymodal distribution of chain lengths. The amylopectin of mutant leaves—containing only the B isoform—has a distribution of chain lengths fro m 5 to 40 glucoses that is virtually indistinguishable from the wild-type. The extent to which the A and B isoforms play distinct roles in amylopectin synthesis in vivo remains unclear.

13.4.1.4 Debranching Enzyme

- Studies of mutants of cereals and Chlamydomonas that are deficient in debranching enzyme activity have given rise to the proposal that these enzymes, as well as SBEs, are important in determining the branching pattern of amylopectin. The developing endosperms of sugaryl maize and rice accumulate a highly-branched, water-soluble glucan, phyto-glycogen, in place of much of their starch. Both endosperms have reduced activities of debranching enzyme, and the gene at the sugary] (sul) locus of maize has been shown to encode a debranching enzyme of the isoamylase class. Mutations at the sta7 locus of Chlamydomonas cause the production of phytoglycogen instead of starch, and are associated with the loss of a debranching enzyme. It is proposed on the basis of these phenotypes that debranching enzyme may act together with starch-branching enzyme to determine the degree of branching of amylopectin in a broad range of starch-synthesising species.

- Ball and colleagues put forward a model for the synthesis of amylopectin in which trimming by debranching enzyme of a highly-branched, 'pre-amylopectin', synthesized by starch synthases and SBEs, is an essential step. It has proved difficult to test the model rigorously, in part because of lack of detailed knowledge ofthe occurrence and properties of debranching enzymes and in part because the sui mutations have multiple pleiotropic effects on enzymes of starch synthesis in the endosperm. The mutation in maize lies in an isoamylase gene, the mutant endosperm has lower activity of a second class of debranching enzyme, pullulanase (also called limit-dextrinase or R enzyme), and alterations in activities of several other enzymes It is thus not possible to attribute the accumulation of phytoglycogen in endosperms unambiguously to the loss of isoamylase activity. New light has been shed on the involvement of isoamylase in starch synthesis in general, and in starch synthesis in leaves in particular, by the discovery of an Arabidopsis mutant that contains phytoglycogen as well as starch in its leaves. This mutant (dbel) is analogous to the sul mutants in that isoamylase activity is dramatically reduced or absent in its leaves, and there is evidence that the mutation lies in an isoamylase gene.

◉ Unlike the sul mutants, the dbel mutant is not deficient in pullulanase activity, and there are no major effects on other enzymes of starch synthesis. Phytoglycogen accumulates in mutant leaves at the same time and in the same chloroplasts as starch containing amylopectin of essentially identical structure to that in wild-type leaves. This phenotype is not obviously compatible with the trimming model proposed by Ball and colleagues. The simplest expectation from the model is that lack of debranching enzyme would give phytoglycogen only, and reduced debranching activity would give a single type of polymer with a structure intermediate between that of phytoglycogen and amylopectin. To account for the simultaneous accumulation of phytoglycogen and amylopectin in chloroplasts of the dbel mutant, Zeeman et al. (1998a) propose that isoamylase is not directly involved in amylopectin synthesis, but rather plays a role in preventing elaboration by starch synthase and SBE of small, soluble glucans in the stroma. They argue that such glucans—for example maltose and maltotriose—can potentially be elongated by starch synthase and then branched by SBE. In the normal plastid, this elaboration of soluble glucan is prevented by a suite of stromal starch-metabolising enzymes such as amylases, phosphorylase and isoamylase. In an isoamylase-deficient mutant, branched soluble glucan can be synthesized in the stroma. Its synthesis competes with that of amylopectin at the surface of the granule forthe starch synthase and SBE available in the plastid, resulting in a reduction in the rate of amylopectin synthesis and the accumulation of phytoglycogen. Although the dbel mutant illustrates that isoamylase plays an important role in carbohydrate metabolism in the chloroplast during starch synthesis, we emphasize that both the trimming model and the alternative model proposed by Zeeman and colleagues to explain the nature of its involvement remain speculative. Much further work is required to elucidate the precise role of the enzyme.

13.4.2 Regulation of Starch Synthesis

- The fact that AGPase is allosterically regulated by 3-PGA and Pi has led to the idea that changes in the levels of these metabolites in the chloroplast may be the primary means by which transitory starch synthesis is regulated. It is argued that when the rate of photosynthesis is low relative to the demand for sucrose, the ratio of 3-PGA to Pi will be low, hence AGPase will be inhibited and the flux of carbon through the pathway of starch synthesis will be relatively restricted. Newly assimilated carbon will thus be preferentially diverted to sucrose biosynthesis. Conversely, when the rate of photosynthesis is high relative to the demand for sucrose, the ratio of 3-PGA to Pi will be high, AGPase will be activated, and the flux through the pathway of starch synthesis will be relatively high. Thus newly-assimilated carbon will be partitioned more towards starch under conditions of sucrose sufficiency. This hypothesis is attractive because it would be consistent with a role for starch synthesis as an overflow as has been demonstrated in many physiological studies. It is less clear how this mechanism would ensure that starch synthesis was sufficient to store enough carbohydrate for the following night period as has been observed in several plant species.
- There could be additional complexity such as diurnal changes in the activation status of sucrose-phosphate synthase that ensure a minimum production of starch during the photoperiod. There are several lines of evidence for the hypothesis that regulation of starch synthesis in vivo proceeds primarily through changes in the 3-PGA to Pi ratio. First, there are good theoretical reasons for believing that the ratio of 3-PGA to Pi in the chloroplast will change as the ratio of assimilation to sucrose synthesis changes. When the rate of assimilation exceeds the rate of sucrose synthesis, import of Pi and export of triose phosphate from the chloroplast via the TPT is expected to be restricted, leading to an increase in the 3-PGA to Pi ratio in the chloroplast. Conversely, whenthe rate of assimilation is lower than the rate of sucrose synthesis then the high rate of generation of Pi in the cytosol from sucrose synthesis is expected to result in an increased export of triose phosphate from the chloroplast and a relatively low 3-PGA to Pi ratio in the stroma.
- Reports of stromal phosphate concentrations, particularly the reduction in stromal phosphate concentration observed following feedback limitation of photosynthesis are consistent with these models. Further support is provided by the strong evidence that signal metabolite fructose 2,6-bisphosphate is involved in linking the biosynthesis of sucrose in the cytosol to the 3-PGA to Pi ratio in the chloroplast (see Stitt, 1990 for a review). This is discussed in accompanying chapters. We wish only to highlight the link between sucrose sufficiency and the 3-PGA to Pi ratio. Studies of spinach leaves show that towards the end of the

photoperiod, when the assimilation rate exceeds the rate of sucrose biosynthesis, there is a restriction in sucrose-phosphate synthase activity and a consequent rise in the hexose phosphate levels. The rise in fructose 6-phosphate stimulates fructose 6-phosphate 2-kinase and inhibits fructose 2,6-bisphosphatase. The subsequent increase in fructose 2,6-bisphosphate inhibits fructose 1,6-bisphosphatase, which in turn leads to an increase in 3-PGA and triose phosphate in the cytosol.

- The increased concentration of phosphorylated compounds in the cytosol probably leads to a restriction in the availability of Pi in the chloroplast. When Pi becomes limiting in the chloroplast, the stromal ATP concentration falls, and the rate of reduction of 3-PGA to triose phosphate is decreased, thus leading to a rise in the stromal 3-PGA to Pi ratio and a stimulation of starch synthesis. Recently, further evidence for the importance of fructose 2,6-bisphosphate has come from studies with transgenic tobacco plants that contain elevated levels of this signal metabolite. The authors Chapter 9 Transitory Starch Metabolism 217 found an increased partitioning of carbohydrate towards starch in plants with elevated fructose 2,6- bisphosphate levels and proposed that this switch in partitioning was mediated through the 3-PGA to Pi ratio. Second, studies with isolated chloroplasts confirm that changes in the ratio of 3-PGA to Pi can alter the rate of starch synthesis., A decrease in the supply of Pi to isolated chloroplasts caused an increase in the partitioning of assimilated carbon into starch. It was proposed that the low concentration of Pi led to a decrease in the ATP/ADP ratio and hence to a restriction of 3-PGA reduction. An increase in the 3-PGA to Pi ratio in the chloroplast would then lead to an activation of AGPase and starch synthesis. Third, circumstantial evidence for the importance ofthe 3-PGA to Pi ratio in regulating starch synthesis in vivo has come from studies where the activity of the TPT has been altered through genetic manipulation.
- The TPT is the major protein in the chloroplast inner envelope and catalyzes a strict 1:1 counter-exchange of substrates: it has similar affinities for triose phosphates, 3-PGA and Pi. It therefore plays a central role in the regulation ofthe 3-PGA to Pi ratio and the communication of this ratio between cytosol and stoma. Transgenic lines ofpotato and tobacco have been generated in which TPT activity is reduced through the expression of antisense RNA. A 20-30% reduction in TPT activity in potato led to an increase in leaf starch by a factor of 300% while in tobacco an 80% reduction in activity led to an increase in leaf starch of only 10%. In both cases it would be expected that restricted TPT activity would increase the stromal 3-PGA to Pi ratio and hence stimulate starch synthesis. The fact that the extent of the increase in starch in response to a decrease in TPT activity is different in potato and tobacco indicates that factors other than simply the 3-PGA to Pi ratio are likely to be involved in regulating starch synthesis in some species. Fourth, the importance of the AGPase in determining the rate of

starch synthesis has been clearly shown by mutant and transgenic plants in which the activity of AGPase is reduced. A reduction of 50% in AGPase activity in a mutant Arabidopsis led to 23 and 39% reductions in starch synthesis under low and high light, respectively. Antisense inhibition of AGPase in potato using a leaf-specific promoter revealed a good positive correlation between the AGPase activity and the accumulation of starch in the leaf. The case for the modulation of starch synthesis through the allosteric regulation of AGPase by the 3-PGA to Pi ratio is convincing. However, it should not be assumed that the AGPase is the only site at which control is exerted over the rate of starch synthesis. The development of a sound theoretical approach to the analysis of metabolic fluxes has led to an increased realization that every step in an enzymic sequence contributes to some extent to the determination of flux.

- The role of each enzyme in controlling flux can be summarised as a flux control coefficient which is the fractional change in flux given by a fractional change in this enzyme activity. The larger the flux control coefficient of a particular enzyme, the more important this enzyme is in controlling the flux. By definition the flux control coefficients of a linear sequence of reactions without branch points sum to 1. It should therefore be the aim of a biochemist to determine all the flux control coefficients in a given pathway, although such an approach is beset by practical problems and has not yet been achieved for an entire pathway in plants. In the case of transitory starch synthesis some initial estimates of flux control coefficients have been made using data from a pea mutant deficient in starch-branching enzyme,Arabidopsis mutants deficient in plastidial PGM and AGPase, and a Clarkiaxantiana mutant defective in plastidial phosphoglucoisomerase. The calculated flux control coefficients indicate that AGPase dominates the control of starch synthesis in both high and low light. The flux control coefficients of phosphoglucoisomerase, PGM and SBE in high light were significantly greater than in low light. This analysis must be regarded as preliminary and incomplete because it is an amalgamation of data from different species and does not include an estimate ofthe flux control coefficients for the starch synthases. However, it is valuable in indicating that the balance of control in the pathway can be different according to the environmental conditions. Despite the technical problems associated with the deter-mination of flux control coefficients in plants, it is to be hoped that further progress will be made in this direction in the future. 218 V. The Degradation of Transitory Starch A. Enzymes Implicated in the Degradation of Transitory Starch

- Despite extensive investigation, the precise nature of the pathway of transitory starch degradation remains unclear. Many enzyme activities that could contribute to degradative pathways have been described and characterised, and cDNAs encoding some of these enzymes have been cloned. However, the precise function of many of degradative enzymes in vivo is not known, and in several cases it is not clear whether the degradative activities are localised in the chloroplast.

13.4.2.1 Amylase

Amylases are widespread in higher plants. They can be divided into two types: endoamylases which can catalyze the hydrolysis of the a(1-4) linkages at any point within amylose and amylopectin except immediately adjacent to all-6) branch points, and exoamylases which cleave maltose groups only from the non-reducing end of a(1-A)glucan chains. In plants the endoamylase and exoamylase activities are normally called a-amylase (EC 3.2.1.1) and 13- amylase (EC 3.2.1.2), respectively.

a. a-amylase

Multiple isoforms of a-amylase are present in the leaves of many species of plant. Although at least some of these isoforms are not present in the chloroplast, others are chloroplastic and could therefore be involved in transitory starch degradation. Leaves of sugar beet, for example, contain five forms of a-amylase separable on zymograms, of which one is extrachloroplastic and four are exclusively or at leastpartly chloroplastic. Reports of the nature of the chloroplastic a-amylases differ from one species to another. Those of mesophyll and bundle-sheath cells of pearl millet (Pennisetum americanum) leaves are reported to be of the heat-stable, calcium dependent type extensively characterised from cereal endo-sperms. Calcium-dependent a- occurrence in chloroplasts requires further investigation. Amylases have also been reported from the leaves of sugar beet and poplar, but the enzyme was extrachloroplastic in the former case and localization studies were not performed in the latter case. The a-amylase of spinach chloroplasts, in contrast, is reported to be of the heat-labile, calcium-independent type first described from Arum spadix. Activity of a-amylase was not detected in studies of the chloroplasts of pea and barley leaves. However, in these cases the assays were conducted under conditions in which only heat-stable, calcium dependent a-amylases would be detected. The possibility that heat-labile, calcium-independent a-amylases are of widespread. There are two recent reports of amylases from Arabidopsis leaves which may be involved in chloroplastic starch degradation. First, activity of an isoform of endoamylase with a pH optimum of 7 is reported to vary diurnally in a manner consistent with a role in starch degradation at night.

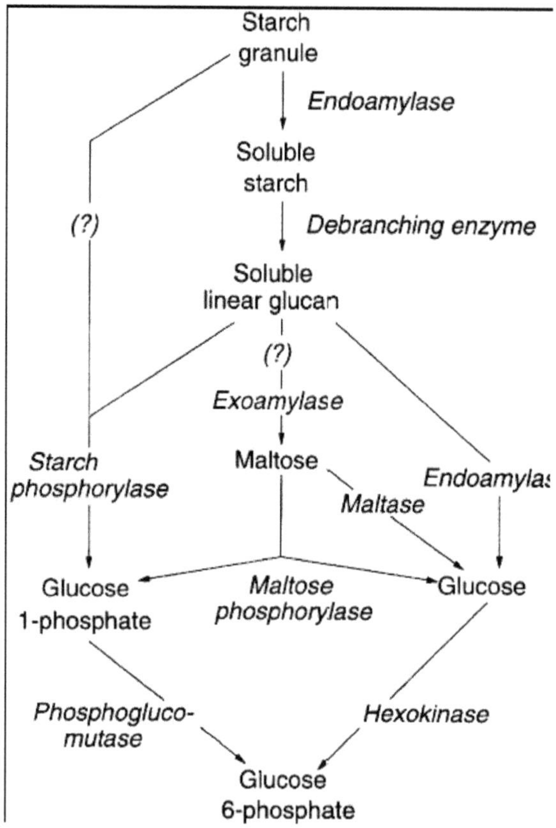

Fig. 4: Enzymes and intermediates implicated in the degradation of transitory starch.

However, ithas not been established whether this activity is in the chloroplast. Second, evidence for the involvement of a different endoamylase in starch degradation has been provided by the discovery of a mutant (sex4) which apparently specifically lacks one isoform of the enzyme. The mutant is deficient in starch degradation at night. Although this endoamylase has been shown to be chloroplastic, its activity does not vary on a diurnal basis. The possible roles of these two endoamylases are discussed further in Sections VB and VC below.

b. J3-amylase

Aamylase is present in leaves, and is often discussed in the context of starch degradation in plants. However, it now seems unlikely that this enzyme plays a role in degradation of transitory starch. Rigorously-conducted cell-fractionation studies have shown that the enzyme is not associated with the chloroplasts in pea leaves, and that it is associated with the vacuole rather than the chloroplast in protoplasts derived from pea and wheat leaves. A report that /3-amylase is present in pea chloroplasts may be based on preparations of chloroplasts with significant vacuolar contamination. Extensive studies of the 0-amylase of Arabidopsis leaves—where this enzyme accounts for over 80% of the total amylolytic activity—suggest that it is not chloroplastic but is likely to be vacuolar and perhaps localised within the vascular bundle.

13.4.2.2 Debranching Enzyme

Debranching enzymes catalyze the hydrolysis of a(1-6) linkages in starch. Two classes of these enzymes have been described from higher plants. Pullulanase [also called limit dextrinase or R-enzyme: amylopectin 6—glucanohydrolase (EC 3.2.1.41)] can hydrolyze the yeast glucan pullulan in addition to starch polymers, whereas isoamylase (EC 3.2.1.68) cannot. Pullulanases have been shown to be present in the leaves of several species of plant. Both extrachloroplastic and chloroplastic isoforms occur in pea leaves, and chloroplasts of spinach and Viciafaba contain pullulanase activity. Isoamylases have been reported from kernels of maize, pea embryos, leaves of Arabidopsis and potato tubers. Possible roles for debranching enzymes in the synthesis of starch are discussed in Section IVA.4 above.

13.4.2.3 a-glucosidase

a-glucosidase (EC 3.2.1.20), also called maltase, hydrolyzes maltose and malto-oligosaccharides to glucose. a-glucosidases often display wide substrate preferences, which may differ considerably from one source of the enzyme to another. For example, some are specific for a(1-4) linkages and short malto-oligosaccharides, whereas others will hydrolyze a(1-6) and other a-glucosidic linkages in larger substrates. Thus, depending on their characteristics, these enzymes could act solely on the products of amylolytic degradation of starch to ensure their complete conversion to glucose, or could play a wider role in starch degradation by attacking larger, a(1-4), a(1-6) glucans. There are conflicting reports on the occurrence and subcellular localization of a-glucosidase in leaves. Activity was not detected in studies of pea leaves and chloroplasts of pea and spinach. However, identified two apoplastic and one chloroplastic isoform from pea seedlings. The chloroplastic isoform had a pH optimum of 7.0 and a IS„ for maltose of 7.2 mM. Beers and colleagues suggested that failure to detect a-glucosidase: activity in previous studies was due to inappropriate extraction and assay conditions. Thepotential contribution ofthis enzyme to starch degradation in leaves generallythus remains to be established.

13.4.2.4. Maltose Phosphorylase

Little is known about the occurrence of maltose phosphorylase (EC 2.4.1.8) in higher plants. The enzyme is reported to be present in pea chloroplasts (Kruger and ap Rees, 1983b), but we know of no other information on its occurrence in leaves. Maltose phosphorylase catalyzes the interconversion of maltose with glucose 1-phosphate and glucose, and 220 Richard N. Trethewey and Alison M. Smith could potentially contribute to the conversion to sugars of the products of amylolytic degradation of starch.

13.4.2.5 Starch Phosphorylase

Starch phosphorylase (a-glucan phosphorylase, EC 2.4.1.1) is of wide occurrence in higher plants. The enzyme catalyzes the phosphorolytic cleavage of a glucose residue from the non-reducing end of a glucan chain, to form glucose 1-phosphate. Plastidial and extraplastidial isoforms ofthe enzyme (referred to as type I and type II phosphorylases) have been demonstrated in many plant organs, including leaves. The properties and primary sequences of the two types are distinct and highly conserved. In potato, in which the enzyme has been very extensively studied, there are two plastidial isoforms, encoded by the pho la and pho lb genes. The Phola product accounts for most of the plastidial activity in the tuber, and the Pholb product for most of the activity in chloroplasts. Two plastidial isoforms have also been identified in pea. Plastidial isoforms of phosphorylase, including those of chloroplasts, prefer malto-oligosaccharides to larger, branched glucans as substrates. The smallest malto-oligosaccharide that will act as a substrate is maltotetraose. There is reason to believe that the reaction catalyzed by the enzyme in the chloroplast is effectively irreversible in the direction of glucose 1-phosphate synthesis. Kruger and ap Rees (1983b) found that the apparent equilibrium constant (0.3-0.7) was very different from the mass action ratio of glucose 1-phosphate: Pi in the chloroplast (7.9 x 10-3) calculated from estimates of stromal volume and metabolite contents in spinach leaves. These properties are consistent with a potential role for the enzyme in the conversion to hexose phosphate of the products of amylolytic degradation of starch. Partially-purified starch phosphorylase from pea chloroplasts exhibits some regulatory properties that could be of relevance to a role in the degradation of transitory starch at night. Its activity is twice as great at pH 7.2 as at pH 8.0, and it is inhibited by 60% in the presence of 4 mM ADPglucose.

13.4.2.6 D-Enzyme

D-enzyme (4-a-D-glucanotransferase: EC 2.4.1.25) catalyzes a disproportionating' reaction: A (l -4)glucan(a) + a(1-4)glucan() a(1-4) glucantR, bH + glucose The enzyme has been studied in relatively few plant organs, but has been shown to be chloroplastic in pea and Arabidopsis leaves. The cDNA for D-enzyme from potato tubers encodes a putative plastid transit peptide. D-enzyme could thus potentially convert short glucans generated by degradation oftransitory starch into longer substrates more suitable for amylases and phosphorylase. A recent study of the properties of a D-enzyme from potato following its expression in E. coli reveals that the enzyme can also catalyze the formation of cyclic glucans from amylose. The existence of such cyclic glucans in plants has not been demonstrated. Large reductions in the activity of D-enzyme brought about by the expression of antisense RNA in potatoes reduced the growth rate ofthe plant, but the precise role played by the enzyme in the metabolism of starch in vivo remains to be established.

13.4.2.7 Hexokinase

The entry into intermediary metabolism of glucose released from starch requires the formation of glucose 6-phosphate via a hexokinase (EC 2.7.1.1). There is continuing debate about whether this phosphorylation occurs within the chloroplast or in the cytosol.

The ability of chloroplasts from several species of plant, including pea and spinach, to metabolise exogenously-supplied glucose suggests a chloroplastic location for hexokinase. However, cell-fractionation experiments indicated that hexokinase activity is outside the chloroplast envelope in pea and spinach. It seems likely that hexokinase is actually located on the chloroplast envelope, but definitive proof that this is the case is lacking.

13.4.2.8 R1 Protein

A novel protein implicated in the process of starch degradation in leaves has recently been identified in potato. The protein (named R1) was first identified from the matrix of tuber starch granules. An immuno-screening approach led to the isolation of a cDNA encoding a large protein of 1464 amino acids with no obvious sequence similarity to any other known protein. No enzymatic activity could be assigned to the protein expressed from the cDNA. Expression of antisense RNA for R1 in potatoes had several important effects on the nature and content of starch in the plant. First, there was a reduction in the phosphate content of tuber starch although the amount of starch in the tuber was not affected. Second, leaves contained more starch than those of untransformed controls: complete mobilization of starch during extended periods of darkness was not achieved. Third, a significant reduction in cold sweetening in cold-stored tubers was found in conjunction with an enhanced retention of starch after two months of storage. It is difficult to deduce the precise role of R1 from these complex phenotypic effects. However, expression of the R1 protein in E. coli led to a significant increase in the level of phosphorylation of the glycogen made by the bacterium. The authors proposed that the primary function of R1 is in starch phosphorylation, and that a lowering of the phosphate content of the starch of the transgenic potatoes leads by an unknown secondary mechanism to reduced susceptibility to degradation. Further investigation, including the development of an assay, will be required to resolve the role of Rl.

13.4.3 The Pathway of Transitory Starch Degradation in vivo

- There is considerable confusion over which of the above enzymes are important in the degradation of transitory starch. The enzymes can be grouped into pathways around a hydrolytic or a phosphorolytic cleavage of a(1-4) linkages, but it is not clear whether one of these two pathways operates in vivo or whether both occur simultaneously. We shall consider the evidence available for transitory starch about the nature ofthe initial attack on the starch granule, and the fate of products of this initial attack. There is some evidence that both starch phosphorylase and hydrolytic enzymes may be able to attack intact starch granules from leaves. Endo-amylases from several sources and a putative chloroplastic a-glucosidase from pea leaves have been reported to attack isolated starch granules. Starch phosphorylase from pea chloroplasts and poplar wood was also reported to attack intact starch granules, but Steup et al. (1983) were unable to demonstrate degradation of starch granules by this enzyme. All of these in vitro experiments on the degradation of starch granules must be interpreted with caution since they are prone to several problems. First, the properties of the degradative enzymes may be altered during purification. Second, the properties of the starch granules may be altered during isolation. In particular, newly-synthesized, less crystalline material—which would be the substrate for degradative enzymes in vivo—may be lost. Third, some studies have used granules from a different plant source from the degradative enzymes, or have not taken care to isolate granules at the end of a photoperiod when degradation normally commences in vivo. Finally, the possibility that isolated granules may be contaminated with degradative enzymes has not been considered in all studies.

- The only direct evidence for the involvement of a specific starch-degrading enzyme in starch degra-dation in leaves comes from studies of the sex4 mutant of Arabidopsis. This mutant is deficient in starch degradation at night. Although some degradation occurs, the rate is much lower than in wild-type leaves, and as a result starch accumulates to high levels as the leaf develops. The mutant lacks one, chloroplastic isoform of endoamylase (termed A2), identified by native gel electrophoresis of extracts of wild-type leaves. Other enzymes of starch synthesis and degradation, including debranching enzymes, phosphorylase and amylases, are apparently unaffected by the mutation. Although the gene at the sex4 locus has not been identified, it is likely that the reduced rate of starch degradation in sex4 leaves is directly attributable to the loss of the A2 endoamylase. The properties of this enzyme have not yet been established. Studies with isolated chloroplasts have provided some evidence that the hydrolytic and phosphorolytic pathways can both operate in the degradation of transitory starch. Stitt and ap Rees (1980) labeled the starch in chloroplasts isolated from pea shoots by incubating the chloroplasts with “C-glucose in the light. During the subsequent dark period, 14C lost from the starch appeared predominantly in maltose but also in sugar phosphates and 3-PGA. In experiments with pea chloroplasts, the presence of maltose phosphorylase and the dependence of maltose synthesis on exogenous Pi led to the suggestion that maltose was synthesized from glucose 1-phosphate formed via phosphorolytic degradation of starch. Taken as a whole, therefore, experi-ments with isolated starch granules and chloroplasts have thus far failed to define the pathway of starch degradation.

13.4.4 Regulation of the Degradation of Transitory Starch

- Given the uncertainty over the pathway of starch degradation, it is impossible to draw any firm conclusions about the regulation of the process. Net degradation occurs predominantly during the night, although some studies have reported that net starch degradation can also occur at the beginning and end of the photoperiod. One possible model is that the rate of starch degradation in leaves is constant, and that changes in the amount of starch are brought about solely by changes in the rate of synthesis. However, observations about the timing and rate of degradation of transitory starch indicate this is not the case, and that the process must be subject to tight regulation. First, radioactive labeling experiments have provided no evidence for starch turnover in illuminated leaves during the normal photoperiod. Second, the rate of degradation is subject to variation during the dark period. Whereas degradation of transitory starch can occur immediately following the light/dark transition in some species, in others there is a lag at the beginning of the dark period before degradation commences. In studies of spinach and soybean leaves, sucrose content started to fall immediately after the light/dark transition but there was a short lag before starch degradation occurred.

◉ In barley leaves, starch degradation did not occur until sucrose content fell below a threshold value, nine hours after the start of the dark period. A correlation between the starch content of the leaf and the rate of degradation in the dark has been observed in studies of soybeans grown under different light intensities, and mutant Arabidopsis lines with differing starch contents. Taken as a whole these experiments suggest that the rate of starch degradation in leaves may be sensitive to the diurnal cycle, the starch content, and the demand for sucrose in the dark. The only obvious point at which starch degradation could be regulated is at the reaction catalyzed by starch phosphorylase. The pH optimum, the sensitivity to inhibition by ADPglucose and the dependence upon pi of starch phosphorylase would be expected to result in an increase in its activity after a light-dark transition, when stromal pH becomes more alkaline, Pi concentration probably rises and the ADPglucose concentration falls.

◉ The potential importance ofthe effects of Pi concentration on the activity of starch phosphorylase is demonstrated by experiments with isolated chloroplasts, in which the rate of phos-phorolytic starch degradation is sensitive to the exogenous Pi concentration. The dependence of starch phos-phorylase activity on Pi concentration provides a means by which starch degradation in the leaf may be linked to the demand for sucrose. It is believed that the fall in sucrose content at the onset of the dark period leads to an increase in cytosolic and stromal Pi concentration. This would be expected to lead to a higher activity of starch phosphorylase, and hence a greater rate of starch degradation. Although the effects of Pi concentration on starch phosphorylase may contribute to the regulation of starch degradation, it is clear that other mechanisms must also operate. First, although the activity of starch phosphorylase in the chloroplasts of some species is sufficient to account for the rate of starch degradation, in others it is not. In Arabidopsis leaves, only 4% of the starch phosphorylase activity is plastidial, andthis plastidial activity is probably lower than the rate of starch degradation in the dark. Second, removal of exogenous Pi from isolated spinach chloroplasts in the dark results in a shift from phosphorolytic to apparently amylolytic starch degradation and has little effect on the total rate of degradation. Third, a regulatory mechanism based on changes that occur during a light dark transition cannot explain those circum-stances in which starch is degraded in the light.

- It seems probable that amylolytic enzymes as well as starch phosphorylase must play a role in the regulation of transitory starch degradation. As discussed above, there is good evidence that amylolytic activities occur inside chloroplasts and that, at least in Arabidopsis, they are required for starch degradation. However, apart from a sharp pH optimum around pH 6, there are few reports of properties of chloroplastic amylases that suggest how their activity might be modulated in vivo. There are reports that the total amylolytic activity in chloroplasts of spinach and Vicia faba is subject to diurnal variation, but these measurements must be treated with caution because the possibility of diurnal variation in the recovery of activity during the extraction procedure was not ruled out. The extractable activity ofa specific isoform of endoamylase with a pH optimum of 7 has recently been shown to vary on a diurnal basis in Arabidopsis leaves. Activity – detected on native gels – was seen only in leaves harvested in the dark, whereas activity of other starch-hydrolysing enzymes did not vary on a diurnal basis. This amylase has not been shown to be chloroplastic, and its involvement in starch degradation remains to be established.

13.4.5 Export of Starch Products Degradation from Chloroplasts

- The synthesis of sucrose is by far the major fate for the carbon released by degradation of transitory starch. For example, in sugar beet leaves the rates of starch breakdown and sucrose synthesis at night have been estimated as 0.23 and 0.16 pg C cm^{-2} per min respectively. Transitory starch degradation at night must be precisely integrated with cytosolic metabolism, and in particular with sucrose biosynthesis. The nature of the export mechanisms for carbon from the chloroplast are of central importance to this integration. Recent experiments indicate that the triosephosphate translocator (TPT)—responsible for the export of carbon from the chloroplast during the day—is not the main route of export of carbon produced by the degradation of transitory starch at night. Reduction of the activity of the TPT through the expression of antisense RNA in potato plants results, as expected, in increased accumulation of transitory starch during the day, but also causes increased mobilization of transitory starch and export of sucrose from the leaves at night. In tobacco plants with reduced activities of the TPT, starch degradation is promoted not only at night but also at the end of the light period.

- Although leaves of transgenic plants contained three times more starch than those of control plants after six hours of the light period, the starch content of the transgenic leaves actually fell towards the end of the light period and transgenic and control leaves contained similarly low amounts of starch by the end of the dark period. It seems very unlikely from these results that the products of starch degradation are exported via the TPT. The results described above strongly indicate that sucrose synthesis at night is supported by export from the chloroplast of metabolites other than triose phosphate. Three further observations lend support to this idea. First, mutants ofFlaveria linearis with very low activities of cytosolic fructose bisphos-phatase and transgenic potato plants with large reductions in cytosolic fructose bisphosphatase brought about by the expression of antisense RNA partitioned more of their photosynthate into starch and less into sucrose during the day than did wild-type plants. This result is consistent with a restricted ability to convert triose phosphate to sucrose in the mutant and transgenic plants. However, during the night the rate of starch mobilization in leaves of mutant and transgenic plants was twice as great as in wild-type plants, indicating that carbon exported from the chloroplast at night was not in the form of triose phosphates. Second, radioactive feeding experiments with wild-typeArabidopsis leaves showed that there is a net glycolytic flux at night: if carbon export from the chloroplast were at the level oftriose phosphates there would have to be a net gluconeogenic flux. Third, leaves of several species have a high concentration of fructose 2,6- bisphosphate at night. This should ensure that the cytosolic fructose 1,6-bisphosphatase is substantially inhibited, and thus prevent a gluconeogenic flux from triose phosphate to sucrose.
- These experiments suggest strongly that carbon for sucrose synthesis must be exported from the chloroplast as a sugar or a sugar phosphate. A direct demonstration of a TPT- independent route for export from the chloroplast has been provided by investigations into an Arabidopsis mutant, sexl , that has severely reduced rates of starch degradation. The mutation results in a four- to five-fold accumulation of starch, and a reduced growth rate. The mutant contained appreciable activities of all of the enzymes putatively involved in starch degradation, but physiological analysis clearly indicated that it was defective in the conversion of starch to sucrose at night. Uptake studies on intact chloroplasts isolated from mutant and wild type plants revealed that the mutant was deficient in glucose uptake. This work suggested strongly that the products of starch degradation in Arabidopsis chloroplasts are exported via a glucose transporter. Evidence that this may also be the case in tobacco leaves comes from transgenic plants with reduced activity of the TPT.
- As described above, these plants show increased rates of starch degradation at night and at the end of the light period. This phenomenon is accompanied by a

three-fold increase relative to control plants in the capacity for glucose transport by isolated chloroplasts. A glucose transporter has previously been characterised in spinach chloroplasts. It catalyzes a facilitated diffusion with highest affinity for D-glucose (K,20 mM). It also catalyzed similar rates of D-xylose, D-mannose, and L-arabinose transport and was inhibited by phloretin, a known inhibitor of glucose transport in erythrocytes. The existence of maltose transporters in the chloroplast envelopes of spinach provides a further potential route for the export of carbon from starch degradation. Maltose is transported into spinach chloroplasts via a mechanism independent of glucose transport, indicating that there are separate maltose and glucose transporters. Carbon could also potentially be exported as hexose phosphates. Rates of uptake of hexose phosphates by isolated chloroplasts are usually relatively low. However, isolated pea chloroplasts appeared to export hexose phosphates in the dark and glucose-6-phosphate-dependent starch synthesis was substantially induced in spinach and potato chloroplasts isolated from leaves that had been fed glucose for several days through the petiole. It remains an open question whether maltose or hexose phosphate transporters play a role in exporting the products of transitory starch degradation.

13.4.6 Conversion of Starch to Sucrose at Night

- The demonstration that carbon can cross the chloroplast envelope either as triose phosphate or as sugars gives rise to the possibility that the products of hydrolytic and phosphorolytic degradation might be exported in different forms and thus have different fates in the cytosol. It is tempting to speculate that hydrolytic degradation gives rise to sugar residues that are exported to support sucrose biosynthesis, whereas phosphorolytic degradation gives rise ultimately to triose phosphate, which is exported from the chloroplast via the TPT to sustain dark respiration. Such a division could allow the process of starch degradation to be independently regulated according to the demand for sucrose synthesis and darkrespiration.
- Export of free sugar from the chloroplast at night would minimise the requirement for ATP in the stroma to support starch degradation and would be consistent with the apparent absence of hexokinase from chloroplasts. Much further experimental work on the nature and regulation of starch degradation in the chloroplast is clearly required to allow the validity of this hypothesis to be tested.
- For example, as discussed above, nothing is known about regulatory properties of starch-hydrolysing enzymes that would allo w their activities to be modulated to meet a demand for sucrose synthesis. The hypothesis would be invalidated if there is a significant rate of export of hexose phosphate from the chloroplast at night, or if chloroplasts contain significant activity of maltose phosphorylase (Section V.A.4, V.B), since this would allow the products of phosphorolytic degradation to be used in sucrose synthesis.

13.5 AT A GLANCE

- Transitory starch is of central importance to many plants, serving as a store of carbohydrate for periods when photosynthesis is not possible. There is a reasonable understanding of the enzymes involved in the synthesis of transitory starch, although it remains unclear exactly how the various isoforms involved combine to give rise to the complex and poorly-understood structure of the starch granule. The pathway of starch degradation has yet to be firmly established. Many enzymes with potential roles in starch degradation have been identified in leaves, but in most cases there is no convincing evidence that they are involved in transitory starch degradation, or even localised in the chloroplast. Without a full understanding of the enzymic machinery that mediates starch degradation the regulation of the process can only be guessed at. It is to be hoped that substantial progress can be made on these open questions through the careful analysis of transgenic plants altered in the activities of one or more of the enzymes involved in transitory starch metabolism.
- Starch is a carbohydrate reserve used by many higher plants. Transitory starch is synthesized in the chloroplasts of higher plants both as an overflow for newly-assimilated carbon and as a reserve of carbohydrate for periods of darkness when photosynthesis is not possible. Starch is present in the chloroplast as insoluble granules consisting of two types of glucose polymer: amylopectin and amylose. The pathway of synthesis has been well defined and involves the action of phosphoglucomutase, ADPglucose pyrophosphorylase, starch synthases, and branching enzymes. However, there are multiple isoforms of starch synthases and branching enzymes and the extent to which different isoforms play distinct roles in determining starch structure is not clear. The simultaneous action of degradative enzymes may also play a role in determining the structure of starch during its synthesis. It is likely that starch synthesis is regulated through the allosteric control of ADPglucose pyrophosphorylase.
- Many enzymes have been implicated in the process of transitory starch degradation although few have been confirmed to play a role in vivo. It is not clear whether the initial attack on the starch granules is hydrolytic—catalyzed by amylases—or phosphorolytic—catalyzed by starch phosphorylase. The route by which sugar is released from shorter soluble glucan chains also remains unclear. There is physiological evidence that the process of starch degradation is subject to regulation. However, given the uncertainty about the mechanism of starch degradation, it is not possible to develop hypotheses about its regulation. The products of degradation may be exported from the chloroplast as sugars or as triose phosphate, but there is evidence that sucrose synthesis in the cytosol at night is supported by the export from the chloroplast of sugars rather than triose phosphate.

RELATED TERMINOLOGY

Adenosine triphosphate: An adenine molecule, or a **nucleotide**, attached to three linearly connected phosphate groups ($-H_2PO_4R$, where R is a functional group). The breaking of chemical bonds between the 2nd and 3rd phosphate groups provides most of the chemical energy used by a cell. Most of the ATP in a cell is made in the mitochondria, the cell's powerhouse. ATP is a **coenzyme** (the -ase in the name gives it away) and a strong reducing agent, or electron donor, that acts as the principal energy carrier in the cell. Donating the terminal phosphate group, or the phosphate group on the end, from ATP causes the release of a large amount of energy. ATP basically shuffles energy around to support metabolism and a bunch of super important cellular processes, like photosynthesis.

ADP - ADP stands for adenosine diphosphate, a product of the Calvin cycle that is used in the light-dependent reactions.

Antenna: A pigment with the primary function of capturing the energy from photons and transferring that energy to other pigments within the photosystem. Most chlorophyll molecules function as antennae, with relatively few of the hundreds of chlorophyll molecules carrying out photochemistry in the reaction center. Carotenoids also function as antennae, but additionally play an important role in quenching triplet chlorophyll and singlet oxygen molecules that would cause damage to the plant cell.

ATP - ATP stands for adenosine triphosphate. ATP is a major energy molecule in cells. ATP and NADPH are products of the light-dependent reactions in plants. ATP is used in reduction and regeneration of RuBP.

Autotroph: An organism that can use light energy and the **photosynthetic process** to produce organic food (read: containing carbon and hydrogen) from inorganic molecules.

Autotrophs - Autotrophs are photosynthetic organisms which convert light energy into the chemical energy they need to develop, grow, and reproduce.

$C_6H_{12}O_6$) glucose:(A glucose (sugar) molecule is the product of each light-independent reaction that occurs in the stroma of the chloroplast during photosynthesis. Glucose is made by joining two of the three-carbon sugar products of the Calvin cycle into a six-carbon sugar. Glucose can be made into larger sugars (sucrose) or carbohydrates such as starch and cellulose. Sugars exit the leaf through the phloem, traveling to the roots for storage or to other parts of the plant, where they're used as energy to fuel the plant's activities.

Chlorophyll: The thylakoid membrane contains the pigment chlorophyll, which is stored in protein complexes called photosystems I and II. These chlorophyll molecules absorb light energy from the sun, which the cell converts into chemical energy through a series of chemical reactions.

Chloroplast: Chloroplasts are specialized, membrane-bound organelles, located in the cytoplasm of plant cells, that carry out photosynthesis. Plant cells can have anywhere from one to a hundred chloroplasts, which move around the cell and can divide to replicate themselves. Surrounded by two membranes, chloroplasts contain DNA, ribosomes, enzymes, grana, and a thick fluid called stroma. Chloroplasts are also involved in the plant's synthesis of fatty acids and amino acids and in its immune response.

Chloroplast: An organelle within the plant cell in which the photosynthetic reactions are compartmentalized. The thylakoid membranes within the chloroplast are the site of the photosynthetic pigments and electron transfer components used to make energy from photosynthesis. The non-membrane space within the chloroplast is called the stroma; this is where photosynthetic energy is used to convert CO_2 into sugars.

CO_2Carbon Dioxide:(Carbon dioxide is a reactant of photosynthesis, and six carbon dioxide molecules are needed to produce one glucose molecule. Carbon dioxide molecules move from the surrounding air, entering a plant's leaf through the stomata in the epidermis. Once inside the leaf, these gas molecules move through the spongy mesophyll, entering the chloroplasts of palisade mesophyll cells. Within the stroma of the chloroplast, enzymes and the ATP, NADPH, and light energy harvested during the light-dependent reactions drive the light-independent reactions (also known as dark reactions or the Calvin cycle) that convert carbon dioxide into sugars.

Conjugated Double Bonds: Double bonds on alternating pairs of carbon molecules in a linear or cyclic organic molecule. An example is as follows -C=C-C=C-C=C- The electrons in conjugated double bonds form a resonance structure that is smeared over the double bond system. These electrons are responsible for light absorption.

Electron Transfer: The transfer of electrons between a series of components in a controlled fashion. Examples include the mitochondrial electron transfer chain involved in oxidative phosphorylation and the chloroplast electron transfer chain carrying out photophosphorylation. Electrons are transferred from molecules with a higher free energy to those with a lower free energy.

Electron transport Chain: A specific process used in the **light-dependent reactions** of **photosynthesis**. During the electron transport chain, **electrons** are repeatedly transferred from a high-energy **electron donor** (**nicotinamide adenine dinucleotide phosphate**, or **NADPH**), to an **electron acceptor**, such as O_2. This process is coupled to the transfer of H^+ ions, or **protons**, across a membrane. Essentially, the transfer of **electrons** across the **thylakoid membrane** in the **chloroplast** drives a **proton gradient** that is used to produce **chemical energy** in the form of **adenosine triphosphate** (**ATP**).

Electron: A negatively charged (-1) subatomic particle. An electron is part of an atom.

Endosymbiosis: The theory that explains how mitochondria and chloroplasts became organelles of other cells. Endosymbiosis is the condition of living within the body or cells of another organism. There is evidence that, millions of years ago, the ancestors of mitochondria and chloroplasts, two organelles ("mini organs") were actually prokaryotic organisms that entered into **endosymbiotic** relationships with eukaryotic cells. We are supposing that they liked each other just a *little* too much.

G3P G3P stands for glucose-3-phosphate. G3P is an isomer of PGA formed during the Calvin cycle

Glucose ($C_6H_{12}O_6$) - Glucose is the sugar that is the product of photosynthesis. Glucose is formed from 2 PGAL›s.

Granum- A granum is a stack of thylakoids (plural: grana)

Granum: Each chloroplast has from ten to a hundred grana (sing. granum), connected to each other by lamellae. Each granum is a stack of disc-shaped thylakoids, where the light-dependent reactions of photosynthesis occur.

Guard Cells: Each stoma is surrounded by two guard cells, specialized epidermal parenchyma cells that open and close the pore. These guard cells regulate transpiration, photosynthesis, and respiration, facilitating gas exchange by opening the stomata and preventing excess water loss by closing them.

H_2O (Water): Water is a reactant of photosynthesis, and six water molecules are needed for each set of photosynthesis reactions. These water molecules move from the soil surrounding the plant, entering the roots through the root hairs and traveling through the xylem in the roots and stem to a leaf. Once inside the leaf, the water molecules move into the chloroplasts inside the palisade mesophyll cells. During the light-dependent reactions that occur in the thylakoids of the chloroplast, the water molecules are split to convert NADP+ to NADPH, to help drive the production of ATP, and to produce oxygen, a byproduct of photosynthesis.

Heterotroph: An organism that consumes **organic** matter created by **autotrophs**. Autotrophs produce organic compounds from inorganic molecules using either **photosynthesis** or **chemosynthesis** (read: chemical reactions using inorganic molecules). Heterotrophs cannot fix carbon and only *use* organic carbon to grow .Heterotrophs cannot make their own organic compounds like plants can. We saw that light bulb go on in your head ;don‹t try to hide it. Yes, humans, and 95% of all other organisms, are **heterotrophic**.

Inner Membrane: Chloroplasts are surrounded by outer and inner phospholipid membranes, which are separated by a narrow intermembrane space. The inner membrane surrounds the stroma, forming a selectively permeable barrier that allows molecules to move into and out of the chloroplast.

Isoprenoids: Also known as terpenes, this is a diverse class of biomolecules. They have in common a five carbon branched molecule as the basic building block.

Leaf: The epidermis is a single layer of epidermal cells that covers all the parts of plants, including roots, stems, leaves, and flowers. Unlike the other plant parts, leaves have two epidermal layers, the upper and lower epidermis, which surround the inner mesophyll. In the epidermis, small pores called stomata allow leaves to exchange gases with the surrounding air. During photosynthesis, carbon dioxide (a reactant) enters the leaf through the stomata and oxygen (a byproduct) exits through the stomata.

Light harvesting complexes (photosystems complexes) - A photosystem (ps) complex is a multi-protein unit in the thylakoid membrane that absorbed light to serve as energy for reactions

Light reactions (light dependent reactions) - The light dependent reactions are chemical reactions requiring electromagnetic energy (light) that occur in the thylakoid membrane of the chloroplast to convert light energy into chemical forms ATP and NAPDH.

Light Reactions, or Light-Dependent reactions: The first stage of photosynthesis, where light energy from the Sun is captured, and with a little help from our good buddy, water (H_2O), changed into chemical energy in the forms of ATP and NADPH. The light-dependent reactions take place in the thylakoid membrane of the chloroplast in plants. They comprise two photosystems, named Photosystem I and Photosystem II.

Light- Light is a form of electromagnetic radiation; the shorter the wavelength the greater amount of energy. Light supplies the energy for the light reactions of photosynthesis.

Light-Dependent Reactions: During the light-dependent reactions that occur in the thylakoids of the chloroplast, the water molecules are split to convert NADP+ to NADPH and to produce oxygen, a byproduct that is released through the leaf's stomata into the surrounding air. Additionally, ATP is formed using ATP synthase, ADP, and a free phosphate.

Light-Independent Reactions (Calvin cycle): In the light-independent reactions (also known as dark reactions or the Calvin cycle) that occur within the stroma of the chloroplast, enzymes, ATP, and NADPH drive the conversion of carbon dioxide into sugars. Three carbon dioxide molecules enter each round of the Calvin cycle, and it takes two rounds to produce a glucose molecule. Two rounds of the Calvin cycle will produce two G3Ps that the plant cell converts into one glucose molecule, or it can use them to make sucrose and other organic molecules. ADP and NADP+ exit the Calvin cycle and return to the thylakoids for the next set of light-dependent reactions.

Light-Independent Reactions, or Dark Reactions: The second stage of **photosynthesis**, where **carbohydrates** from carbon dioxide, or CO_2, are produced using the energy forms (**ATP** and **NADPH**) generated in the first stage of photosynthesis, aka the **light-dependent reactions**. The **light-independent reactions** occur in the **stroma** of the **chloroplast** in **plants.**

Limiting Factor - Carbon dioxide level, temperature and / or light intensity affects the rate of photosynthesis; if it is in short supply it affects plant growth. By increasing these limiting factors it will increase the rate of photosynthesis and plant growth.

Lumen - The lumen is the region within the thylakoid membrane where water is split to obtain oxygen. The oxygen diffuses out of the cell, while the protons remain inside to build positive electrical charge inside the thylakoid.

Lumen: The inner membrane space of a chloroplast. The lumen is also called the **thylakoid lumen.** Protons are pumped into the lumen during the **light-dependent reactions** of **photosynthesis.**

Mesophyll Cell - A mesophyll cell is a type of plant cell located between the upper and lower epidermis that is the site for photosynthesis

Nadph - Nadph is a high-energy electron carrier used in reduction

Nadph: Nicotinamide adenine dinucleotide phosphate, which is a strong reducing agent, or electron donor, as well as a coenzyme. Nadp+ is the common notation for the nonreduced form, and nadph is the name once it has been reduced, or gained an electron. Nadph acts as a carrier of electrons and is used extensively in the synthesis of biological molecules.

O_2 Oxygen: Oxygen molecules are a byproduct of photosynthesis. They are produced during the light-dependent reactions that occur in the thylakoids of the chloroplast. This oxygen gas is released through the leaf's stomata into the surrounding air.

Outer Membrane: Chloroplasts are surrounded by outer and inner phospholipid membranes, which are separated by a narrow intermembrane space. The outer membrane is the external layer, which maintains the chloroplast's shape and protects its inner structures.

Oxidation -Oxidation refers to the loss of electrons

Oxidized: A state of a molecule once it has lost an electron.

Oxidizing Agent: A molecule that accepts electrons and *oxidizes* the molecule that it accepts electrons from.

Oxygen (O_2): Oxygen is a gas that is a product of the light-dependent reactions

P-680: The reaction center of photosystem ii which best absorbs light at a 680 nm wavelength. P-680 is a chlorophyll a molecule.

P-700: The reaction center of photosystem i which best absorbs light at a 700 nm wavelength. P-700 is a chlorophyll a molecule.

Palisade Mesophyll - the palisade meophyill is the area of the mesophyll cell without many air spaces

Pgal - Pgal is an isomer of pga formed during the calvin cycle.

Phloem: Vascular bundles contain xylem and phloem—continuous tubes that transport water, nutrients, and other substances throughout the plant's roots, stem, and leaves. Positive hydrostatic pressure in the phloem moves dissolved sugars and organic compounds from the leaves downward to the stem and roots in a process called translocation.

Photomorphogenesis: The development of a plant cell or organ that is triggered by exposure to light. For example, if a corn plant is germinated in the dark, it will produce an etiolated shoot, lacking chlorophyll or the ability to carry out photosynthesis. Exposing an etiolated shoot to light will initiate development of pigment biosynthesis and the production of cellular components necessary for photosynthesis.

Photons: Photons are units of light energy from the sun that can power photosynthesis in plant cells. Within the chloroplast's thylakoid membranes, pigments—including chlorophyll—absorb photons, initiating the light-dependent reactions of photosynthesis.

The Basic Unit of Light. Although light has some properties consistent with a wave, light from the sun is in individual packages. Each photon is a particle of electromagnetic radiation traveling with the speed of light (3×10^8 m sec^{-1}). The energy in each photon is dependent on the frequency with which the electromagnetic field vibrates.

Photophosphorylation: The coupling of photosystems i and ii with an electron transfer chain that moves electrons from water (which is oxidized to form o_2) to nadp$^+$) which is reduced to form nadph). The transfer of electrons between photosystem ii and photosystem i releases energy, which is conserved in the form of a trans-membrane proton gradient and used to synthesize atp.

Photorespiration: The process where **ribulose-1,5-bisphosphate carboxylase oxygenase (rubisco)**, an enzyme, catalyzes the attachment of oxygen (o2) to the **5-carbon ribulose-1,5-bisphosphate (rubp)** in the **stomata** of plant leaves. The end products are **phosphoglycerate (pga)** and carbon dioxide (co2). Photorespiration is thought to reduce the efficiency of photosynthesis in some plants but is still actively being studied. We'll keep you posted.

Photosynthesis - Photosynthesis is the process by which organisms convert light energy into chemical energy (glucose).

Photosynthesis: Photosynthesis is the process plants use to make their own food, which fuels all their activities. During photosynthesis, light energy from the sun converts groups of six carbon dioxide molecules and six water molecules (the reactants) into one glucose (sugar) molecule and six oxygen molecules (the products).

Photosystem - A photosystem (ps) is a cluster of chlorophyll and other molecules in a thylakoid that harvest the energy of light for photosynthesis.

Photosystem: An array of pigment-protein complexes and electron transfer components that function together to harvest light energy, transfer the energy to photochemical reaction centers, and move the excited electrons in a controlled fashion to produce usable biochemical energy. Each photosystem contains hundreds of chlorophyll and carotenoid molecules functioning as antennae, while only a few chlorophyll molecules are employed in the reaction centers.

Pigment - A pigment is a colored molecule. A pigment absorbs specific wavelengths of light. Chlorophyll absorbs blue and red light and reflects green light, so it appears green.

Proton Gradient: The movement of protons from the **lumen** to the **stroma**, or from high to low concentration, in **chloroplasts**. The proton gradient is exploited to generate **atp** and **nadph**.

Protoporphyrin: An intermediate in the biosynthetic pathways for chlorophyll and heme. This molecule is a strong photosensitizer and causes much cellular damage if it accumulates in tissues exposed to light.

Quenching: The process of causing the de-excitation of a molecule from an excited singlet or triplet state back down to the ground state. Quenching of chlorophyll is an important protective role of carotenoids. If a triplet molecule is not rapidly quenched, it can react with oxygen to generate singlet oxygen. Singlet oxygen can cause much cellular damage.

Rate of photosynthesis - Is how quickly a plant is photosynthesising. Factors such as carbon dioxide levels, temperature and / or light intensity affects the rate of photosynthesis.

Reaction Center: A specialized component of the photosystem that actually carries out the photochemistry with the excitation energy harvested by antenna pigments.

Reducing Agent: A molecule that can donate electrons. A reducing agent *reduces* the molecule that it donates electrons to.

Reduction - Reduction refers to the gain of electrons. It often occurs in conjunction with oxidation.

Reduction: The opposite of oxidation. When a molecule acquires one or two electrons it becomes more reduced.

Rubisco - Rubisco is an enzyme that bonds carbon dioxide with rubp

Rubisco: Ribulose-1,5-bisphosphate carboxylase oxygenase, or the enzyme used in the **light-independent reactions** to fix carbon dioxide (co_2) to the 5-carbon ribulose-1,5-bisphosphate (rubp). Stoma, or stomate (plural stomata): the opening(s) or pore(s) in a leaf that allow the passing of gases into and out of the leaf.

Singlet States: An orbital to which an electron may move after being excited by a photon. Singlet state orbitals require that the excited electron have a spin opposite to that of the spin on the electron remaining in the ground state.

Spongy Mesophyll: in the spongy mesophyll, parenchyma cells are loosely arranged, with spaces between them, and their shape tends to be spherical or stellate. This structure allows the leaf to receive carbon dioxide from the air and to release oxygen and water vapor into the air.

Stomata: Stomata (singular: stoma) are small epidermal pores that allow gases and water vapor to move between interior leaf structures and the surrounding air. During photosynthesis, carbon dioxide (a reactant) enters the leaf through the stomata and oxygen (a byproduct) exits through the stomata.

Stroma: The inner area of a **chloroplast** where the **light-independent reactions** occur. The stroma is the water-based, fluid-filled area of the **chloroplast**

Stroma: The stroma is the thick fluid that fills the space between the inner membrane and the grana of a chloroplast. The light-independent reactions (also known as dark reactions or the calvin cycle) of photosynthesis occur in the stroma. These reactions require carbon dioxide molecules (reactants) from the air surrounding the plant, as well as atp and nadph, energy-storing molecules that are produced by the light-dependent reactions that occur in the thylakoids.

Tetrapyrrole: A class of molecules characterized by a series of four, five-membered rings containing four carbon atoms and one nitrogen. Tetrapyrrole molecules include hemes, chlorophylls, vitamin b_{12} and phytochrome.

Thylakoid membranes: A series of sealed membrane compartments found inside the chloroplast of plant cells. The membranes contain the photosynthetic apparatus and all of the chlorophyll in the leaf. Their structure as closed 'sacks' permits functional 'inside' and 'outside' spaces within the chloroplast across which photophosphorylation can generate the proton gradient necessary for atp synthesis.

Thylakoid - The thylakoid is a disc-shaped portion of chloroplast, found in stacks called grana.

Thylakoids: Chloroplasts contain stacks of disc-shaped thylakoids, where the light-dependent reactions of photosynthesis occur. Each thylakoid consists of a thylakoid membrane and an inner lumen. The thylakoid membrane contains the pigment chlorophyll, which is stored in protein complexes called photosystems i and ii. These chlorophyll molecules absorb light energy from the sun, which the cell converts into chemical energy through a series of chemical reactions.

Transpiration Stream - Is the uninterrupted flow of water through the xylem of the plant, from the roots up to the leaves where it will evaporate.

Transpiration - Is the process of water through the xylem of the plant, from the roots up to the leaves where it changes to vapor and is released to the atmosphere.

Xylem: Vascular bundles contain xylem and phloem—continuous tubes that transport water, nutrients, and other substances throughout the plant's roots, stem, and leaves. The evaporation of water at the leaf's stomata (transpiration) creates negative hydrostatic pressure in the xylem that moves water and dissolved minerals from the roots upward to the stem and leaves.

REFERENCE

Adler K, ArkonaC, Manteuffel R and Siiss K-H (1993) Electron-microscopical localization of chloroplast proteins by immunogold labelling on cryo-embedded spinach leaves. Cell Biol Intern 17: 213-220.

Allen JF, Alexciev K and Hakansson G (1995) Photosynthesis: Regulation by redox signaling. Curr Biol 5: 869-872.

Altekar W and Rajagopalan R (1990) Ribulose bisphosphate carboxylase activity in halophilic archaebacteria. Arch Microbiol 153: 169-174.

Anderson LE (1987) Ribose-5-phosphate isomerase and ribulose5-phosphate kinase show apparent specificity for a specific ribulose 5-phosphate species. FEBS Lett 212: 45-48

Anderson LE and Pacold I (1972) Chloroplast and cytoplasmic enzymes. IV. Pea leaffruc tose 1,6-diphosphate aldolase. Plant Physio149: 393-397.

Anderson LE, Goldhaber-Gordon IM, Li D, Tang X-Y, Xiang M and Prakash N (1995) Enzyme-enzyme interaction in the chloroplast: glyceraldehyde-3-phosphate dehydrogenase, triose phosphate isomerase and aldolase. Planta 196: 245-255.

Anderson LE, Huppe HC, Li AD and Stevens FJ (1996) Identification of a potential redox-sensitive interdomain disulfide in the sedoheptulose bisphosphatase of Chlamydomonas reinhardtii. Plant J 10: 553-560.

Andersson I (1996) Large structures at high resolution: the 1.6 A crystal structure of spinach ribulose-1,5-bisphosphate carboxylase/oxygenase complexed with 2-carboxyarabinitol bisphosphate. J Mol Biol 259: 160-174.

Andralojc PJ, Keys AJ, Martindale W, Dawson GW and Parry MAJ (1996) Conversion of D-hamamelose into 2-c arboxy-Darabinitol and 2-carboxy-D-arabinitol 1-phosphate in leaves of Phaseolus vulgaris L. J Biol Chem 271: 26803-26809.

Andrews TJ, Hudson GS, Mate CJ, Von Caemmerer 5, Evans JR and Arvidsson YBC (1995) Rubisco: The consequences of altering its expression and activation in transgenic plants. J Exptl Bot 46: 1293-1300.

Antia NJ (1967) Comparative studies on aldolase activity in marine planktonic algae and their evolutionary significance. J Phycol 3: 81-84.

Ashton AR (1998a) A simple procedure for purifying the major chloroplast fructose-1,6-bisphosphatase from spinach (Spinacia oleracea) and characterization of its stimulation by subfemtomolar mercuric ions. Arch Biochem Biophys 357: 207224.

Ashton AR (1998b) Sedoheptulose-1,7-bisphosphate phosphatase activity of chloroplast fructose-1,6-bisphosphatase and sedoheptulose-1,7-bisphosphate in stromal extracts from chloroplasts of spinach (Spinacia oleracea). Austr J Plant Physiol 25: 531-537.

Assail N-E, Martin, W and Loiseaux-de Goer S (1991) Evolution of the Rubisco operon from prokaryotes to algae: Structure and analysis of the rbcS gene of the brown alga Pylaiella &torahs. Plant Mol Biol 17: 853-863.

Avilan L, Gontero B, Lebreton S and Ricard J (1997) Memory imprinting effects in multienzyme complexes. I. Isolation, dissociation and reassociation of a pho sphoribulokinaseglyceraldehyde-3-phosphate dehydrogenase complex from Chlamydomonas reinhardtii chloroplasts. Eur J Biochem 246: 74-84.

Baalmann E, Backhausen JE, Kitzmann C and Scheibe R (1994) Regulation of NADP-dependent glyceraldehyde 3-phosphate dehydrogenase activity in spinach chloroplasts. Bot Acta 107: 313-320

Baalmann E, Backhausen JE, Rak C, Vetter S and Scheibe R (1995) Reductive modification and nonreductive activation of purified spinach chloroplast glyceraldehyde-3-phosphate dehydrogenase. Arch Biochem Biophys 324: 201-208.

Baalmann E, Scheibe R, CerffR and Martin W (1996) Functional studies of chloroplast glyceraldehyde-3-phosphate dehydrogenase subunits A and B expressed in Escherichia coli: Formation of highly active A4 and B4 homotetramers and evidence that aggregation of the 134 complex is mediated by the B subunit carboxyterminus. Plant Mol Biol 32: 505-514.

Babadzhanova MA and Bakaeva NP (1987) Quaternary structure and some properties of ribophosphate isomerase from the leaves ofArabidopsis thaliana ofEnkheim race and its mutants triplex and 58/15. Biokhimiya 52: 146-153.

Backhausen JE, Vetter 5, Baalmann E, Kitzmann C and Scheibe R (1998)N.ADtdependent malate dehydrogenase and glyceraldehyde-3-phosphate dehydrogenase isoenzymes play an important role in dark metabolism of various plastid types. Planta 205: 359-366.

Baier D and Latzko E (1975) Properties and regulation of C- 1 - fructose-1,6-diphosphatase from spinach chloroplasts. Biochim Biophys Ac to 396: 141-148.

Bailey RW and Staehelin LA (1968). The chemical composition of isolated cell walls of Cyanidium caldarium. J Gen Microbiol 54: 269-276.

Banks FM, Driscoll SP, Parry MAJ, Lawlor DW, Knight JS, Gray JC and Paul MJ (1999) Decrease in phosphoribulokinase activity by antisense RNA in transgenic tobacco. Relationship between photosynthesis, growth, and allocation at different nitrogen levels. Plant Physiol 119: 1125-1136.

Bassham and Calvin (1957) The path of carbon in photosynthesis. Prentice-Hall, Englewood Cliffs

Beh M, Strauss G, Huber R, Stetter KO and Fuchs G (1993) Enzymes of the reductive citric acid cycle in the autotrophic eubacterium Aquifeacpyrophilus and in the archaebacterium Thermoproteus neutrophilus. Arch Microbiol 160: 306-311 Berger C and Feierabend J (1967) Plastidenentwicklung and Bildung von Photosynthese-Enzymen in etiolierten Roggenkeimlingen. Physiol Veg 5: 109-122.

Bernacc hia G, Schwan G, Lottspeich F, Salamini F and Bartels D (1995) The transketolase gene family of the resurrection plant Craterostigma plantagineum: differential expression during the rehydration phase. EMBO J 14: 610-618.

Bernstein BE, Michels PA and Hol and WG (1997) Synergistic effects of substrate-induced conformational changes in phosphoglycerate kinase activation. Nature 385: 275-278.

Bertsch U, Schlicher TB, Schroder I and Soil J (1993) Sequence of mature phosphoglycerate kinase from spinach chloroplasts. Plant Physiol 103: 1449-1450.

Besford RT (1990) The greenhouse effect: Acclimation of tomato plants growing in high CO2, relative changes in Calvin cycle enzymes. J Plant Physiol 136: 458-463.

Beuf L, Bedu 5, Durand MC and Joset F (1994) A protein involved in co-ordinated regulation of inorganic carbon and glucose metabolism in the facultative photoautotrophic cyanobacterium Synechocystis PCC6803. Plant Mol Biol 25: 855-864.

Biesecker G, Harris JI, Thierry JC, Walker JE and Wonacott AJ (1 977) Sequence and structure of D-glyceraldehyde3-phosphate dehydrogenase from Bacillus stearothermophilus. Nature 266: 328-333.

Bilger W, Fisahn J, Brummet W, Kossmann J and Willmitzer L (1995) Violaxanthin cycle pigment contents in potato and tobacco plants with genetically reduced photosynthetic capacity. Plant Physiol 108: 1479-1486.

Bilker M, Schiinemann D and Borchert 5 (1998) Enzymic properties and capacities of developing tomato (Lycopersicon esculentum L.) fruit plastids. J Exp Bot 49: 681-691.

Blom N and Sygusch J (1997) Product binding and role of the C-terminal region in class I D-fructose 1,6-bisphosphate aldolase. Nat Struct Biol 4: 36-39.

Boldt R, Burner T and Schnarrenberger C (1992) Repression of the plastidic isoenzymes of aldolase, 3-phosphoglycerate kinase, and triosephosphate isomerase in the barley mutant `albostrians.' Plant Physiol 99: 895-900.

Boldt R, Koshuchova S , Gross W, B Omer T and Schnarrenberger C (1997) Decrease in glycolate pathway enzyme activities in plastids and peroxisomes of the albostrians mutant of barley (Hordeum vulgare L.). Plant Sci Lett 124: 33-40.

Boldt R, Pelzer-Reith B, Burner T and Schnarrenberger C (1994) Aldolases in barley (Hordeum vulgare L.): Properties and repression of the plastid enzyme in the plastome mutant `albostrians.' J Plant Physiol 144: 282-286.

Bommer D, Schfferjohann J and Bowien B (1996) Identification of cbbBc as an additional distal gene of the chromosomal cbb CO_2 fixation operon fromRaistonia eutropha. Arch Microbiol 166: 245-251.

Bradbeer JW (1969) The activities of the photosynthetic carbon cycle enzymes of greening bean leaves. New Phytol 68: 233245.

Bradbeer JW (1971) Plastid development in primary leaves of Phaseolus vulgaris. The effects of short blue, red, far-red and white light treatments on dark-grown plants. J Exp Bot 22: 382-390.

Bradbeer JW and B Omer T (1978) Activities of glyceraldehydephosphate dehydrogenase (N AD P+) and phosphoribulokinase in two barley mutants deficient in chloroplast ribosomes. In: Akoyunoglou G and Argyroundi-Akoyunoglou JH (eds) C hloroplas t Devel opmen t, pp 727-732.

Bradbeer JW, Atkinson YE, Borner T and Hagemann R (197 9) Cytoplasmic synthesis of plastid polypeptides may be controlled by plastid synthesized RNA. Nature 279: 816-817.

Bradbeer JW, Gyldenholm AO, Wallis ME and Whatley FR (1969) Studies on the biochemistry of c hloroplast development. Progr Photosynth Res 1: 272-279.

Brandes HK, Hartman FC, Lu T-YS and Larimer FW (1996a) Efficient expression of the gene for spinach phosphoribulokinase in Pichia pastoris and utilization of the recombinant enzyme to explore the role of regulatory cysteinyl residue by site-directed mutagenesis. J Biol Chem 271: 64906496.

Brandes HK, Larimer FW and Hartman FC (1996b) The molecular pathway for the regulation of phosphoribulokinase by thioredoxinf J Biol Chem 271: 3333-3335.

Brandt P and Wilhelm C (1990) The light-harvesting system of Euglena gracilis during the cell cycle. Planta 180: 293-296.

Breazeale VD, Buchanan BB and Wolosiuk RA (197 8) Chloroplast sedoheptulose 1,7-bisphosphatase: Evidence for regulation by the ferredoxin/thioredoxin system. Z Naturforsch 33c: 521-528.

Briiggemann W, Klaucke S and Maas-Kantel K (1994) Longterm chilling of young tomato plants under low light. V. Kinetic and molecular properties of two key enzymes of the Calvin cycle in Lycopersicon esculentum Mill, and L. peruvianum Mill. Planta 194: 160-168.

Bringloe DH, Rao SK, Dyer TA, Raines CA and Bradbeer JW (1996) Differential genee xpression of chloroplast and cytosolic phosphoglycerate kinase in tobacco. Plant Mol Biol 30: 637640.

Brinkmann H and Martin W (1996) Higher plant chloroplast and cytosolic 3-phosphoglyceratekinases: A case of end osymbiotic gene replacement. Plant Mol Biol 30: 65-75

Brooks K and Criddle RS (1966) Enzymes of the carbon cycle of photosynthesis I. Isolation and properties of spinach chloroplast aldolase. Arch Biochem Biophys 117: 650-659.

Buchanan BB (1980) Role of ligh tin the regulation of chloroplast enzymes. Ann Rev Plant Physiol 31: 341-374.

Buchanan BB (1991) Regulation ofCO2 assimilation in oxygenic photosynthesis: the ferredoxin/thioredoxin system. Arch Biochem Biophys 288: 1-9.

Bukowiecki AC and Anderson LE (1974) Multiple forms of aldolase and triosephosphate isomerase in diverse plant species. Plant Sci Lett 3: 381-386.

Bult CJ, White 0, Olsen GJ, Zhou L, Fleischmann RD (35 other coauthors) and Venter JC (1996) Complete genome sequence of the methanogenic Archeon, Methanococcus jannaschii. Science 273: 1058-1073.

Burner T, Schumann B and Hagemann R (1976) Biochemical studies on plastid ribosome-deficient mutant of Hordeum vulgare. In: BucherT, Neupert W,Sebald W and Wern S(eds) Genetics and Biogenesis ofChloroplasts and Mitochondria, pp 42-48. Elsevier/North-Holland Biomedical Press, Amsterdam Bovarnick JG, Schiff JA, Freedman Z and Egan JM (1974) Events surrounding the early development of Euglena chloroplasts: Cellular origins of chloroplast enzymes in Euglena. J Gen Microbiol 83: 63-71.

Cadet F and Meunier J-C (1988a) Spinach chloroplast (Spinacia oleracea) sedoheptulose-1,7-bisphosphatase. Biochem J 253: 243-248.

Cadet F and Meunier J-C (1988b) pH and kinetic studies of chloroplast sedoheptulose-1,7-bisphosphatase from spinach (Spinacia oleracea). Biochem J 253: 249-254.

Cadet F, Meunier JC and Ferte N (1987) Isolation and purification of chloroplastic spinach (Spinacia oleracea) sedoheptulose1,7-bisphosphatase. Biochem J 241: 71-74.

Camara B, Hugueny P, Bouvier F, Kuntz M and Moneger R (1995) Biochemistry and molecular biology of chromoplast development. Int Rev Cytol 163: 175-247.

Cerff R (1973) Glyceraldehyde 3-phosphate dehydrogenases and glyoxylate reductase. I. Their regulation under continuous red and far red light in the cotyledons of Sinapis alba L. Plant Physiol 51: 76-81.

Cerff R (1978a) Glyceraldehyde-3-phosphate dehydrogenase (NADP) from Sinapis alba: steady state kinetics. Phytochemistry 17: 2061-2067.

Cerff R (1978b) Glyceraldehyde-3-phosphate dehydrogenase (NADP) from Sinapis alba L. NAD(P)-induced conformation changes of the enzyme. Eur J Biochem 82: 45-53.

Cerff R and Kloppstech K (1982) Structural diversity and differential light control of mRNAs coding for angiosperm glyceraldehyde-3-phosphate dehydrogenases. Proc Natl Ac ad Sci USA 79: 7624-7628.

Cerff R and Quail PH (1974) Glyceraldehyde 3-phosphate dehydrogenases and glyoxylate reductase. II. Far red light-dependent development of glyceraldehyde 3-phosphate dehydrogenase isoenzyme activities in Sinapis alba cotyledons. Plant Physiol 54: 100-104.

Chan RL, Keller M, Canaday J, Weil J-H and Imbault P (1990) Eight small subunits of Euglena ribulose-1,5-bisphosphate carboxygenase/oxygenase are translated from a large mRNA as a polyprotein. EMBO J 9: 333-338.

Chao 5, Raines CA, Longstaff M, Sharp PJ, Gale MD and Dyer TA (1989) Chromosomal location and copy number in wheat and some of its close relatives of genes for enzymes involved in photosynthesis. Mol Gen Genet 218: 423-430.

Charles SA and Halliwell B (1980) Properties of freshly purified and thiol-treated spinach chloroplast fructose bisphosphatase. Biochem J 185: 689-693.

Charlet T, Moore BD and Seemann JR (1997) Carboxyarabinitol 1-phosphate phosphatase from leaves of Phaseolus vulgaris and other species. Plant Cell Physiol 38: 511-517.

Chen YR , Hartman FC, Lu TY, Larimer FW (1998) D-Ribulose5-phosphate 3-epimerase: cloning and heterologous expression of the spinach gene, and purification and characterization of the recombinant enzyme. Plant Physiol 118: 199-207.

Chen Z, Silva H and Klessig DF (1993) Active oxygen species in the induction of plant systemic acquired resistance by salicylic acid. Science 262: 1883-1886.

Chory J, Chatterjee M, Cook RK, Elich T, Fankhauser C, Li J, Nagpal P, Neff M, Pepper A, Poole D, Reed J and Vitart V (1996) From seed germination to flowering, light controls plant development via the pigment phytochrome. Proc Natl Acad Sci USA 93: 12066-12071.

Clasper 5, Chelvaraj an RE, Easterby JS and Powls R (1994) Isolation of multiple dimeric forms of phosphoribulokinase from an alga and a higher plant. Biochim Biophys Acta 1209: 101-106.

Clasper 5, Easterby JS and Powls R (1991) Properties of two high-molecular-mass forms of glyceraldehyde-3-phosphate dehydrogenase from spinach leaf, one of which also possesses latent phosphoribulokinase activity. Eur J Biochem 202: 12391246.

Clermont 5, Corbier C, Mely Y, Gerard D, Wonacott A and Branlant G (1993) Determinants of coenzyme specificity in glyceraldehyde-3-phosphate dehydrogenase: Role of the acidic residue in the fingerprint region of the nucleotide binding fold. Biochemistry 32: 10178-10184.

Conley TR and Shih M-C (1995) Effects of light and chloroplast functional state on expression of nuclear genes encoding chloroplast glyceraldehyde-3-phosphate dehydrogenase in hypocotyl (hy) mutants and wild-type Arabidopsis thaliana. Plant Physiol 108: 1013-1022.

Conley TR, Park S-C, Kwon H-B, Peng H-P and Shih M-C (1994) Characterization of cis-acting elements in light regulation of the nuclear gene encoding the A subunit of chloroplast isozyme of glyceraldehyde-3-phosphate dehydrogenase from Arabidopsis thaliana. Mol Cell Biol 14: 25252533.

Cooper SJ, Leonard GA, McSweeney SM, Thompson AW, Naismith JH, Qamar 5, Plater A, Berry A and Hunter WN (1996) The crystal structure of a class II fructose-1,6-bisphosphate aldolase shows a novel binuclear metal-binding active site embedded in a familiar fold. Structure 4: 1303-1315.

Criqui MC, Durr A, Marbach J, Fleck J and Jamet E (1992) How are photosynthetic genes repressed in freshly-isolated mesophyll protoplasts of Nicotiana sylvestris? Plant Physiol Biochem 30: 597-601.

Delwiche CF and PalmerJD (1996) Rampant horizontal transfer and duplication of Rubisco genes in eubacteria and plastids. Mol Biol Evol 13: 873-882.

Dennis ES, Gerlach WL, Walker JC, Lavin M and Peacock WJ (1988) Anaerobically regulated aldolase gene of maize. J Mol Biol 202: 759-767.

Derocher EJ, Quigley F, Mache R and Bohnert HJ (1993) The six genes of the Rubisco small subunit multigene family from Mesembryanthemum crystallinum, a facultative CAM plant. Mol Gen Genet 239: 450-462.

Dewdney J, Conley TR, Shih M-C and Goodman HM (1993) Effects of blue and red light on the expression of the nuclear genes encoding chloroplast glyceraldehyde-3-phosphate dehydrogenase of Arabidopsis thaliana. Plant Physiol 103: 1115-1121.

Dockerty A and Merret MJ (1979) Isolation and enzymatic characterization of Euglena proplastids. Plant Physiol 63: 468-473.

Donath M, Mendel R, Cerff R and Martin W (1995) Introndependent transient expression of the maize GapA 1 gene. Plant Mol Biol 28: 667-676.

Douglas SE and Dunford DG (1989) The small subunit of ribulose-1,5-bisphosphate carboxylase is plastid-encoded in the chlorophyll c-containing alga Cryptomonas phi. Plant Mol Biol 13: 13-20.

Dunford RP, Catley MA, Raines CA, Lloyd JC and Dyer TA (1998) Purification of active chloroplast sedoheptulose-1,7- bisphosphatase expressed in Escherichia coli. Protein Expr Purif 14: 139-145.

Eckardt NA, Snyder GW, Portis AR Jr and Ogren WL (1997) Growth and photosynthesis under high and low irradiance of Arabidopsis thaliana antisense mutants with reduced ribulose1,5-bisphosphatecarboxylase/oxygenase activase content. Plant Physiol 113: 575-586.

Effects °flight intensity on photosynthetic carboxylativephase enzymes and chlorophyll synthesis in greening leaves of Hordeum vulgare L. Plant Physiol 41: 913-918.

Ehleringer JR, Sage RF, Flanagan LB and Peaivy RW (1991) Climate change and the evolution ofe4 photosynthesis. Trends Ecol Evol 6: 95-99.

Eklund H, Gleason FK and Holmgren A (1991) Structural and functional relations among thioredoxins of different species. Proteins: Struct Funct Gen 11: 13-28.

ErasoJM and Kaplan S (1994) prrA, aputative response regulator involved in oxygen regulation of photosynthesis gene expression in Rhodobacter sphaeroides. J Bacteriol 176: 3243.

Ernes MJ and Fowler MW (1978) Location of transketolase and transaldolase in apical cells of pea roots. Biochem Soc Transactions 6: 203-205.

Escoubas JM, Lomas M, LaRoche J and Falkowski PG (1995) Light intensity regulation of cab gene transcription is signaled by the redox state of the plastoquinone pool. Proc Natl Acad Sci USA 92: 10237-10241.

Evans JR, Von Caemmerer 5, Setchell BA and Hudson GS (1994) The relationship between CO2 transfer conductance and leafanatomy in transgenic tobacco with a reduced content of Rubisco. Austr J Plant Physiol 21: 475-495.

Fagan T, Hastings JW and Morse D (1998) The phylogeny of glyceraldehyde-3-phosphate dehydrogenase indicates lateral gene transfer from cryptomonads to dinoflagellates. J Mol Evol 47: 633-639.

Fahey RC, Hunt JS and Windham GC (1977) On the cysteine and cystine content ofproteins. Differences between intracellular and extracellular proteins. J Mol Evol 10: 155-160.

Falcone DL and Tabita FR (1991) Expression of endogenous and foreign ribulose-1,5-bisphosphate carboxylase/oxygenase (R uB i sC 0) genes in a RuBi sCO dele ti on m utan t of Rhodobacter sphaeroides. J Bacteriol 173: 2099-2108.

Falkenstein E, von Schaewen A and Scheibe R (1994) Full-length cDNA sequences for both ferredoxin-thioredoxin reductase subunits from spinach (Spinaciaoleracea L.) Biochim Biophys Acta 1185: 252-254.

Faske M, Holtgrefe 5, Ocheretina 0, Meister M, Backhausen JE and Scheibe R (1995) Redox equilibria between the regulatory thiols of light/dark-modulated chloroplast enzymes and dithiothreitol: Fine-tuning by metabolites. Biochim Biophys Acta 1247: 135-142.

Feierabend J (1966) Enzymbildung in Roggenkeimlingen warend der Umstellung von heterotrophem auf autotrophes Wachstum. Planta 71: 326-355.

Feierabend J (1979) Role of cytoplasmic protein synthesis and its coordination with the plastidic protein synthesis in the biogenesis of chloroplasts. Ber Deutsch Bot Ges 92: 553-574 Feierabend J (1986) Investigation of the site of synthesis of chloroplastic enzymes of nitrogen metabolism by the use of heat-treated 70S ribosome-deficient rye leaves. Physiol Plant 67: 145-150.

Feierabend J and Pirson A (1966) Die Wirkung des Lichts and die Bildung von Photosyntheseenzymen in Roggenkeimlingen. Z Pflanzenphysiol 55: 235-245.

Feierabend J and Schrader-Reichhardt U (1976) Biochemical differentiation of plastids and other organelles in rye leaves with a high-temperature-induced deficiency of plastid ribosomes. Planta 129: 133-145.

FeierabendJ and Gringel G (1983) Plant transke tol ase: subcellular distribution, search for multiple forms, site of synthesis. Z Phanzenphysiol 110: 247-258.

Ferri G, Stoppini MC, Meloni ML, Zapponi MC and Iadarola M (1990) Chloroplast glyceraldehyde-3 -phosphate dehydrogenase (NADP): Amino acid sequence of the subunits from isoenzyme I and structural relationship with isoenzyme II. Biochem Biophys Acta 1041: 36-42.

Fichtner K, QuickWP, Schulze ED, Mooney HA, Rodermel SR, Bogorad L and Stitt M (1993) Decreased ribulose-1,5-bisphosphate carboxylase-oxygenase in transgenic tobacco transformed with `antisense' rbcS. Planta 190: 1-9.

Fickenscher K and Scheibe R (1988) Limited proteolysis of inactive tetrameric chloroplast NADP-malate dehydrogenase produces active dimers. Arch Biochim Biophys 260: 711-779.

Fisahn J, Kossmann J, Matzke G, Fuss H, BilgerW and Willmitzer L (1995) Chlorophyll fluorescence quenc hing and violaxanthin deepoxidation of FBPase antisense plants at low light intensities and low temperatures. Physiol Plant 95: 1-10.

Fischer KH and Latzko E (1979) Chloroplastribulose-5-phosphate kinase: light-mediated activation and detection of both soluble and membrane-associated activity. Biochem Biophys Res Commun 89: 300-306.

Flechner A, DreBen U, WesthoffP, HenzeK, SchnarrenbergerC and Martin W (1996) Molecular characterization of transketolase (EC 2.2.1.1) active in the Calvin cycle of spinach chloroplasts. Plant Mol Biol 32: 475-484.

Flechner A, Gross W, Martin W and Schnarrenberger C (1999) Chloroplast class I and class II aldolases are bifunctional for fructose-1,6-bisphosphate and sedoheptulose-1,7-bisphosphate cleavage in the Calvin cycle. FEBS Lett 447: 200-202.

Fluhr R, Moses P, Morelli G, Coruzzi G and Chua N-H (1986) Expression dynamics of the pea rbcS multigene family and organ distribution of the transcripts. EMBO J 5: 2063-2072.

Fothergill-Gilmore LA and Michels PAM (1993) Evolution of glycolysis. Progr Biophys Mol Biol 59: 105-238.

Fourcroy P (1986) Phytochrome control of gene expression in radish (Raphanus sativus) seedlings: III. Evidence for a rapid control of the ribulose-1,5-bisphosphate carboxylase small subunit gene expression by red light. Plant Sci 44: 183-190 FrederickJF (1968) Thepolyglucoside and enzymes of Cyanidiurn caldarium. Phytochemistry 7, 1573-1576.

Fridlyand LE, Backhausen JE, Holtgrefe 5, Kitzmann C and Scheibe R (1997) Quantitative evaluation of the role of 3- phosphoglycerate reduction in chloroplasts. Plant Cell Physiol 38:1177-1186.

Fritz CC, Wolter FP, Schenkemeyer V, HergetT and Schreier PH (1993) The gene family encoding the ribulose-(1,5)- bisphosphatecarboxylase/oxygenase (Rubisco)small subunit of potato. Gene 137: 271-274.

Frommer WB and Sonnewald U (1995) Molecular analysis of carbon partitioning in solanaceaous species. J Exp Bot 46: 587-730.

fructose during catabolism of sucrose and mannitol in halophilic archaebacteria. Arch Microbiol 158: 356-363.

Furbank RT, Chitty JA, Von Caemmerer S and Jenkins CLD (1996) Antisense RNA inhibition of RbcS gene expression reduces rubisco level and photosynthesis in the C4 plant Flaveria bidentis. Plant Physiol 111: 725-734.

Gardemann A, Stitt M and Heldt HW (1983) Control of CO2 fixation. Regulation of spinach ribulose-5-phosphate kinase by stromal metabolite levels. Bioc him Biophys Acta 722: 5160.

Gatenby AA and Viitanan PV (1994) Structural and functional aspects of chaperonin-mediated protein folding. Ann Rev Plant Physiol Plant Mol Biol 45: 469-491.

Geiger DR and Servaites JC (1994) Diurnal regulation of photosynthetic carbon metabolism in C3 plants. Annu Rev Plant Physiol Plant Mol Biol 45: 235-256.

Gerbling KP , Steup M and Latzko E (1986) Fructose 1,6-bisphosphatase form B from Synechococcus leopoliensis hydrolyzes both fructose and sedoheptulose bisphosphate. Plant Physiol 80: 716-620.

Gibson JL and Tabita FR (1997) Analysis of the cbbXYZ operon in Rhodobacter sphaeroides. J Bacteriol 179: 663-669 .

Gilbert HF (1984) Redox control of enzyme activities by thiol/ disulfide exchange. Methods Enzymol 107: 330-351 Gilmartin PM, Memelink J, Hiratsuka K, Kay SA and Chua NH (1992) Characterization of a gene encoding a DNA binding protein with specificity for a light-responsive element. Plant Cell 4: 839-849.

Gilmartin PM, Sarokin L, Memelink J and Chua N-H (1990) Molecular light switches for plant genes. Plant Cell 2: 369378.

Goloubinoff P, Christeller JT, Gatenby AA and Lorimer GH (1989) Reconstitution of active dimeric ribulose bisphosphate c arboxylase from an unfolded state depends on two c haperonin proteins and Mg-ATP. Nature 342: 884-889.

Gontero B, C6rdenas ML and Ricard J (1988) A functional five-enzyme complex of chloroplasts involved in the Calvin cycle. EurJ Biochem 173: 437-443.

Gontero B, Guidici-Orticoni MT and Ricard J (1994) The modulation of enzyme reaction rates within multi-enzyme complexes. 2. Information transfer within a Chloroplast multi-enzyme complex containing ribulose bisphosphate carboxylaseoxygenase. EurJ Biochem 226: 999-1006.

Gontero B, Mulliert G, Rault M, Giudici-Orticoni MT and Ric ard J (1993) Structural and functional properties of a mul ti-enzyme complex from spinach chloroplasts: 2. Modulation of the kinetic properties of enzymes in the aggregated state. Eur J Biochem 217: 1075-1082.

Gottlob-McHugh SG, Sang wan RS, Blakely SD, Venleberghe GC, Ko K, Turpin DH, Plaxton WC, Mild BL and Dennis DT (1992) Normal growth of transgenic tobacco plants in the absence of cytosolic pyruvate kinase. Plant Physiol 100: 820825.

Graham D, Grieve AM and Smillie R (1968) Phytochrome as the primary p ho to regulator of the sy nthesis of Calvin cycle enzymes in etiolated pea seedlings. Nature 218: 89-90.

Gray JC, Paul MJ, Barnes SA, Knight JS , Loynes A, Habash D, Parry MAJ and Lawlor DW (1995) Manipulation of phosphoribulokinase and phosphate translocator activities in transgenic tobacco plants. J Exp Bot 46: 1309-1315.

Grisson FE and Kahn JS (1974) Glyceraldehyde-3-phosphate dehydrogenases from Euglena gracilis. Purification and physical and chemical characterization. Arch Biochem Biophys 171: 444-458.

Gross W and SchnarrenbergerC (1995) Heterotrophic growth of two stains of the acidothermophilic red alga Galdieria sulphuraria. Plant Cell Physiol 36: 633-638.

Gross W, Bayer MG, SchnarrenbergerC, Gebhart UB, Maier TL and Schenk HEA (1994) Two distinct aldolases of class II type in the cyanoplasts and in the cytosol of the alga Cyanaphora paradoxa. Plant Physiol 105: 1393-1398.

Gross W, Lenze D, Nowitzki U, Weiske J, Schnarrenberger C (1999) Characterization, cloning, and evolutionary history of the Chloroplast and cytosolic class I aldolases of the red alga Galdieria sulphuraria. Gene 230: 7-14.

Grotjohann N (1997) Activation of Chloroplast fructose-1,6- bisphosphatase of Pisum sativum by mole mass change. Bot Acta 110: 323-327.

Gunasekera D and Berkowitz GA (1993) Use of transgenic plants with ribulose-1,5-bisphosphate carboxylase/oxygenase antisense DNA to evaluate the rate limitation of photosynthesis under water stress. Plant Physiol 103: 629-635.

Gutteridge S and Gatenby AA (1995) Rubisco synthesis, assembly, mechanism and regulation. Plant Cell 7: 809-819.

Haake V, Zrenner R, Sonnewald U and Stitt M (1998) A moderate decrease of plastid aldolase activity inhibits photosynthesis, alters the levels of sugars and starch, and inhibits growth of potato plants. Plant J 14: 147-157.

Hagemann R and Scholz F (1962) Ein Fall geninduzierter Mutationen des Plasmotypus bei Gerste. Der Ziichter 32: 5058.

Hahn D, Bennoun P and Kiick U (1996) Altered expression of nuclear genes encoding chloroplast polypeptides in non-photosynthetic mutants of Chlamydomonas reinhardtii: Evidence for post-transcription regulation. Mol Gen Genet 252:362-370.

Hahn D, Kaltenbach C and Kiick U (1998) The Calvin cycle enzyme sedoheptulose-1,7-bisphosphatase is encoded by a light-regulated gene in Chlamydomonas reinhardtii. Plant Mol Biol 36: 929-934.

Harbinson J and Foyer CH (1991) Relationships between the efficiencies of Photosystem I and II and stromal redox state in CO2-free air. Plant Physiol 97: 41-49.

Harbinson J, Genty B and Foyer CH (1990) Relationship between photosynthetic electron transport and stromal enzyme activity in pea leaves. Plant Physiol 94: 545-553.

Hariharan Tara, Johnson PJ and Cattolico RA (1998) Purification and characterization of phosphoribulokinase from the marine chromophytic alga Heterosigma carterae. Plant Physiol 117: 321-329.

Harris GC and Koniger M (1997) The 'high' concentrations of enzymes within the chloroplast. Photosynth Res 54: 5-23 Harrison DHT, Runquist JA, Holub A and Miziorko HM (1998) The crystal structure of phosphoribulokinase from Rhodobacter sphaeroides reveals a fold similar to that of adenylate kinase. Biochemistry 37: 5074-5085.

Harrison EP, Willingham NM, Lloyd JC and Raines CA (1998) Reduced sedoheptulose-1,7-bisphosphatase levels in transgenic tobacco lead to decreased photosynthetic capacity and altered carbohydrate accumulation. Planta 204: 27-36.

Hatch MD (1997) Resolving C4 photosynthesis: Trials, tribulations and other unpublished stories. AustJ Plant Physiol 24: 413-422.

He Z, Von Caemmerer 5, Hudson GS, Price GD, Badger MR and Andrews TJ (1997) Ribulose-1,5-bisphosphate carboxylase/ oxygenase activase deficiency delays senescence of ribulosel,5-bisphosphate carboxylase/oxygenase but progressively impairs its catalysis during tobacco leaf development. Plant Physiol 115: 1569-1580.

Heber U and Walker D (1992) Concerning a dual function of coupled cyclic electron transport in leaves. Plant Physiol 100: 1621-1626.

Heber U, Hallier UW and Hudson MA (1967) Untersuchungen zur intralzelluthren Verteilung von Enzymen und Substraten in der Blattzelle. II. Lokalisation von Enzymen des reduktiven und dem oxidativem Pentosephosphat-Zyklus in den Chloroplasten und Permeabilitat der Chloroplasten-Membran gegeniiber Metaboliten. Z Naturforschg 22b: 1200-1215.

Heber U, Pon NG and Heber M (1963) Localization of carboxydismutase and triosephosphate dehydrogenases in chloroplasts. Plant Physiol 38: 355-360.

Hensel R, Laumann 5, Lang J, Heumann H and Lottspeich F (1987) Characterisation of two D-glyceraldehyde-3 -phosphate dehydrogenases from the thermophilic archaebacterium Thermoproteus tenax. EurJ Biochem 170: 325-333.

Henze K (1997) Molekulare Studien zu den Enzymen des CalvinZyklus aus Spinacia oleracea. Ph.D. thesis. Technische Universitat Braunschweig.

Henze K, Morrison HG, Sogin ML, Muller M (1998) Sequence and phylogenetic position of a class II aldolase gene in the amitochondriate protist, Giardia lamblia. Gene 222: 163-168.

Hermoso R, Castillo M, Chueca A, Lazaro JJ, Sahrawy M and Lopez-Gorge J (1996) Binding site on pea chloroplast fructose1,6-bisphosphatase involved in the interaction with thioredoxin. Plant Mol Biol (1996 30: 455-465).

Hess WR, Linke B and Borner T (1997) Impact of plastid differentiation on transcription of nuclear and mitochondrial genes. In: Schenk HEA, Herrmann RG, KW Jeon, Muller NE, Schwemmler W (eds) Eukaryotism and Symbiosis, pp 233242.

Hess WR, Miller A, Nagy F and Borner T (1994) Ribosome-deficient plastids affect transcription of light-induced nuclear genes: Genetic evidence for a plastid-derived signal. Mol Gen Genet 242: 305-312.

Hess WR, Schendel R, Borner T and Rudiger (1991) Reduction in mRNA level for two nuclear encoded light regulated genes in the barley mutant `albostrians' is not correlated with phytochrome content and activity. J Plant Physiol 138: 292298.

Himmo SD, Thomson M and Gubler CJ (1988) Isolation of transketolase from human erythrocytes. Prep Biochem 18: 261-276.

Hirasawa M, Schiirmann P, Jacquot J-P, Manieri W, Jacquot P, Keryer E, Hartman FC and Knaff DB (1999) Oxidation-reduction properties of chloroplast thioredoxins, ferredwdnthioredoxin reductase, and thioredoxinf-regulated enzymes. Biochemistry 38: 5200-5205.

Hoefgen R, Streber WR and Pohlenz HD (1995) Antisense gene expression as a tool for evaluating molecular herbicide targets. Pesticide Science 43: 175-177.

Holfgrefe 5, Backhausen JE, Kitzmann C, and Scheibe R (1997) Regulation of steady-state photosynthesis in isolated intact chloroplasts under constant light: Responses of carbon fluxes, metabolite pools and enzyme-activation states to changes of electron pressure. Plant Cell Physiol 38: 1207-1216.

Houlne G and Schantz R (1988) Characterization of cDNA sequences for LHCI apoprotein in Euglena gracilis: The mRNA encodes a large precursor containing several consecutive polypeptides. Mol Gen Genet 213: 479-486.

Houlne G and Schantz R (1993) Expression of polyproteins in Euglena. CM Rev Plant Sci 12: 1-17

Hove-Jensen B and Maigaard M (1993) Escherichia coli rpiA gene encoding ribose phosphate isomerase A. J Bacteriol 175: 5628-5635.

Hovenkamp-Obbema R and S tegwee D (1974) Effect of chloramphenicol on the developmentofproplastidsinEuglenagracilis. I. Thesynthesisofribulosediphosphate carboxylase, NADP-linked glyceraldehyde-3-phosphate dehydrogenase and aminolaevulinate dehydratase. Z Pflanzenphysiol 73: 430438.

Hudson GS, Evans JR, Von Caemmerer 5, Arvidsson YBC and Andrews TJ (1992) Reduction of ribulose-1,5-bisphosphate carboxylase/oxygenase content by antisense RNA reduces photosynthesis in transgenic tobacco plants. Plant Physiol 98: 294-302.

Hughes J, Lamparter T, Mittmann F, Hartmann E, Gartner W, Wilde A and Borner T (1997) A prokaryotic phytochrome. Nature 386: 663-663.

Ireland HMM and Bradbeer JW (197 1) Plastid development in primary leaves of Phaseolus vulgaris. The effects of D-threoand L-threo-chloramphenicol on the light-induced formation of enzymes of the photosynthetic carbon pathway. Planta 96: 254-261.

Ishii M, Miake T, Satoh T, Sugiyama H, KodamaT and Igarashi Y (1996) Autotrophic carbon dioxide fixation in Acidianus brierleyi. Arch Microbiol 166: 368-371.

Jac quotJ-P, Lopez-Jaramillo J, Miginiac-Maslow M, Lemaire 5, Cherfils J, Chueca A and Lopez-Gorge J (1997a) Cysteine-153 is required for redox regulation of pea chloroplast fructose1,6-bisphosphatase. FEBS Lett 401: 143-147.

Jacquot J-P, Lopez-Jaramillo J, Chueca A, Cherfils J, Lemaire 5, Chedozeau B, Miginiac-Maslow M, Decottignies P, Wolosiuk RA and Lopez-Gorge J (1995) High-level expression of recombinant pea chloroplast fructose-1,6-bisphosphatase and mutagenesis of its regulatory site. Eur J Biochem 229: 675681.

Jebanathiraj ahJA and Coleman JR (199 8) Association of carbonic anhydrase with a Calvin cycle enzyme complex in Nicotiana tabacum. Planta 204: 177-182.

Jiang CZ and Rodermel SR (1995) Regulation of photosynthesis during leaf development in RbcS antisense DNA mutants of tobacco. Plant Physiol 107: 215-224.

Jiang CZ, Quick WP, Aired R, Kliebenstein D and Rodermel SR (1994) Antisense RNA inhibition of Rubisco activase expression. Plant J 5: 787-798.

Jones PG, Lloyd JC and Raines CA (1996) Glucose feeding of intact wheat plants represses the expression of a number of Calvin cycle genes. Plant Cell Environ 19: 231-236.

Joshi BM and Tabita FR (1996) A global two component signal transduction system that integrates the control of photosynthesis, carbon dioxide assimilation, and nitrogen fixation. Proc Natl Acad Sci USA 93: 14515-14520.

Kacser H and Bums JA (1973) The control of flux. Symp Soc Exp Biol 27: 65-104.

Kagaya K, Nakamura H, Hidaka 5, Ejiri S-I and Tsutsumi K (1995) The promoter from the rice nuclear gene encoding chloroplast aldolase confers mesophyll-specific and light-regulated expression in transgenic tobacco. Mol Gen Genet 248: 668-674.

KaldenhoffR, Henningsen U and Richter G (1994) Gene activation in suspension-cultured cells of Arabidopsis thaliana during blue-light-dependent plantlet regeneration. Planta 195: 182-187.

Kamber L and Feller U (1998) Influence of the activation status and of ATP on phosphoribulokinase degradation. J Exp Bot 49:139-144.

Kaneko T, Sato 5, Kotani H, Tanaka A, Asamizu E, Nakamura Y, Miyajima N, Hirosawa M, Sugiura M, Sasamoto 5, Kimura T, Hosouchi T, Matsuno A, Muraki A, Nakazaki N, Naruo K, Okumura 5, Shimpo 5, Takeuchi C, Wada T, Watanabe A,

Yamada M, Yasuda M and Tabata S (1996b) Sequence analysis of the genome of the unicellularcyanobacterium Synechocystis sp. strain PCC6803. II. Sequence determination of the entire genome and assignment of potential protein-coding regions (supplement). DNA Res 3: 185-209.

Karlan AW and Russell GK (197 6) Aldolase levels in wild type and mutant Euglena gracilis. J Protozool 23: 176-179.

Karmali A, Drake AF and Spencer N (19 83) Purification, properties and assay of D-ribulose-5-phosphate 3-epimerase from human erythrocytes. Biochem J 211: 617-623.

Karpinski S, Esc obarC, Karpinska B, Creissen G andM ulline aux PM (1997) Photosynthetic electron transport regulates the expression of cytosolic ascorbate peroxidase genes in Arabidopsis during excess light stress. Plant Cell 9: 627-640.

Keeling PJ and Doolittle WF (1997) Eukaryotic triosephosphate isomerase originated in the mitochondrion. Proc Natl Acad Sci USA 94: 1270-1275.

Keller CJ and Huffaker RC (1967) Evidence for in vivo light-induced synthesis of ribulose-l,5-diP carboxylase and phosphoribulokinase in greening barley leaves. Plant Physiol 42: 1277-1283

Kelley PM and Freeling M (1984) Anaerobic expression of maize fructose-1,6-bisphosphate aldolase. J Biol Chem 259: 14180-14183.

Kelley PM and Tolan PR (1986) The complete amino acid sequence for the anaerobically induced aldolase derived from cDNA clones. Plant Physiol 82: 1076-1080.

Kern D, Kern G, Neef H, Tittmann K, Killenberg-Jabs M, Wilmer C, Schneider G and Hubner G (1997) How thiamine diphosphate is activated in enzymes. Science 275: 67-70 .

Kersanach R, Brinkmann H, Liaud M-F, Zhang D-X, Martin W and Cerff R (1994) Five identical intron positions in ancient duplicated genes of eubacterial origin. Nature 367: 387-389 .

Kingston-Smith AH, Harbinson J, Williams J and Foyer CH (1997) Effect of chilling on carbon assimilation, enzyme activation, andphotosynthetic electron transportin the absence ofphotoinhibition in maize leaves. Plant Physiol 114: 10391046.

Kishore R, Muchhal U and Schwartzbach 5 (1993) The presequence of Euglena LHCPII, a cytoplasmically synthesized chloroplast protein, contains a functional endoplasmatic reticulum-targeting domain. Proc Natl Acad Sci USA 90: 11845-11849.

Kitayama M, Togasaki RK (1995) Purification and cDNA isolation ofchloroplastic phosphoglycerate kinase from Chlamydornonas reinhardtii. Plant Physiol 107: 393-400.

Kloppstech K (1997) Light regulation of photosynthetic genes Physiol Plant 100: 739-747

Knaff DB and Hirasaw a M (1 991) Ferredoxin-dependent chloroplast enzymes. Biochim Biophys Acta 1056: 93-125 .

Knowles VL, McHugh 5, Hu Z, Dennis DT, Mild BL and Plaxton WC (1998) Altered growth of transgenic tobacco lacking leaf cytosolic pyruvate kinase. Plant Physiol 116: 4551.

Koch KE (1996) Carbohydrate-modulated gene expression in plants. Ann Rev Plant Physiol Plant Mol Biol 47: 509-540 .

Kohler U, Donath M, Mendel RR, Cerff R and Hehl R (1996b) Intron-specific stimulation of anaerobic gene expression and splicing efficiency in maize cells. Mol Gen Genet 251: 252258.

Kohler U, Mendel RR, Cerff R and Hehl R (1996a) A promoter for strong and ubiquitous anaerobic gene expression in tobacco. PlantJ 10: 175-183.

Kohlhoff M, Dahm A and Hensel R (1 996) Tetrameric triosephosphate isomerase from hyperthermophilic Archaea. FEBS Lett 383: 245-250.

Koksharova 0, Schubert M, Shestakov S and Cerff R (1998) Genetic and biochemic al evidence for distinctkey functions of two highly divergent GAPDH genes incatabolic and anaboloic carbon flow of the cyanobacterium Synechocystis sp. PCC6803. Plant Mol Biol 3 6: 183-194.

Kopke-Secondo E, Molnar I and Schnarrenberger C (1990) Isolation andcharacterization ofthe cytosolic andchloroplastic 3-phosphoglycerate kinase from spinach leaves. Plant Physiol 93: 40-47.

Koroleva OA, Farrar JF, Tomos AD and Pollock CJ (1998) Carbohydrates in individual cells of epidermis, mesophyll, and bundle sheath in barley leaves with changed export or photosynthetic rate. Plant Physiol 118: 1525-1532.

Kossmann J, Mueller-Roeber B, Dyer T A, Raines C A, 5 onnewald U and Willmitzer L (1992) Cloning and expression analysis of the plastidic fructose- 1,6-bisphosphatase coding sequence from potato: Circumstantial evidence for the import of hexoses into chloroplasts. Planta 188: 7-12.

Kossmann J, Sonnewald U and Willmitzer L (1994) Reduction of the chloroplastic fructose-1,6-bisphosphatase in transgenic potato plants impairs photosynthesis and plant growth. Plant J 6: 637-650.

Kozaki A and Takeba G (1996) Photorespiration protects C3 plants from photooxidation. Nature 384: 557-560.

Kramer 5, Stitt M and Heldt HW (1988) Mitochondrial oxidative phosphorylation participating in photosynthetic metabolism of a leaf cell. FEBS Lett 226: 352-356.

Kramer DM, Wise RR, Frederick JR, Alm DM, Hesketh JD, Ort DR and Crofts AR (1990) Regulation of coupling factor in field-grown sunflower: A redox model relating coupling factor activity to the activities of other thioredoxin-dependent chloroplast enzymes. Photosynth Res 16: 213-222.

Krapp A, Chaves MM, David MM, Rodrigues ML, Pereira JS and Stitt M (1994) Decreased ribulose-1,5-bisphosphate carboxylase/oxygenase in transgenic tobacco transformed with `antisense' rbcS: VIII. Impact on photosynthesis and growth in tobacco growing under extreme high irradiance and high temperature. Plant Cell Environ 17: 945-953.

Krapp A, Hoffmann B, Schaefer C and Stitt M (1993) Regulation of the expression of rbcS and other photosynthetic genes by carbohydrates: A mechanism for the sink regulation of photosynthesis. Plant J 3: 817-828.

Krapp A, Quick WP and Stitt M (1991) Ribulose-1,5-bisphosphate carboxylase-oxygenase, other Calvin cycle enzymes, and chlorophyll decrease when glucose is supplied to mature spinach leaves via the transpiration stream. Planta 186: 58-69.

Krishnan G and Altekar W (1991) An unusual class I (Schiff base) fructose-1,6-bisphosphate aldolase from the halophilic archaebacterium Haloarculavallismortis. EurJ Biochem 195: 343-350

Kruger I and Schnarrenberger C (1983) Purification, subunit structure and immunochemical properties of fructosebisphosphate aldolases from spinach and corn leaves. Eur J Biochem 136: 101-106.

Kruger I and Schnarrenberger C (1985) Development ofcytosol and chloroplast aldolases during germination ofspinach seeds. Planta 164: 109-114.

Kruse A, Fieuw 5, Heineke D and Miiller-Rober B (1998) Antisense inhibition of cytosolic NADP-dependent isocitrate dehydrogenase in transgenic tobacco plants. Planta: 205: 8291.

Kurzok H-G and Feierabend J (1984) Comparison of a cytosolic and a chloroplast triosephosphate isomerase isoenzyme from rye leaves. I. Purification and catalytic properties. Biochim Biophys Acta 788: 214-221.

Kurzok H-G and Feierabend J (1986) Comparison of the development and site ofsynthesis ofacytosolic and a chloroplast isoenzyme of triosephosphate isomerase in rye leaves. J Plant Physiol 126: 207-212.

Kusian B and Bowien B (1997) Organization and regulation of ebb CO2 assimilation genes in autotrophic bacteria. FEMS Microbiol Rev 21: 135-155.

Kusian B, Yoo JG, Bednarski R and Bowien B (1992) The Calvin cycle enzyme pentose-5-phosphate 3-epimerase is encoded within the cfx operons of the chemoautotroph Alcaligenes eutrophus. J Bacteriol 174: 7337-7344.

Kwon H-B, Park S-C, Peng H-P, Goodman HM, Dewdney J and 5 hih M-C (1994) Identification of a light responsive region of the nuclear gene encoding the B subunit of chloroplast glyceraldehyde-3-phosphate dehydrogenase fromArabidopsis thaliana. Plant Physiol 105: 357-367.

Lam E and Chua N-H (1990) GT-1 binding site confers light responsive expression in transgenic tobacco. Science 248: 471-474.

Lambert DH, Bryant DA, Stirewalt VL, Dubbs JM, Stevens SE Jr and Porter RD (1985) Gene map for the Cyanophora paradoxa cyanelle genome. J Bacteriol 164: 659-664.

Latzko E and Gibbs M (1968) Distribution and activity of enzymes of the reductive pentose phosphate cycle in spinach leaves and in chloroplasts isolated by different methods. Z Pflanzenphysiol 59: 184-194.

Latzko E and Gibbs M (1969) Enzyme activities of the carbon reduction cycle in some photosynthetic organisms. Plant Physiol 44:295-300.

Latzko E and Kelly GJ (1979) Enzymes of the reductive pentose phosphate cycle. Encyclopedia Plant Physiol (New Series) 6: 239-250.

Lauerer M, Saftic D, Quick WP, Labate C, Fichtner K, Schulze E-D, Rodermel SR, Bogorad L and Stitt M (1993) Decreased ribulose-1,5-bisphosphatec arboxylase-oxygenase in transgenic tobacco transformed with `antisense' rbcS. Planta 190: 332345.

Lebherz HG, Leadbetter MM and Bradshaw RA (1984) Isolation and characterization of the cytosol and chloroplast form of spinach leaf fructose diphosphate aldolase. J Biol Chem 259: 1011-1017.

Lebreton 5, Gontero B, Avilan L and Ricard J (1997) Memory imprinting effects in multienzymecomplexes. II. Kineticsofthebienzymecomplexfrom Chlaznydomonas reinhardtii and hysteretic activation of chloroplast oxidized phosphoribulokinase. Eur J Biochem 246: 85-91.

Leegood RC (1990) Enzymes of the Calvin cycle. Methods Plant Biochem 3: 15-37.

Li L-A and Tabita FR (1994) Transcription control of ribulose bisphosphate carboxylase/ oxygenase activase and adjacent genes mAnabaena species. J Bacteriol 176: 6697-6706.

Li L-A and Tabita FR (1997) Maximum activity of recombinant ribulose 1,5-bisphosphate carboxylase/oxygenase of Anabaena sp. strain CA requires the product of the rbcXgene. J Bacteriol 179: 3793-3796.

Liaud MF, Brandt U, Scherzinger M and Cerff R (1997) Evolutionary origin of cryptomonad microalgae: Two novel chloroplast/cytosol-specific GAPDH genes as potential markers of ancestral endosymbiont and host cell components. J Mol Evol 44 :S28-37.

Lloyd J and Farquhar GD (1994) 13C discrimination during CO2 assimilation by the terrestrial biosphere. Oecologia 99: 201215.

Lloyd JC, Horsnell PR, Dyer TA and Raines CA (1991a) Structure and sequence of a wheatphosphoribulokinase gene. Plant Mol Biol 17: 167-168.

Lloyd JC, Raines CA, John UP and Dyer TA (1991b) The chloroplast FBPase gene of wheat: Structure and expression of the promoter in photosynthetic and meristematic cells of transgenic tobacco plants. Mol Gen Genet 225: 209-216.

Loferer H and Hennecke H (1994) Protein disulfide oxidoreductases in bacteria. Trends Biochem Sci 19: 169-171 Longstaff M, Raines CA, McMorrow EM, Bradbeer JW and Dyer TA (1989) Wheat phosphoglycerate kinase: Evidence for recombination between the genes for the chloroplastic and cytosolic enzymes. Nucl Acid Res 17: 6569-6580.

Lopez-Jaramillo J, Chueca A, JacquotJ-P, Hermoso R, Lazaro JJ, Sahrawy M and Lopez-Gorge J (1997) High-yield expression of pea thioredoxin m and assessment of its efficiency in chloroplast fructose-1,6-bisphosphatase activation. Plant Physiol 114: 1169-1175.

Lorimer GH and Miziorko H (1980) Carbamate formation of the &amino group of a lysyl residue as the basis for the activation of ribulose-1,5-bisphosphate carboxylase by CO2 and Mg2'. Biochemistry 19: 5321-5328.

Macioszek J Anderson JB and Anderson LE (1987) Isolation of chloroplastic phosphoglycerate kinase/glyceraldehyde3-phosphate couple. Plant Physiol 94: 291-296.

MacioszekJ and Anderson LE (1987) C hanging kinetic properties of the two-enzyme phosphoglycerate kinase/NADP-linked glyceraldehyde-3-phosphate dehydrogenase couple from pea chloroplasts during photosynthetic induction. Bioc him Biophys Acta 892: 185-190.

Maier U-G (1992) The four genomes of the cryptomonad alga Pyrenomonas sauna. BioSystems 28: 69-74.

Malhorta OP, Kumar A and Tikoo K (1987) Isolation and quaternary structure of a complex of glyceraldehyde 3-phosphate dehydrogenase and phosphoglycerate kinase. Ind J Biochem Biophys 24: 16-20.

Mann CC (1999) Genetic engineers aim to soup up crop photosynthesis. Science 283: 314-316.

Marcus A (1960) Photocontrol of formation of red kidney bean leaf triphosphopyridine nucleotide linked triosephosphate dehydrogenase. Plant Physiol 35: 126-128.

Marcus F and Harrisch PB (1990) Amino acid sequence of spinach chloroplast fructose-1,6-bisphosphatase. Arch Biochem Biophys 279: 151-157.

Markos A, Miretsky A and Milner M (1993) A glyceraldehyde3-phosphate dehydrogenase with eubacterial features in the amitochondriate eukaryote Trichomonas vaginalis. J Mol Evol 37: 631-643.

Marques IA, Ford DM, Muschinek G and Anderson LE (1987) Photosynthetic carbon metabolism in isolated pea chloroplasts: metabolite levels and enzyme activities. Arch Biochem Biophys 252: 458-466.

Marsh JJ and Lebherz HG (1992) Fructose-bisphosphate aldolases: an evolutionary history. Trends Biochem Sci 17: 110-113.

Martin W and Herrmann RG (1998) Gene transfer from organelles to the nucleus: How much, what happens and why? Plant Physiol 118: 9-17.

Martin W and Schnarrenberger C (1997) The evolution of the Calvin cycle from prokaryotic to eukaryotic chromosomes: A case stu dy of functional redundancy in ancient pathways through endosymbiosis. Curr Genet 32: 1-18.

Martin W andMiillerM (1998) The hydrogen hypothesis for the first eukaryote. Nature 392:37-41

Martin W, Brinkmann H, Savona C and CerffR (1993) Evidence for a chimaeric nature of nuclear genomes: Eubacterial origin of eukaryotic glyceraldehyde-3-phosphate dehydrogenase genes. Proc Natl Acad Sci USA 90: 8692-8696.

Martin W, Henze K, Kellermann J, Flechner A and Schnarrenberger C (1996b) Microsequencing and cDNA cloning of the Calvin cycle/OPPP enzyme ribose-5-phosphate isomerase (EC 5.3.1.6) from spinach chloroplasts. Plant Mol Biol 30: 795-805.

Martin W, Mustafa A-Z, Henze K and SchnarrenbergerC (1996a) Higher plant chloroplast and cytosolic fructose-1,6-bisphophosphatase isoenzymes: Origins via duplication rather than prokaryote-eukaryote divergence. PlantMol Bio132:485-491.

Martin W, Stoebe B, Goremykin V, Hansmann 5, Hasegawa M, and Kowallik KV (1998) Gene transfer to the nucleus and the evolution ofchloroplasts. Nature 393:162-165.

Masle J, Hudson GS and Badger MR (1993) Effects of ambient CO2 concentration on growth and nitrogen use in tobacco (Nicotiana tabacum) plants transformed with an antisense gene to the small subunit of ribulose-1,5-bisphosphate carboxylase/oxygenase. Plant Physiol 103: 1075-1088.

Mate CJ, Hudson GS, Von Caemmerer 5, Evans JR and Andrews TJ (1993) Reduction of ribulose bisphosphate carboxylase activase levels in tobacco (Nicotiana tabacum) by antisense RNA reduces ribulo se bisph osp hate c arbo xylase c arbamylatio n and impairs photosynthesis. Plant Physiol 102: 1119-1128.

McFadden G, Gilson PR, Hoffmann CJB, Adcock GJ and Maier U-G (1994) Evidence that an amoeba acquired a chloroplast by retaining part of an engulfed eukaryotic alga. Proc Natl Acad Sci USA 91: 3690-3694.

McFadden GI, Gilson PR, Douglas SE, Cavalier-Smith T, Hofmann CJ and Maier UG (1997) Bonsai genomics: Sequencing the smallest eukaryotic genomes. Trends Genet 13:46-49.

McMorrow EM and BradbeerJW (1990) Separation, purification, and comparative properties of chloroplast and cytoplasmic phosphoglycerate kinase from barley leaves. Plant Physiol 93: 374-383.

McNeil PH and Walker DA (1981) The effect of magnesium and other ions on the distribution of ribulose 1,5-bisphosphate carboxylase in chloroplast extracts. Arch Biochem Biophys 208:184-188.

Meijer WG, Van Den Bergh ERE and Smith LM (1996) Induction of the gap-pgk operon encoding glyceraldehyde-3-phosphate dehydrogenase and 3-phosphoglycerate kinase of Xanthobacter flavus requires the LysR-type transcriptional activator CbbR J Bacteriol 178: 881-887.

Melkonian M (1996) Systematics and evolution of the algae: Endocytobiosis and the evolution of the major algal lineages. Prog Bot 57: 281-311.

Meyer-Gauen G, Herbrandt H, Pahnke J, Tholl D, Cerff R and Martin W (1998) Gene structure, expression in Escherichia coli and biochemical properties of the NADt-dependent glyceraldehyde-3-phosphate dehydrogenase from Pinus sylvestris chloroplasts. Gene 209: 167-174.

Meyer-Gauen G, Schnarrenberger C, Cerff R and Martin W (1994) Molecular characterization of a novel, nuclear-encoded, NADtdependent glyceraldehyde-3-phosphate dehydrogenase in plastids of the gymnosperm Pinus sylvestris L. Plant Mol Biol 26: 1155-1166.

Michalowski CB, Derocher EJ, Bohnert HJ and Salvucci ME (1992) Phosphoribulokinase from ice plant: Transcription, transcripts and protein expression during environmental stress. Photosyn Res 31: 127-138.

Michels 5, Rogalaska E and BranlantG (1996) Phosphate binding sites in phosphorylating glyceraldehyde-3-phosphate dehydrogenase from Bacillus stearothermophilus. Eur J Biochem 235: 641-647.

Michels PAM, Marchand M, Kohl L, Allert 5, Vellieux FMD, Wierenga R and Opperdoes F (1991) The cytosolic and glycosomal isozymes of glyceraldehyde-3-phosphate dehydrogenase in Trypanosoma brucei have a distant evolutionary relationship. Eur J Biochem 198: 421-428.

Miffier B, Ziegler I and Ziegler H 81969) Lichtinduzierte reversible Aktivitatssteigerung der Glyceraldehyd-3-phosphat-Dehydrogenase in Chloroplasten. EurJ Biochem 9: 101-106.

Miles AJ, Potts SC, Willingham NM, Raines CA and Lloyd JC (1993) A light- and developmentally-regulated DNA-binding interaction is common to the upstream sequences of the wheat Calvin cycle bisphosphatase genes. Plant Mol Biol 22: 507516.

Mills JD and Mitchell P (1982) Modulation of coupling factor ATPase activity in intact chloroplasts. Reversal of thiol modulation in the dark. Biochim Biophys Acta 679: 75-83.

Mo Y, Harris BG and Gracy RW (1973) Triosephosphate isomerase and aldolases from light and dark-grown Englena gracilis. Arch Biochem Biophys 157: 580-587.

Mojzsis SJ, Arrhenius G, McKeegan KD, Harrison TM, Nutman AP and Friend CR (1996) Evidence for life on Earth before 3,800 million years ago. Nature 384: 55-59.

Moorhead BG and Plaxton WC (1990) Purification and characterization ofcytosolic aldolase from carrot storage root. Biochem J 269: 133-139.

Mora-Garcia 5, Rodriguez-Suarez R and Wolosiuk RA (1998) Role of electrostatic interactions on the affinity of thioredoxin for target proteins. Recognition of chloroplast fructose-1, 6bisphosphatase by mutant Escherichia coli thioredoxins. J Biol Chem 273: 16273-16280.

Morse D, Salois P, Markovic P and Hastings JW (1995) A nuclear-encoded form II RuBisCO in dinoflagellates. Science 268:1622-1624.

MUllerB (1972) A labile CO2-fixing enzyme complex in spinach chloroplasts. Z Naturforsch 27b: 925-932.

Muschak M, Hoffmann-Benning 5, Fuss H, Kossmann J, Willmitzer L, and Fisahn J (1997) Gas exchange and ultrastructural analysis of transgenic potato plants expressing mRNA antisense construct targeted to the cp-fructose-1,6- bisphosphate phosphatase. Photosynthetica 33: 455-465.

Nagashima H and Fukuda I (1983) Floridosides in unicellular hot spring algae. Phytochemistry 22, 1949-1951.

Nakamura H, Satoh W, Hidaka 5, KagayaY, Ejiri S and Tsutsumi K. (1996) Genomic structure of the rice aldolase isozyme C-1 gene and its regulation through a Ca2 +-mediated protein kinasephosphatase pathway. Plant Mol Biol 30: 381-385.

Nakamura H, Tokairin Y, Tamayama 5, Kon 5, Hidaka 5, Ejiri and Tsutsumi K (1997) Cell type- and positionally specific regulation of the aldolase P gene expression in rice seedlings. Biosci Biotechnol Biochem 61: 256-262.

Neuhaus HE, Batz 0, Thorn E and Scheibe R (1993) Purification of highly intact plastids from various heterotrophic plant tissues: Analysis of enzymic equipment and precursor dependency for starch biosynthesis. Biochem J 296: 395-401.

Nicholson 5, Esterby JS and Powls R (1987) Properties of a multimeric protein complex from chloroplasts possessing potential activities of NADPH-dependent glyceraldehyde3-phosphate dehydrogenase and phosphoribulokinase. Eur J Biochem 162: 423-431.

Nikkola M, Lindqvist Y and Schneider G (1994) Refined structure of transketolase from Saccharomyces cerevisiae at 2.0 A resolution. J Mol Biol 238: 387-404.

Nilsson U, Meshalkina L, Lindqvist Y and Schneider G (1997) Examination of substrate binding in thiamin diphosphatedependent transketolase by protein crystallography and site-directed mutagenesis. J Biol Chem 272: 1864-1869.

Nishimura M and Beevers H (1981) Isoenzymes of sugar phosphate metabolism in endosperm of germinating castor beans. Plant Physiol 67: 1255-1258.

Nishizawa AN and Buchanan BB (1981) Enzyme regulation in C4 photosynthesis: Purification and properties of the thi ore do)rin -lin ke d fructo se bisph o sphatase and se doheptulose bisphosphatase from corn leaves. J Biol Chem 256: 61196126.

Nowitzki U, Westhoff P, Henze K, Schnarrenberger C and Martin W (1995) Cloning of the amphibolic Calvin cycle/ OPPP enzyme D-ribulose-5-phosphate 3-epimerase (EC 5.1.3.1) from spinach chloroplasts: Functional and evolutionary aspects. Plant Mol Biol 29: 1279-1291.

Ocheretina 0 and Scheibe R (1994) Cysteines of chloroplast NADP-malate dehydrogenase form mixed disulfides. FEBS Lett 355: 254-258.

Ocheretina 0, Hamecker J, Rother T, Schmid R and Scheibe R (1993) Effects of N-terminal truncations upon chloroplast NADP-malate dehydrogenase from pea and spinach. Biochim Biophys Acta 1163: 10-16.

Oelmueller R and Mohr H (1985) Specific action of blue light on phytochrome-mediated enzyme syntheses in the shoot of milo (Sorghum vulgare). Plant Cell Environ 8: 27-32.

of the genomeof the unicellular cyanobacterium Synechocystis sp. strain PCC6803. II. Sequence determination of the entire genome and assignment of potential protein-coding regions. DNA Res 3: 109-136.

Ort DR and Oxborough K (1992) In situ regulation of chloroplast coupling factor activity. Ann Rev Plant Physiol Plant Mol Biol 43: 269-291.

Osmond CB (1997) C4 photosynthesis: Thirty or forty years on. Aust J Plant Physiol 24: 409-412

Ostrem JA, Vernon DM and Bohnert HJ (1990) Increased expression of a gene coding for NAD:glyceraldehyde3-phosphate dehydrogenase during the transition from C-3 photosynthesis to crassulacean acid metabolism in Mesembryanthemum crystallinum. J Biol Chem 265: 3497-3502.

Otto S and Feierabend J (1989) Enzymes of starch and sugar phosphate metabolism in achlorophyllous ribosome-deficient plastids from high-temperature-grown rye leaves. Physiol Plant 76: 65-73.

Pacold ME and Anderson LE (1975) Chloroplast and cytosolic enzymes. VI. Pea leaf 3-phosphoglycerate kinases. Plant Physiol 55: 168-171.

Pacold ME, Anderson LE, Li D and Stevens FJ (1995) Redox sensitivity and light modulation of enzyme activity in the rhodophytes Gracilaria tikvahiae and Chondrus crispus. J Phycol 31:297-301.

Paoli GC, Morgan NS, Tabita FR and Shively JM (1995) Expression of the cbbLcbbS and cbbM genes and distinct organization of the cbb Calvin cycle structural genes of Rhodobacter capsulatus. Arch Microbiol 164: 396-405.

Park SC, Kwon HB and Shih MC (1996) Cis-acting elements essential for light regulation of the nuclear gene encoding the A subunit of chloroplast glyceraldehyde 3-phosphate dehydrogenase in Arabidopsis thaliana. Plant Physiol 112: 1563-1571.

Parry MAJ, Andralojc PJ, Parmar 5, Keys AJ, Habash D, Paul MJ, Alred R, QuickWP and Servaites JC (1997) RegulatiorD f Rubisco by inhibitors in the light. Plant Cell and Environment 20: 528-534.

Paul MJ, Knigh tJS, Habash D, Parry MAJ, Lawlor DW, Barnes SA, Loynes A and Gray JC (1995) Reduction in phosphoribulokinase activity by antisense RNA in transgenic tobacco: Effect on CO_2 assimilation and growth in low irradiance. Plant J 7: 535-542.

Paul MJ, Knight JS, Habash D, Parry MAJ, Lawlor DW, Barnes SA, Loynes A and Gray JC (1995) Reduction in phosphoribulokinase activity by antisense RNA in transgenic tobacco: Effect on CO_2 assimilation and growth in low irradiance. Plant J 7: 535-542.

Pelzer-Reith B, Freund 5, Schnarrenberger C, Yatsuki H and Hori K (1995) The plastid aldolase gene from Chkunydomonas reinhardtii: Intron/exon organization, evolution, and promoter structure. Mol Gen Genet 248: 481-486.

Pelzer-Reith B, Wiegand S and Schnarrenberger C (1994b) Plastid class I and cytosol class II aldolase of Euglena gracilis. Purification and characterization. Plant Physiol 106: 11371144.

Pettersson G and Ryde-Pettersson U (1989) Dependence of the Calvin cycle activity on kinetic parameters for the interaction of non-equilibrium cycle enzymes with their substrates. Eur J Biochem 186: 683-687.

PfannschmidtT, Nilsson A, and Allen JF (1999) Photosynthetic control of chloroplast gene expression. Nature 397: 625-628.

Plaumann M, Pelzer-Reith B, Martin W and Schnarrenberger C (1997) Cloning of fructose-1,6-bisphosphate aldolases from Euglena gracilis: Multiple recruitment of class I aldolase to chloroplasts and eubacterial origin of eukaryotic class II aldolase genes. Curr Genet 31: 430-438.

Plaxton WC (1996) The organization and regulation of plant glycolysis. Ann Rev PlantPhysiol Plant Mol Biol 47 :185-214.

Pohlmeyer K, Paap BK, Soil J and Wedel N (1996) CP12: a small nuclear-encoded chloroplast protein provides novel insights into higher-plant GAPDH evolution. Plant Mol Biol 32: 969978.

Porter MA, Milanez 5, Stringer CD and Hartman FC (1986) Purification and characterization of ribulose-5-phosphate kinase from spinach. Arch Biochem Biophys 245: 14-23.

Porter MA, Stringer CD and Hartman FC (1988) Characterization of the regulatory thioredoxin site of phosphoribulokinase. J Biol Chem 263: 123-129.

Portis AR (1992) Regulation of ribulose-1,5-bisphosphate carboxylase/oxygenase activity. Ann Rev Plant Physiol Plant Mol Biol 43: 415-437.

Pratt LH (1995) Phytochromes: Differential properties, expression pattern and molecular evolution. Photochem Photobiol 61: 10-21.

Price GD, Evans JR, Von Caemmerer 5, Yu JW and Badger MR (1995) Specific reduction of chloroplast glyceraldehyde3-phosphate dehydrogenase activity by antisense RNA reduces CO_2 assimilation via a reduction in ribulose biphosphate regeneration in transgenic tobacco plants. Planta 195: 369378.

Price GD, von Caemmerer , Evans JR, Yu J-W, Lloyd J, Oj a V, Kell P, Harrison K, Gallagher A and Badger MR (1994) Specific reduction of chloroplast carbonic anhydrase activity by antisense RNA in transgenic tobacco plants has a minor effect on photosynthetic CO_2 assimilation. Planta 193: 331340.

PrilB B, Meyer HE and HolldorfAW (1993) Characterization of the glyceraldehyde-3-phosphate dehydrogenase from the extremely halophilic arc haebacterium Haloarculavallismortis. Arch Microbiol 160: 5-11.

Pupillo P and Faggiani R (1979) Subunit structure of three glyceraldehyde-3 -phosphate dehydrogenases of some flowering plants. Arch Biochem Biophys 194: 581-592.

Pupillo P and Giuliani-Piccari G (1975) The reversible depolymerization of spinach chloroplast glyceraldehyde-3phosphate dehydrogenase. Interactions with nucleotides and dithiotreitol. EurJ Biochem 51: 475-482.

Purcell M, Mabrouk YM and Bogorad L (1995) Red/far-red and blue light-responsive regions of maize rbcS-m3 are active in bundle sheath and mesophyll cells, respectively. Proc Natl Acad Sci USA 92: 11504-11508.

Qian Y and Tabita FR (1996) A global signal transduction system regulates aerobic and anaerobic CO_2 fixation in Rhodobacter sphaeroides. J Bacteriol 178: 12-18.

Quick WP, Fichtner K, Schulze ED, Wendler R, Leegood RC, Mooney H, Rodermel SR, Bogorad L and Stitt M (1992) Decreased ribulose-1,5-bisphosphate carboxylase-oxygenase in transgenic tobacco transformed with `antisense' rbcS: IV. Impact on photosynthesis in conditions of altered nitrogen supply. Planta 188: 522-531.

Quick WP, Schurr U, Scheibe R, Schulze ED, Rodermel SR, Bogorad L and Stitt M (1991) Decreased ribulose-1,5-bisphosphate carboxylase-oxygenase in transgenic tobacco transformed with `antisense' rbcS: I. Impact on photosynthesis in ambient growth conditions. Planta 183: 542-554.

Quigley F, Martin W and Cerff R (1988) Intron conservation across the prokaryote-eukaryote boundary: Structure of the nuclear gene for chloroplast glyceraldehyde-3-phosphate dehydrogenase from maize. Proc Natl Acad Sci USA 85: 2672-2676.

Ragsdale SW (1991) Enzymology of the acetyl-CoA pathway of CO_2 fixation. Crit Rev Biochem Mol Biol 26: 261-300.

Raines CA, Lloyd JC, Longstaff M, Bradley D and Dyer TA (1988) Chloroplast fructose-1,6-bisphosphatase: the product of a mosaic gene. Nucleic Acid Res 16: 7931-7942.

Raines CA, Longstaff M, Lloyd JC and Dyer TA (1989) Complete coding sequence of wheat phosphoribulokinase: developmental and light-dependent expression of the mRNA. Mol Gen Genet 220: 43-48.

Rajagopalan R and Altekar W (1994) Characterisation and purification of ribulose-bisphosphate carboxylase from heterotrophically grown halophilic archaebacterium, Halofercvc mediterranei. Eur J Biochem221: 863-869.

Rathnam CKM (1978) Malate and dihydroxyacetone phosphate-dependent nitrate reduction in spinach leaf protoplasts. Plant Physiol 62: 220-223.

Rault M, Guidici-Orticoni M-T, Gontero B and Ricard J (1993) Structural and functional properties of a multi-enzyme complex from spinach chloroplasts. I. Stoichiometry of the polypeptide chains. EurJ Biochem 217: 1065-1073.

Rawal N, Kelkar SM and Altekar W (1988) Ribulose 1,5-bisphosphate dependent CO_2 fixation in the halophilic archaebacterium, Hcdobacterium mediterranei. Biochem Biophys Res Commun 156: 451-456.

Reddy PS, Ramanjulu 5, Sudhakar C and Veeranjaneyulu K (1998) Differential sensitivity of stomatal and non-stomatal components NaC1 or Na2SO4 salinity in horsegram, Macrotyloma uniflorum (Lam.). Photosynthetica 35: 99-105 Reinbothe 5, Reinbothe C and Parthier B (1991a) Glucose repression of chloroplast development in Euglena gracilis. Plant Physiol Biochem 29: 297-307.

Reinbothe 5, Reinbothe C, Krauspe R and Parthier B (199 lb) Changing gene expression during dark-induced dedifferentiation in Euglena gracilis. Plant Physiol Biochem 29: 309318.

Ricard B, Rivoal J and Pradet A (1989) Rice cytosolic glyceraldehyde 3-phosphate dehydrogenase contains two subunits differentially regulated by anaerobiosis. Plant Mol Biol 12: 131-139.

Ricard J, Giudici-Orticoni MT and Gontero B (1994) The modulation of enzyme reaction rates within multi-enzyme complexes: 1. Statistical thermodynamics of information transfer through multi-enzyme complexes. Eur J Biochem 226: 993-998.

Roberts DL, Runquist JA, Miziorko HM and Kim JJ (1995) Crystallization and preliminary X-raycrystallographic analysis ofphosphoribulokinase from Rhodobacter sphaeroides. Protein Sci 4: 2442-2443.

Robinson BH and Chun K. (1993) The relationships between transketolase, yeast pyruvate decarboxylase and pyruvate dehydrogenase of the pyruvate dehydrogenase complex. FEBS Lett 328: 99-102.

Robinson JM, Smith MG and Gibbs M (1980) Influence of hydrogen peroxide upon carbon dioxide photoassimilation in the spinach chloroplast. Plant Physiol 65: 755-759.

Robinson SP and Walker DA (1981) Photosynthetic carbon reduction cycle. In: Hatch MD and Boardman NK (eds) Biochemistry of Plants, Vol 8, Photosynthesis, pp 193-236. Academic Press, New York.

Rodermel SR, Haley J, Jiang CZ, Tsai CH and Bogorad L (1996) A mechanism for intergenomic integration: Abundance of ribulose bisphosphate carboxylase small-subunit protein influences the translation of the large-subunit mRNA. Pine Natl Acad Sci USA 93: 3881-3885.

Rodriguez-Suarez RJ and Wolosiuk RA (1995) High level expression in Escherichia coil, purification and properties of fructose-1,6-bisphosphatase from rapeseed (Brassica napus) leaves. Photosynth Res 46: 313-322.

Rodriguez-Suarez RJ, Mora-Garcia S and Wolosiuk RA (1997) Characterization of cysteine residues involved in the reductive activation and the structural stability of rapeseed (Brassica napus) chloroplast fructose-1,6-bisphosphatase. Biochem Biophys Res Commun 232: 388-393.

Rowan R, Whitney SM, Fowler A and Yellowlees D (1996) Rubisco in marine symbiotic dinoflagellates: form 11 enzymes in eukaryotic oxygenic phototrophs encoded by a nuclear multigene family. Plant Cell 8: 539-553.

Rozario C, Morin L, Roger A, Smith M and Milner M (1996) Primary structure and phylogenetic relationships of glyceraldehyde-3-phosphate dehydrogenase genes of free-living and parasitic diplomonad flagellates. J Euk Microbiol 43:330-340.

Runquist JA, Narasimhan C, Wolff CE, Koteiche HA and Miziorko HM (1996) Rhodobacter sphaeroides phosphoribulokinase: Binary and ternary complexes with nucleotide substrate analogs and effectors. Biochemistry 35: 1504915056.

Russel DA and Sachs MM (1989) Differential expression and sequence analysis of the maize glyceraldehyde-3-phosphate dehydrogenase gene family. Plant Cell 1: 793-803.

Rutner AC (1970) Spinach 5-phosphoribose isomerase. Purification and properties of the enzyme. Biochemistry 9: 178-184.

Sahrawy M, Chueca A, Hermoso R, Lazaro JJ and Lopez-Gorge J (1997) Directed mutagenesis shows that thepreceding region of the chloroplast fructose-1,6-bisphosphatase regulatory sequence is the thioredoxin docking site. J Mol Biol 269: 623630.

Sainis JK and Harris GC (1986) The association of ribulose1,5-bisphosphate carboxylase-oxygenase with phosphoribuloisomerase and phosphoribulokinase. Biochem Biophys Res Commun 139: 947-954.

Sainis JK, Merriam K and Harris GC (1989) The association of D-ribulose-1,5-bisphosphate carboxylase/oxygenase with phosphoribulokinase. Plant Physiol 89: 368-374.

SainisJ K. and Srinivasan VT (1993) Effect of the state of water as studied by pulsed NMR on the function of RUBISCO in a multienzyme complex. J Plant Physiol 142: 564-568.

Sarokin LP and Chua N-H (1992) Binding sites for two novel phosphoproteins, 3AF5 and 3AF3, are required for rbcS-3A expression. Plant Cell 4: 473-483.

SasajimaK, YonedaM (1974) Simpleproceduresforthepreparation ofD-ribose-5-phosphate ketol-isomerase, Dribulose-5-phosphate 3-epimerase and D-sedoheptulose1,7-bisphosphate : D-glyceraldehyde-3-phosphate glycoaldehydetransferase. Agr Biol Chem 7: 1297-303.

Sassenrath-Cole GF, Pearcy RW and Steinmaus S (1994) The role of enzyme activation state in limiting carbon assimilation under variable light conditions. Photosynth Res 41: 295-302

Scagliarini 5, Trost P and Pupillo P (1998) The non-regulatory isoform of NAD(P)-glyceraldehyde-3-phosphate dehydrogenase from spinach chloroplasts. J Exp Bot 49: 1307-1315.

Scagliarini 5, Trost P, Pupillo P and Valenti V (1993) Light activation and molecular mass changes of NAD(P)- glyceraldehyde 3-phosphate dehydrogenase of spinach and maize leaves. Planta 190: 313-319.

Schaefer DG and Zryd JP (1997) Efficient gene targeting in the moss Physcomitrella patens. Plant J 11: 1195-1206.

Schafer 5, Barkowski C and Fuchs G (1986) Carbon assimilation by the autotrophic thermophilic archaebacterium Thermoproteus neutrophilus. Arch Microbiol 146: 301-308.

Scheibe R (1 990) Light/dark modulation: Regulation of chloroplast metabolism in a new light. Bot Acta 103: 327-334 Scheibe R (1991) Redox-modulation of chloroplast enzymes. A common principle for individual control. Plant Physiol 96: 1-3 Scheibe R (1995) Light-modulation of stromal enzymes. In: MA Madore and WJ Lucas (eds) Carbon Partitioning and Sucrose-Sink Interactions in Plants, pp 13-22. American Society of Plant Physiologists, Rockville.

Scheibe R and Beck E (1994) The malate valve: Flux control at the enzymic level. In: Schulze ED (ed) Flux Control in Biological Systems, pp 3-11. Academic Press, San Diego Scheibe R, Kampfenkel K, Wessels R and Tripier D (199 1) Primary structure and analysis of the location of the regulatory disulfide bond of pea chloroplast NADP-malate dehydrogenase. Biochim Biophys Acta 1076: 1-8.

Scheibe R, Baalmann E, Backhausen JE, Rak C and Vetter S (1996) C-terminal truncation of spinach chloroplast NADPdependent glyceraldehyde 3-phosphate dehydrogenase prevents inactivation and reaggregation. Biochim Biophys Acta 1296: 228-234.

Schidlowski M (1988) A 3,800 million year isotopic record of life from carbon sedimentary rocks. Nature 333: 313-318 Schmidt G and Lyman H (1974) Photocontrol of chloroplast enzyme synthesis in mutant and wild-type Euglena gracilis. Proc 3rd Intern Congr Photosynthesis 3: 1755-1764.

Schmidt M, Svendsen I and Feierabend J (1995) Analysis of the primary structure of the chloroplast isozyme of triosephosphate isomerase from rye leaves by protein and cDNA sequencing indicates a eukaryotic origin of its gene. Biochim Biophys Acta 1261: 257-264.

Schmidt S, Drumm-Herrel H, Oelmueller R and Mohr H (1987) Time course of competence in phytochrome-controlled appearance of nuclear-encoded plastidic proteins and messenger RNA species. Planta 170: 400-407.

Schnarrenberger C (1987) Regulation and structure of isozymes of sugar phosphate metabolism in plants. In: Rattazzi MC, Scandalios JG and Whitt GS (eds) Isozymes: Current Topics in Biological and Medical Research, Vol. 16, Agriculture, Physiology, and Medicine, pp 223-240. Alan R Liss, New York.

Schnarrenberger C , Flechner A and Martin W (1995) Enzymatic evidence for a complete oxidative pentose phosphate pathway in chloroplasts and an incomplete pathway in the cytosol of spinach leaves. Plant Physiol 108: 609-614.

Schnarrenberger C and KrUger I (1986) Distinction of cytosol and chloroplast fructose-bisphosphate aldolases from pea, wheat and corn leaves. Plant Physiol 80: 301-304.

Schnarrenberger C, Pelzer-Reith B, Yatsuki H, Freund 5, Jacobshagen S and Hori K (1994) Expression and sequence of the only detectable aldolase in Chlamydornonas reinhardtii. Arch Biochem Biophys 313: 173-178.

SchnarrenbergerC,Tetour M and Herbert M (1975) Development and intracellular distribution of the oxidative pentose phosphate cycle in radish cotyledons. Plant Physiol 56: 836-840.

Sganga MW and B auer CE (1992) Regulatory factors controlling photosynthetic reaction center and light-harvesting gene expression in Rhodobacter capsulatus. Cell 68: 945-954.

Sheen J (1990) Metabolic repression of transcription in higher plants. Plant Cell 2: 1027-1038.

Sheen J (1994) Feedback control of gene expression. Photosynth Res 39: 427-438.

Shen B, Jensen RG and Bohnert HJ (1997) Mannitol protects against oxidation by hydroxyl radicals. Plant Physiol 115: 527-532.

Shibata N, Inoue T, Fukuhara K, Nagara Y, Kitagawa R, Harada 5, Kasai N, Uemura K, Kato K, Yokota A and Kai Y (1996) Orderly disposition of heterogeneous small subunits in D-ribulose-1,5-bisphosphate carboxylase/oxygenase from spinach. J Biol Chem 271: 26449-26452.

Shih M-C and Goodman HM (1988) Differential lightregulated expression of the nuclear genes encoding chloroplast and cytosolic glyceraldehyde-3-phosphate dehydrogenase on Nicotiana tabacum. EMBO J 7: 893-898.

Shih M-C, Heinrich P and Goodman HM (1992) Cloning and chromosomal mapping of nuclear genes encoding chloroplast and cytosolic glyceraldehyde-3-phosphate dehydrogenase from Arabidopsis thaliana. Gene 119: 3 17-319.

Shively JM, van Keulen G, Meijer WG (1998) Something from almost nothing: carbon dioxide fixation in chemoautotrophs. Annu Rev Microbiol 52: 191-230.

Sicher RC, Kremer DF and Rodermel SR (1994) Photosynthetic acclimation to elevated CO2 occurs in transformed tobacco with decreased ribulose-1,5-bisphosphate carboxylase/ oxygenase content. Plant Physiol 104: 409-415.

Skrukrud CL, Gordon IM, Dorwin 5, Yuan X-H, Johansson G and Anderson LE (19 91) Purification and characterization of pea chloroplastic phosphoriboisomerase. Plant Physiol 97: 730-735.

Smillie R (1963) Formation and function of soluble proteins in chloroplasts. Can J Bot 41: 123-137.

Smillie RM (1968) The Biochemistry of Euglena In: DE Buetow (ed) The Biology of Euglena, Vol. II, Biochemistry, pp 1-54. Academic Press, New York.

spinach: Purification, microsequencing and cDNA cloning of the chloroplast enzyme. Plant Mol Biol 26: 1961-1973 Henze K, Badr A, Wettern M, Cerff R and Martin W (1995) A nuclear gene of eubacterial origin in Euglena gracias reflects cryptic endosymbioses during protist evolution. Proc Natl Acad Sci USA 92:9122-9126.

Stitt M and Schulze D (1994) Does Rubisco control the rate of photosynthesis and plant growth? An exericise in molecular ecophysiology. PlantCell Environ 17: 465-487.

Structure and function in plant cells. New Phytol 136:543-570 Jang J-C and Sheen J (1997) Sugar sensing in higher plants.

Synthetic bacterium. Proc Natl Acad Sci USA 55: 928-934 Fabry S and Hensel R (1988) Primary structure ofglyceraldehyde3-phosphate dehydrogenase deduced from the nucleotide sequence of the thermophilic archaebacterium Methanothermus fervidus. Gene 64: 189-197.

Ueno O (1998) Induction of Kranz anatomy and C4-like biochemical characteristics in a submerged amphibious plant by abscisic acid. Plant Cell 10: 571-583.

INDEX

V

www.ingramcontent.com/pod-product-compliance
Ingram Content Group UK Ltd.
Pitfield, Milton Keynes, MK11 3LW, UK
UKHW050945290926
15470UKWH00005B/1656

9 781787 150454